Clinical Nuclear Medicine Physics with MATLAB®

T0187648

Series in Medical Physics and Biomedical Engineering

Series Editors
Kwan-Hoong Ng, E. Russell Ritenour,
and Slavik Tabakov

For more information about this series, please visit: https://www.routledge.
com/Series-in-Medical-Physics-and-Biomedical-Engineering/book-series/
CHMEPHBIOENG

Clinical Nuclear Medicine Physics with MATLAB®

A Problem-Solving Approach

Edited by

Maria Lyra Georgosopoulou

CRC Press
Taylor & Francis Group
Boca Raton London New York

CRC Press is an imprint of the
Taylor & Francis Group, an **informa** business

First edition published 2022
by CRC Press
6000 Broken Sound Parkway NW, Suite 300, Boca Raton, FL 33487-2742

and by CRC Press
2 Park Square, Milton Park, Abingdon, Oxon, OX14 4RN

Library of Congress Cataloging-in-Publication Data

Names: Lyra Georgosopoulou, Maria, editor.
Title: Clinical nuclear medicine physics with MATLAB : a problem-solving approach / edited by Maria Lyra Georgosopoulou.
Description: First edition. | Boca Raton : CRC Press, 2021. | Series: Series in medical physics and biomedical engineering | Includes bibliographical references and index.
Identifiers: LCCN 2021006755 (print) | LCCN 2021006756 (ebook) | ISBN 9780367747510 (hardback) | ISBN 9780367756079 (paperback) | ISBN 9781003163183 (ebook)
Subjects: LCSH: MATLAB. | Medical physics. | Nuclear medicine--Data processing.
Classification: LCC R895 .C55 2021 (print) | LCC R895 (ebook) | DDC 610.1/53--dc23
LC record available at https://lccn.loc.gov/2021006755
LC ebook record available at https://lccn.loc.gov/2021006756

ISBN: 9780367747510 (hbk)
ISBN: 9780367756079 (pbk)
ISBN: 9781003163183 (ebk)

Typeset in Times
by KnowledgeWorks Global Ltd.

Contents

Foreword

Clinical Nuclear Medicine Physics with MATLAB®: A Problem-Solving Approach by Professor Maria Lyra Georgosopoulou, PhD, and her team is the third in a collection of textbooks within the *Series in Medical Physics and Biomedical Engineering* dedicated to software used by clinical medical physicists. The first two books in the series are companions to the present one and are dedicated to the use of MATLAB in Radiotherapy and Diagnostic Physics [1,2]. This book will complete the three-book set.

Given the ongoing and accelerating development of medical device technology and user protocols, fulfilling the mission and delivering the full competence profile of the clinical medical physicist has become a daunting task [3,4]. However, well-written software and programming skills can help in a multitude of ways [5-8]. Regrettably, few suitable textbooks are available, with the result that the acquisition of high-level programming skills by students and clinical scientists is often a hit-and-miss affair. Few books include exemplar scripts illustrating application to the clinical milieu whilst didactic approaches are insufficiently comprehensive or low in communicative power. *Clinical Nuclear Medicine Physics with MATLAB®: A Problem-Solving Approach* aims to fill this gap in the use of MATLAB in clinical nuclear medicine.

In this textbook, the university tutor will find structured teaching text and real-world case study examples with which to enhance presentations and to set as learning tasks. On the other hand, the student will find a pedagogically appealing and engaging manuscript for individual study whilst the practicing clinical medical physicist a learning tool for further development of own skills. Professor Georgosopoulou and her team have put their multifarious clinical and MATLAB experience into this book so there is something for everybody here.

Professor Georgosopoulou and her team are clinical nuclear medicine physicists with many years of experience and I sincerely thank them for finding the time within their busy schedules to dedicate to this important educational initiative and to share their nuclear medicine and MATLAB expertise with the readers of this book. I am sure it has not been easy and I appreciate their contribution.

Finally, I would like to wish the readers of this textbook many happy MATLAB programming hours – and to remind them that programming is power! I also invite potential authors with novel ideas regarding textbooks for other software used in medical physics to write to me, they will find a willing ear.

Carmel J. Caruana, PhD, FIPEM
Professor and Head, Medical Physics Department, University of Malta
Past-Chair, Education and Training Committee, European Federation of
Organizations for Medical Physics
Past Associate Editor for Education and Training:
Physica Medica – European Journal of Medical Physics
Past Chair Accreditation Committee 1:
International Medical Physics Certification Board

REFERENCES

1. Dvorak P. (2018) Clinical Radiotherapy Physics with MATLAB: A Problem-Solving Approach, CRC Press.
2. Helmenkamp J., Bujila R., & Poludniowski G. (2020) Diagnostic Radiology Physics with MATLAB: A Problem-Solving Approach, CRC Press.
3. Caruana, C. J., Christofides, S., & Hartmann, G. H. (2014). European Federation of Organizations for Medical Physics (EFOMP) policy statement 12.1: recommendations on medical physics education and training in Europe 2014, Physica Medica – European Journal of Medical Physics, 30 (6), 598.
4. Guibelalde E., Christofides S., Caruana C. J., Evans S., & van der Putten W. (Eds.) (2015). European Guidelines on the Medical Physics Expert (Radiation Protection Series 174). Publications Office of the European Union, Luxembourg: European Commission.
5. Lyra M., Ploussi A., & Georgantzoglou A. (2011). MATLAB as a tool in nuclear medicine image processing, In: Ionescu C. (Ed.) MATLAB – A Ubiquitous Tool for the Practical Engineer InTech, DOI: 10.5772/19999.
6. Ferris M.C., Lim J., & Shepard D.M. (2005). Optimization tools for radiation treatment planning in MATLAB, In: Brandeau M.L., Sainfort F., Pierskalla W.P. (Eds) Operations Research and Health Care. International Series in Operations Research & Management Science, vol 70. Boston, MA: Springer.
7. Nowik P., Bujila R., Poludniowski G., & Fransson A. (2015). Quality control of CT systems by automated monitoring of key performance indicators: a two-year study, J Appl Clin Med Phys. 16 (4): 254–265.
8. Donini B., Rivetti S., Lanconelli N., & Bertolini M. (2014) Free software for performing physical analysis of systems for digital radiography and mammography, Med Phys, 41 (5): 051903.

Contributors

Maria Argyrou
Department of Nuclear Medicine
Athens Medical Center
Athens, Greece

Christos Chatzigiannis
Imaging Centre
Regional Medical Physics Service
Royal Victoria Hospital
Belfast, UK

Antonios Georgantzoglou
Department of Physiology, Development
    and Neuroscience
University of Cambridge
Cambridge, UK

Maria Lyra Georgosopoulou
1st Department of Radiology
    Radiation Physics Unit
School of Medicine
National and Kapodistrian University of
    Athens
Athens, Greece

Nefeli Lagopati
National Technical University of Athens
School of Chemical Engineering
Laboratory of General Chemistry
National and Kapodistrian University of
    Athens
and
Department of Medicine
School of Health Sciences
Molecular Carcinogenesis Group
Athens, Greece

Marios Sotiropoulos
Unité Signalisation radiobiologie et
    cancer
Institut Curie/CNRS/Inserm/Université
    Paris-Saclay
Orsay, France

Stella Synefia
Metropolitan College
Maroussi, Greece

Elena Ttofi
Limassol, Cyprus

1 Introduction

Maria Lyra Georgosopoulou

CONTENTS

1.1 INTRODUCTION – QUALITY IN NUCLEAR MEDICINE

Nuclear Medicine includes measurement or imaging of radioactive samples in a tiny size to a human adult. It is a medical specialty by which we use safe, painless, and cost-effective techniques for the body and organ imaging as well as for treatment of diseases.

Nuclear Medicine Imaging is unique in discovering organ function and structure using radiopharmaceuticals for diagnostic and treatment purposes. It emerges valuable quantitative data as base to diagnosis and therapy and many useful parameters and their relationship are studied.

As it has to deal with radioactivity, all scientists involved in clinical procedures have to be well and continuously educated and must be well aware with new techniques and innovation technology applied. Knowledge of fundamental physiological data is necessary.

Radiation protection, basic protocols, quality control, and dosimetry are parameters and procedures that personnel of Nuclear Medicine must be familiar. New trends by the modern technology in instruments and software bring developments in favor of patients and workers. Children should safely undergo these procedures to evaluate, e.g., bone pain, injuries, infection of kidneys, or bladder function.

Protocols and Guidelines for the inspection of equipment and parameters used by the imaging systems in Nuclear Medicine will drive to the desirable qualified results.

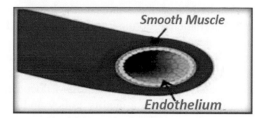

FIGURE 1.1 An example of *high quality*. Innovations in 3D printing lead to implantable high Quality blood vessels. Advanced bioinks for 3D printing are rationally designed materials intended to improve the functionality of printed scaffolds. (Extracted from Gao G. et al., 2019.)

Quality generally is referred to a specific characteristic of an object or to the achievement of an object or a service. The meaning of quality is synonymous with good or excellent. So, the quality of something depends on the criteria that are applied for it and it is the completion of the characteristics of a product that it will satisfy our steady needs; e.g., High Quality example could be the use of bioinks formulated from smooth muscle cells from a human aorta and endothelial cells from an umbilical vein. The artificial blood vessel is an essential tool to save patients suffering from cardiovascular disease. There are products in clinical use made from polymers, but they don't have living cells and vascular functions. The researchers built a high-quality biomimetic blood vessel using three dimensional (3D) printing techniques (Figure 1.1) (Gao G. et al., 2019). The researchers fabricated the printed blood vessel; they used a special algorithm to print multiple concentric layers using bioink. Tissue-engineered biomimetic blood vessels provide a promising route for the construction of durable small-diameter vascular grafts that may be used in future treatments of cardiovascular diseases.

1.2 QUALITY IN NUCLEAR MEDICINE FOCUSES ON THREE MAIN TASKS

(1) Quality Management (QM), (2) Quality Assurance (QA), and (3) Quality Control (QC)

1.2.1 QUALITY MANAGEMENT (QM)

Quality Management covers all activities that organizations use to direct, control, and coordinate quality. These activities include formulating a quality policy and setting quality objectives. They also include quality schedule, quality control, quality assurance, and quality corrections. Continuous updating on modern techniques, such as Artificial Intelligence and Deep learning techniques that differ in the way that information is extracted from the input, helps toward a dynamic education necessary for the optimum results of a patient's health (Cook, G. J. R. and Goh, V., 2019).

Quality Management Systems (QMS) should be maintained with the intent to continuously improve effectiveness and efficiency, enabling Nuclear Medicine to achieve the expectations of its quality policy, satisfy its customers, and improve professionalism. The Quality Management audit methodology in Nuclear Medicine practice is published by International Atomic Energy Agency (IAEA No. 33, Quality Management Audits in Nuclear Medicine Practices, 2015a).

Regular quality audits and assessments are essential for modern Nuclear Medicine services. The entire Quality Management (QM) and audit process has to be systematic, patient oriented, and outcome based. The QM documentation in Nuclear Medicine should contain documents of a quality policy and quality objectives, a quality manual, standards of operating procedures for diagnosis and therapy management, and records of parameters and remarks.

The Quality Management (QM) Service in the Nuclear Medicine department should also take into account multidisciplinary contributions including clinical, technical, radiopharmaceutical, medical physics, and radiation safety procedures. Its effectiveness should be improved according to the requirements of professional, regulatory, standardization, or accrediting bodies. The adoption of internationally recognized audits and protocols as from International Atomic Energy Agency (IAEA), Society of Nuclear Medicine and Molecular Imaging (SNMMI), European Association of Nuclear Medicine (EANM), International Commission on Radiation Units (ICRU), etc. is preferable and offers the necessary homogeneity of results quality (EANM Carrio et al., 2006; EANM Cuocolo et al., 2007).

1.2.2 Quality Assurance (QA)

Quality Assurance is the function of a management system that provides confidence for the fulfillment of specified requirements and it is a necessity in Nuclear Medicine.

Quality Assurance (QA) covers all activities as design, development, production, installation, servicing, and documentation. All the results must be monitored ensuring that appropriate action is taken when performance is outside acceptable ranges. That is: *"fit for purpose"* or *"do it right as the first time"*.

QA standards, protocols, guidelines, preparation of radiopharmaceuticals, and phantoms have been set by world organizations. National Electrical Manufacturers Association (NEMA), International Atomic Energy Agency (IAEA), International Electrotechnical Commission (IEC), Deutsches Institut fur Normung (DIN), European Association of Nuclear Medicine (EANM), Society of Nuclear Medicine and Molecular Imaging (SNMMI), and other scientific organizations have set down guidelines in order to ensure that the appropriate standards in Nuclear Medicine are met in most countries of the world. These guidelines will be acceptable as a common denominator for the benefit of patient care on a world level (IAEA, 2006). Every department should organize its own program according to its needs and workload, in order to be efficient, to keep the instrumentation in good condition, and to prevent any deterioration.

1.2.2.1 Why Do We Need a Quality Assurance (QA) Program in Nuclear Medicine?

We wish to ensure that we get a high-quality, correct, clinical result which addresses the patient's clinical problem with minimal detrimental effects on the patient concerning radiation dose, waiting time, and minimal errors.

Nuclear Medicine product is "Diagnosis and Therapy". This is what we wish to guarantee, and Quality Assurance (QA) reassures that the obtained result is the best, more probable for the available equipment and personnel.

Quality Assurance (QA) in Nuclear Medicine includes:

1. High-performance equipment
2. Staff Education (Clinical Medical Physicists)
3. Image acquisition and processing reliable software
4. Radioisotopes – Radiopharmaceutical Quality

1.2.2.2 Education/Clinical Medical Physicists in Nuclear Medicine

In 2013, a cooperation of European Association of Nuclear Medicine (EANM) and European Federation of Organizations of Medical Physicists (EFOMP) produced the "Curriculum for education and training of Medical Physicists in Nuclear Medicine" to provide a guideline curriculum covering theoretical and practical aspects of education and training for Medical Physicists in Nuclear Medicine (EANM/EFOMP Del Guerra et al., 2013). This joint EANM/EFOMP European guideline curriculum is a further step to harmonize specialized training of Medical Physicists in Nuclear Medicine and is a contribution in Quality Assurance (QA) of the Nuclear Medicine departments.

The growing number and the increasing complexity of the procedures in Nuclear Medicine require clinically qualified Medical Physicists to ensure safe and effective patient diagnosis, treatment, and management; they are health professionals responsible of quality and safety of Nuclear Medicine applications of ionizing radiation. IAEA, Human Health Reports No. 15 describes an algorithm developed for determination of the recommended staffing levels for clinical medical physics services in medical imaging and radionuclide therapy (IAEA, 2018b) endorsed by the International Organization for Medical Physics (IOMP).

1.2.2.3 Radioisotopes – Radiopharmaceutical Quality

Progress in Nuclear Medicine is tightly linked to the development of new radiopharmaceuticals and efficient production of relevant radioisotopes. New radiopharmaceuticals can provide extremely valuable information in the evaluation of many diseases (cancer, brain, or heart diseases). IAEA (2015b) is a practical support for the introduction of new radiotracers, including recommendations on the necessary steps needed to facilitate and speed the introduction of radiopharmaceuticals in clinical use, while ensuring that a safe and high-quality product is administered to the patients at all times. Radiopharmaceuticals are an important tool in Nuclear Medicine for early diagnosis of many diseases and developing effective treatments (IAEA Bulletin, 2019).

The IAEA International Symposium on Trends in Radiopharmaceuticals (ISTR-2019), 28 October–1 November 2019, in Vienna, Austria, provided an international forum for scientists working in the fields of production of radioisotopes and radiopharmaceuticals to discuss the most recent progress in the field. Development, production, and use of diagnostic, therapeutic, and theranostic (used for diagnosis and therapy) radioisotopes and radiopharmaceuticals were analytically discussed (International Symposium on Trends in Radiopharmaceuticals [ISTR], 2019).

By the completion of a Quality Assurance (QA) program, we should also be harmonized with the European Union (EU) legislation against the dangers arising from Ionizing Radiation. Requirements of Council Directives 96/29 and 97/43 of Euratom are "Basic Safety Standards Directive (BSS) on the protection of workers and the general population" and "Medical Exposure Directive (MED), on health protection of individuals in relation to medical exposure". These are included in the new (Euratom, 2014), "Basic Safety Standards Directive Council Directive 2013/59/Euratom".

1.2.3 QUALITY CONTROL (QC)

Quality Control (QC) is a part of Quality Management (QM) focused on fulfilling quality requirements. Quality Control (QC) is referred to the actions we take to achieve, support, and improve the quality.

- *Total Quality Control* is the completeness of quality characteristics of a system.
- *Statistical Quality Control* is the collection and analysis of Total Quality Controls.

Though much has been done to improve the stability of Nuclear Medicine systems performance, their components will drift from the initial specifications, causing a gradual deterioration in performance over a period of time (e.g., fluctuations in the electrical power supply and environmental conditions may cause changes in system performance). Quality Control (QC) of imaging system and radiopharmaceuticals is a necessity for reliable results in Nuclear Medicine.

For best image quality in Nuclear Medicine examination performance, Quality Control (QC) of the imaging system is a necessity. Quality Control (*QC*) of the imaging system requires special choice of:

- System characteristics (e.g., Resolution-Uniformity-Linearity-COR)
- Proper examination time-interval or quite enough counts collection to increase statistics
- Low level of scattered radiation
- Optimum patient position (minimum distance of patient from imaging system)
- Suitable imaging size

Acceptance testing – The first Quality Control (QC) procedure has as scope to test newly installed systems with proper phantoms to provide sufficient data for evaluation of a new system's specification and obtain relevant results. These results should be used as reference data for annual survey of the system. The annual survey of the system performance will be a part of an imaging Quality Assurance (QA) program.

The following examples present a few images obtained by the Quality Control (QC) approach:

- **System characteristics** (Resolution-Uniformity-Linearity; Figure 1.2)
- **Image quality as a function of counts**

FIGURE 1.2 Resolution-Uniformity-Linearity phantom imaging in a planar γ-camera quality control (QC).

Right examination time interval or proper counts collection to increase statistics (Figure 1.3).

- **Low level of scattered radiation (Figure 1.4)**
- **Suitable imaging size**
  The spatial resolution of a digital image is governed by the size of the pixels used to represent the digitized image. The pixel size is smaller as the number of pixels and the matrix size are larger.

Data collection matrix size depends on the number and size of the contained pixels. A pixel represents the smallest sampled 2D element in an image. It has dimensions given along two axes in millimeters, dictating in-plane spatial resolution.

Digital images are characterized by matrix size and pixel depth. Matrix size refers to the number of discrete picture elements (pixels) in the matrix. This affects the degree of spatial detail that can be presented. Larger matrices provide more detail. Matrix sizes used in Nuclear Medicine images typically range from 64 × 64 to 512 × 512 pixels. The larger the matrix size, the smaller the pixels and the more detail that is visible in the image (Figure 1.5) (Radiology Key, 2016).

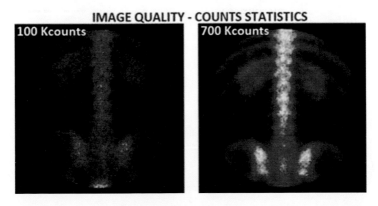

FIGURE 1.3 Choice of termination of image acquisition (time or counts) is crucial for the optimum statistics in improvement of image quality.

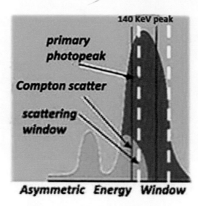

FIGURE 1.4 By determination of an asymmetric energy window, scattered radiation is avoided, thereby improving the quality of acquisition.

A smaller pixel size can display more image detail; though there is no further improvement at a certain point because of resolution limitations of the imaging device.

Most Nuclear Medicine images consist of multidimensional datasets of counts (pixels or counts/voxel). A voxel represents a value on a regular grid in three-dimensional space; in the word *voxel*, [vo] represents volume and [el] represents element; similar formation with [el] for element includes the word *pixel*.

The dataset directly reflects regional concentration of radioactivity. Thus, Nuclear Medicine is a quantitative technique for the detection of molecular interaction of a radio-pharmaceutical with the endogenous target. However, its quantitative power for clinical research can only become productive when there is strict standardization of imaging protocols. Quality Control (QC) and Quality Assurance (QA) procedures of the imaging system must be in use so that optimal quantitative images and data are obtained.

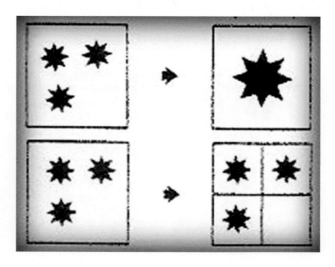

FIGURE 1.5 Matrix size. (Top) Small matrix registers the three events as one. (Bottom) In a larger matrix the three events are recorded clearly.

1.2.3.1   Quality Control of Radiopharmaceuticals

Radiopharmaceuticals are another important quality factor under consideration in Nuclear Medicine. Benefits of high-quality radiopharmaceuticals include accurate diagnosis and secure therapy for the patient.

Principal properties affecting labeling of radiopharmaceuticals are:

- Radioisotope concentration
- Radioisotope purity
- Radiochemical purity
- Specific concentration
- Sterility

Radiopharmaceuticals must be prepared within a reliable Quality Control (QC) system that considers materials, personnel, accurate documentation, and improvement of renewal results. Guidelines and best practices for the Quality Control (QC) of medical radioisotopes and radiopharmaceuticals written by a group of experts with experience of a large range of radiopharmaceuticals are published by the IAEA. These guidelines support professionals in preparing radiopharmaceuticals of good quality and high safety (IAEA, 2018a).

During preparation of radiopharmaceuticals for injection, regulations on radiation protection as well as appropriate rules of working under aseptic conditions ought to be adhered to, strictly. Particular attention must be paid to the prevention of cross-contamination and to waste disposal. A continuous assessment of the effectiveness of the Quality Assurance (QA) system is essential to prove that the procedures applied in the Radiopharmacy section of the Nuclear Medicine department lead to the expected quality (EANM, 2007, Radiopharmacy Committee, cGRPP-guidelines, version2).

1.2.3.2   Quality Control/System Performance Degradation

In Nuclear Medicine Quality Control (QC), sources of system performance degradation must be carefully indicated. Some of them are:

- Collimator damage; photomultiplier tube (PMT) drift; energy peak drift; electronic offsets drift; mechanical COR change; electronic noise; crystal or light coupling degradation; contamination and background; variations with angle of rotation.

The following *images/examples* of system deterioration emerged during the Quality Control (QC) procedures are useful in training:

- **Collimator damage (Figure 1.6)**
- **Photomultiplier tube drift (Figure 1.7)**
- **Photomultipliers out of order (Figure 1.8)**
- **Effect of large radius of rotation (Figure 1.9A)**
  Acquisition of striatum SPECT tomography (Figure 1.9B)
- **Truncation** on a SPECT myocardial perfusion study; a patient example (Figure 1.10) (IAEA, 2003)

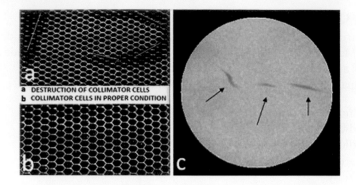

FIGURE 1.6 Collimator destruction during transportation. (a) Distorted cells made of foils. (b) Collimator cells in proper condition. (c) flood source image completed by destructed collimator a. Inhomogeneities on the image caused by destructed collimator cells

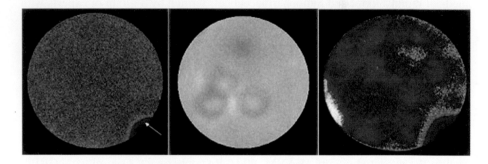

FIGURE 1.7 Uniformity checks of a planar γ-camera by Co57 Flood Source. Left: Peripheral photomultiplier (PMT) defect. Middle: Photomultipliers PMTs voltage deficiency; see rings of low intensity. Right: Non-uniformed image because of wrong energy window selection (20% in Tc-99m photopeak [140 keV] instead of energy window at Co57 photopeak [122 keV] as well as a peripheral photomultiplier (PMT) defect.

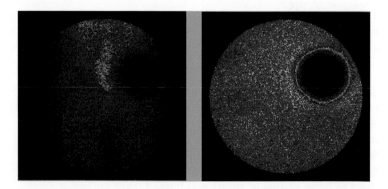

FIGURE 1.8 Left: A cold circular region completing a scintigraphy image before the daily Quality Control (QC). Right: A uniformity control test shows that a group of photomultipliers (PMTs) is out of order. Consequence is the cold region in the patient's scintigraphy image.

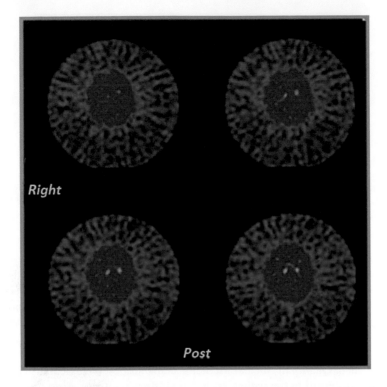

FIGURE 1.9A The patient was extremely claustrophobic and an acquisition by I-123-Ioflupane (DATSCAN) for Striatum study was done with γ-camera heads at full radius. No clear distinction of the imaged striatum. As a result, the obtained tomographic slices were nondiagnostic – very bad quality images. (Images supplied courtesy of GE Healthcare.)

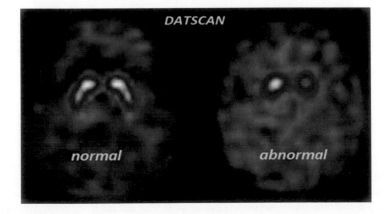

FIGURE 1.9B Properly acquired striatum images by DATSCAN. Left: Diagnosed as normal. Right: Characterized as abnormal.

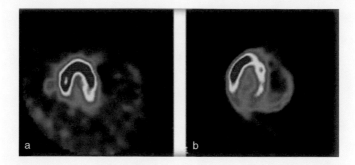

FIGURE 1.10 (a) Transaxial image with no truncation. All 64 rows of the matrix were used for the data. (b) Transaxial slice from severe truncation, which has been simulated by using a row of ±10 pixels from the axis of rotation. (IAEA Quality Control Atlas for Scintillation camera Systems, 2003.)

With truncation, the myocardium is still defined but is severely distorted.

During myocardial perfusion scintigraphy, truncation can occur when a patient is too large to fit entirely within the Field of View (FOV) or is positioned in such a way that a part of the body is outside the FOV. Missing emission data are clinically acceptable as long as the entire heart is kept within the FOV.

1.2.3.3 Abnormal Distribution of the Radiopharmaceutical

Improvement of Quality in Nuclear Medicine demands best biological distribution of Radiopharmaceutical in the patient body by: *Optimum activity dose; the most suitable radiopharmaceutical;* and *proper chemical and physical form.*

Two image examples of the impact of abnormal distribution of radiopharmaceuticals in the Quality of the results are shown in Figure 1.11.

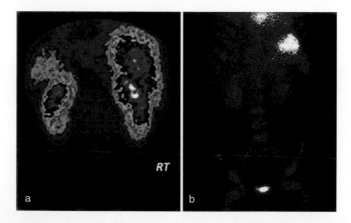

FIGURE 1.11 Degradation of image due to abnormal distribution of radiopharmaceutical. (a) Image of a series of dynamic renal study. Radiopharmaceutical abnormal concentration to liver and spleen is observed. (b) Unsuccessful labeling of Tc-99m and red blood cells. Abnormal concentration in skeleton and other organs is acquired due to presence of high percentage of alumina.

1.2.4   IMAGING PHANTOMS ARE QUALITY CONTROL TOOLS IN NUCLEAR MEDICINE

Phantoms are Quality Control tools in Nuclear Medicine. Three variations of them depending to the scope of use are for:

Calibration – Ideal geometrical characteristics
Imaging – Tissue equivalent of known geometry
Dosimetry – Radioactive sources of standard dimensions

Calibration phantoms are used in procedures employing or measuring radioactive material to evaluate performance. Phantoms often have properties similar to human tissue. Water demonstrates absorbing properties similar to normal tissue; hence, water-filled phantoms are used to map radiation levels (Iturralde P. M., 1990.

Imaging phantom in Nuclear Medicine is a specially designed object that is scanned or imaged to evaluate, analyze, and tune the performance of imaging devices used. A phantom is more readily available and provides more consistent results than the use of a living subject and avoids subjecting a living individual to direct risk. Phantoms are either employed for use in 2D-based imaging techniques in planar γ-camera evaluation and phantoms with desired imaging characteristics have been developed for 3D techniques such as SPECT, PET, CT, MRI, and hybrid imaging methods or modalities.

A phantom used to evaluate an imaging device should respond in a similar manner to how human tissues and organs would act in that specific imaging modality. Single Photon Emission Computed Tomography (SPECT) and Positron Emission Tomography (PET) systems require periodic performance testing. A modular phantom can be used to collect the data for the Quality Control (QC) completeness of SPECT or PET systems. It permits measurements of the resolution of radioactive (hot) lesions in a cold (clear background) field and (cold) lesions in a radioactive (hot) field; uniformity of response to a uniform field, linearity, and a simple way to find the center of rotation. A more sophisticated test suitable for acceptance testing may be done by measuring the Point Spread Function (PSF), from which the Modulation Transfer Function (MTF) may be calculated. Imaging phantoms are used for Quality Control (QC) of diagnostic imaging systems in Nuclear Medicine as well as for training students about used techniques.

Dosimetric phantoms are used to measure absorbed dose at the surface or at a depth within their irradiated volume. Actually they are radioactive sources of standard dimensions.

Some of the phantoms used in Quality Control (QC) procedures of PET, SPECT, and CT systems are shown in Figure 1.12.

AAPM CT performance phantom meets guidelines in American Association of Physicists in Medicine (AAPM) Report No. 1, Performance Evaluation and Quality *Assurance of CT Scanners*. It offers the user a single test object that measures ten

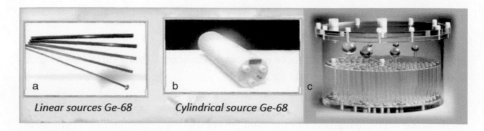

Linear sources Ge-68   Cylindrical source Ge-68

FIGURE 1.12   (a) National Electrical Manufactures Association (NEMA) sensitivity PET phantom. Six concentric aluminum tubes used to detect camera sensitivity in PET. (b) NEMA PET scatter fraction phantom. Made of four sections for ease of carrying/storage. (c) Elliptical PET and SPECT Phantom™ Model ECT/ELP/P Deluxe ECT phantom with elliptical body shape: Use with high spatial resolution SPECT and PET systems for evaluation of data acquisition using noncircular orbit. (Data Spectrum Corporation, http://www.spect.com/products.html.)

distinct CT performance parameters. Quality measurements are also useful in Quality Control of hybrid systems SPECT/CT and PET/CT (Figure 1.13).

By **NEMA IEC PET body phantom** Image Quality (IQ), a test can be completed to simulate whole-body imaging with hot and cold lesions. Body phantom (IQ) is used with different fillable spheres for imaging parameters determination in PET and PET/CT. It is designed in accordance with the recommendations by the International Electro-technical Commission (IEC) and modified by the National Electrical Manufacturers Association (NEMA). It is recommended for use in the evaluation of reconstructed image quality in whole-body PET imaging (Figure 1.14).

FIGURE 1.13   An imaging phantom for determining computed tomography (CT) performance in SPECT/CT and PET/CT Nuclear Medicine imaging systems, AAPM Report No. 1, 1977, CT Performance Phantom. (CIRS Model 610 AAPM CT Performance Phantom, Universal Medical, https://www.universalmedicalinc.com/aapm-ct-performance-phantom.html.)

FIGURE 1.14 NEMA IEC PET Body Phantom Set™, The body phantom with a lung insert and an insert of six spheres with various sizes.

1.2.5 Hybrid Systems Quality Assurance (QA) – Fusion

In addition to stand-alone imaging modalities, integrated dual-modality imaging methods by the hybrid imaging systems SPECT/CT, PET/CT, and PET/MRI are used to provide more diagnostic information.

Single Photon Emission Tomography (SPECT) or Positron Emission Tomography (PET) and Computed Tomography (CT) differ in that SPECT or PET evaluates physiology and CT gives anatomy information.

The clinical application of hybrid technology includes the ability to register or "fuse" a CT or a magnetic resonance imaging (MRI) data with an emission tomography data (SPECT or PET) to add anatomy information to the physiology imaging.

Hybrid systems of SPECT/CT or PET/CT allow the combined assessment of anatomy and function. Furthermore, Integrated PET/MRI can provide complementary functional and anatomical information about a specific organ or body system at the molecular level.

Hybrid SPECT/CT cameras are well accepted for improved attenuation correction and accurate localization. The potential of SPECT/CT imaging for lesion characterization increases by the use of diagnostic CT. Recent technological advances will also enhance the role to be played by quantitative SPECT/CT for dosimetry estimates in individualized radionuclide therapy (Israel. O. et al., 2019).

In (AAPM, 2019c), recommendations for performing acceptance tests and annual physics surveys of γ-camera and SPECT and SPECT/CT systems are made. SPECT and CT spatial alignment and attenuation correction accuracy for optimum image quality are the recommendations for SPECT/CT hybrid system.

The tests described in the AAPM Report No. 177, 2019 are categorized into four sections: Physical Inspection, γ-Camera Planar Tests, SPECT Tests, and SPECT/CT

Tests. The tests include recommendations for both acceptance testing and an annual physics survey.

1.2.5.1 Quality of Performance and Advantages Utilizing SPECT/CT

Image registration is defined as the transfer of two image datasets into one common coordinate system. A distinction can be made between:

- Software-based registration of datasets acquired independently one from each other by two different imaging devices; and
- Hardware-based registration where the two datasets are obtained by hybrid equipment in a single imaging session.

A high-quality SPECT/CT study requires a reliable, well-functioning hybrid system which has met acceptance testing criteria and which is regularly monitored for *Quality of Performance*.

The acquisition and processing protocols must be carefully followed and:

- The technical staff (doctors and Medical Physicists as well) of the Nuclear Medicine department should be well trained to perform and monitor both components of the study according to a well-defined protocol.
- Workstations that allow integrated viewing of the functional and anatomic data should be used for the interpretation of the SPECT/CT images.
- The images must be reviewed for technical and diagnostic *quality* before the patient leaves the department.

Precise image registration in the hybrid imaging study by SPECT/CT makes the interpretation of high signal-to-background (S/B) functional images – combined with better anatomic information – more credible.

Anatomical accuracy of image registration in SPECT/CT hybrid imaging is a high advantage of the system. Increased specificity is achieved through a more precise localization by CT and characterization of functional findings by SPECT.

The use of the CT component of a SPECT/CT patient examination, for correcting the SPECT data for attenuation and scatter is crucial, for proper estimation of radioactivity concentration in specific organs or tissues on a volumetric basis (IAEA, 2008).

1.2.5.2 SPECT/CT Fusion

Skeletal SPECT/CT is the new imaging gold standard when searching for osseous metastases. The improved diagnostic accuracy of SPECT/CT is associated with greater diagnostic confidence.

Two relative cases are shown in Figures 1.15 and 1.16.

Through hybrid imaging by SPECT/CT (Figure 1.17), the Nuclear Medicine expert contributes to better health care by clinical interventions to the individual patient's needs (Israel O. et al., 2019).

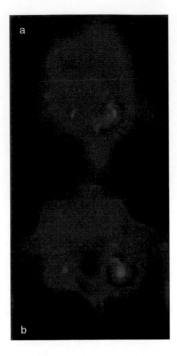

FIGURE 1.15 (a) Fused SPECT/CT image of a lumbar vertebra in a patient with breast cancer is shown. Increased uptake of Tc-99m-MDP into the arthrosis of the facet joint is observed. (b) A similar image in another breast cancer patient is depicted. Although the SPECT appearance of the lesion is quite similar to that in patient A, the CT overlay proves it to be a small osteolysis Fused images also indicate that SPECT/CT is a clinically relevant component of the diagnostic process in patients with non-oncological disease referred for bone scintigraphy. (Extracted from IAEA, TECDOC 1597, 2008.)

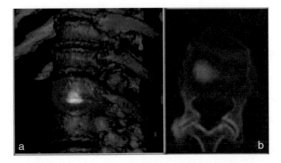

FIGURE 1.16 A 74-year-old patient after recent trauma showing enhanced uptake of Tc-99m-MDP in a vertebral body of the lower thoracic spine. (a) 3D-volume rendering of the SPECT/CT fusion shows that a lesion is in the 12th vertebral body. The inspection of the fused tomogram proves it to be a fracture; moreover, the examination discloses it to be unstable since (b) the posterior cortical is involved, thus motivating immediate surgery. (Extracted from IAEA, TECDOC 1597, 2008.)

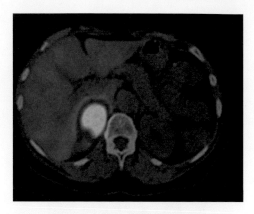

FIGURE 1.17 A 6-year-old child with newly diagnosed neuroblastoma was referred for staging. I-123-mIBG SPECT/CT localizes abnormal tracer uptake to the primary lesion in a large mass in the right adrenal. (Courtesy of Israel O. et al., 2019.)

1.2.5.3 Hybrid PET/CT

A Quality Assurance (QA) program for a hybrid PET/CT system is necessary to complete an initial evaluation of the performance of the system so that a baseline of measurements could be established. Then periodic assessment of system performance on an annual, quarterly, weekly, and daily basis should follow.

1.2.5.3.1 PET/CT Acceptance Testing and Quality Assurance

The American Association of Physics in Medicine (AAPM) published in October 2019 the Report No. 126, which is referred to PET/CT Acceptance Testing and Quality Assurance (AAPM, Report No. 126, 2019b). A rigorous Quality Assurance (QA) program for PET/CT is recommended.

Other international, European, or American organizations have published recommendations in relative reports or guidelines, such as: as International Atomic Energy Agency (IAEA), International Electrotechnical Commission (IEC), National Electrical Manufacturers Association (NEMA), American College of Radiology (ACR), Society of Nuclear Medicine and Molecular Imaging (SNMMI), or European Association of Nuclear Medicine (EANM).

1.2.5.3.2 PET/CT Image Quality by NEMA IEC PET Body Phantom Set™

The evaluation of PET system requires standard and reliable methods to allow the comparison of different PET systems using accepted measurement standards for the system. NEMA provides guidelines to measure the performance of the PET component of a PET/CT system by the body phantom with a lung insert and an insert of six spheres with various sizes (Figure 1.14); NEMA organization has published a series of procedures to evaluate the physical performance of PET systems. This NEMA standard is revised periodically, and the latest update of this publication resulted in the NEMA NU2-2018.

The NEMA NU2-2012 quality control tests have been performed to evaluate this system on site before clinical use (Jha A. K. et al., 2019). Performance measurements of the PET component were made using the NEMA NU2-2012 procedures for spatial

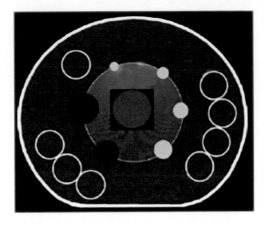

FIGURE 1.18 PET/CT image quality calibrations by NEMA body phantom. Schematically, in this figure, we see the cold, hot spheres and background Region of Interests (ROIs).

resolution, scatter fraction, sensitivity, count rate loss and random coincidence estimation, Noise Equivalent Count Rate (NECR), and *image quality*.

Image quality test for contrast recovery, background variability met all specifications. Overall PET performance of whole-body system was satisfactory and all the performance specifications required by NEMA NU2-2012 were satisfied.

In <u>Positron emission tomography</u> body phantom, the middle pipe represents lungs, six fillable spheres with inner diameters of 10, 13, 17, 2, 28, and 37 mm. Background activity concentration of the body phantom (Figure 1.18), hot lesions, and cold lesions should be filled accordingly.

The analysis can be made on transverse sections (Demir M. et al., 2018) in which a circular region of interest (ROI) is drawn as close as possible to the dense inner diameter of every cold and hot sphere. For background, ROIs of the same size of the hot and cold sphere ROIs, drawn around and are outlined near the edge of the phantom.

1.2.5.4 New Generation of PET/CT

New generation of PET/CT has been introduced using silicon (Si) PM-based technology. In PET/CT devices, conventional analog PMs are replaced by the solid-state technology aiming to improve time resolution, event collection (improvement of system sensitivity), localization, and counting efficiency.

One limiting factor in the development of efficient and accurate PET systems is the sensor technology in the PET detector. There are generally four types of sensor technologies employed: Photomultiplier Tubes (PMTs), Avalanche photodiodes (APDs), Silicon Photon Multipliers (SiPMs), and Cadmium Zinc Telluride (CZT) detectors. PMTs were widely used for PET applications due to their excellent performance parameters of high gain, low noise, and fast timing. However, their sensitivity to magnetic fields turns the interest of technology to solid-state photodetectors applied to PET systems. The two most promising types are SiPMs and CZT (Jiang W. et al., 2019).

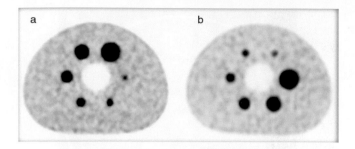

FIGURE 1.19 Transversal images of the phantom obtained from (a) a SiPM-based PET-CT and (b) a PET/CT scanner with a conventional Photon Multipliers (PMs) detector, with a sphere-to-background activity ratio of 4:1. Image acquisitions and reconstructions were performed as in the patient study. (Courtesy by Oddstig J. et al., 2019.)

- *SiPMs* have been employed to both PET/CT and PET/MRI systems due to their high timing resolution, low operation voltage, and their features of compactness and immunity for coexistence with magnetic fields. However, efforts are still needed to improve the performance of SiPMs for lower noise, higher photo-detection efficiency (PDE), and even better timing resolution.

  It is predicted that digital silicon photomultipliers (dSiPMs) might overtake analog SiPMs in future PET systems.
- *Cadmium Zinc Telluride (CZT)* detectors show excellent position and energy resolution, but their PET applications are limited due to their poor timing resolution performance.

  A comparison of measurements performed in analog PET/CT equipped with conventional PMTs with the signal recovery obtained from digital PET systems, in hot sphere inserts of the NEMA/IEC NU2 phantom highlighted the improved *signal-to-noise ratios* achievable with the new digital PET devices compared to conventional ones (Gnesin S. et al., 2020).

To compare the image quality and diagnostic performance of a SiPM-based PET/CT with a PET/CT system with a detector that uses conventional Photo Multipliers, imaging (Figure 1.19) of the National Electrical Manufacturers Association NEMA-IEC body phantom and 16 patients was carried out by (Oddstig J. et al., 2019).

Results indicate that the new generation of PET/CT scanners might provide better diagnostic performance; though further quality studies are needed.

1.2.5.5 Comparison of Quantitative PET/CT and SPECT/CT

Quality Control for quantitative SPECT will require a regular check of agreement between the dose calibrator and reconstructed SPECT values for radioactivity.

For PET systems, cross-calibration is slightly simpler than for SPECT, as in PET all the radiation has the same photon energy and characteristics, 2×511-keV annihilation photons, independent of the radiotracer. Various radiotracers used by SPECT, with different photon energies, require separate calibrations and checking.

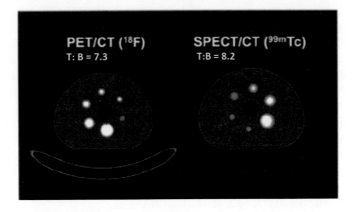

FIGURE 1.20 Comparison of quantitative PET/CT (F-18) and SPECT/CT (Tc-99m) images in IEC body phantom containing six fillable spheres. **T:B** indicates actual target to background ratio for each experiment between concentration of radioactivity in spheres and in general compartment of phantom. Quantitative results are obtained for single slice through phantom corresponding to central section through largest sphere. (Extracted from Bailey D. L. et al., 2013.)

In Figure 1.20, a comparison of the NEMA/IEC phantom quantitative images is shown.

In this example, PET Standardized Uptake Values (SUV) is slightly overestimated by about 8% whereas SPECT images are underestimated by about 11% probably because of Partial-Volume-Effect losses caused by poorer spatial resolution in SPECT than in PET. The absolute differences between the SUVs of SPECT and PET were generally very small.

1.2.5.6 From PET/CT to PET/MRI

PET does not provide clear visualization of the anatomical structure because it has relatively poor spatial resolution. PET/CT has been proposed to overcome this limitation, and this hybrid system provides more clinically useful information than either a PET or a CT stand-alone system.

The inability of simultaneous acquisition and the reduced soft tissue contrast in PET/CT have led to the development of PET/MRI which offers the better soft tissue contrast over CT. CT as synergistic modality offers anatomical details. However, MRI combines structural and molecular imaging (Hu Z. et al., 2014).

PET/CT multimodality imaging provides both metabolic information and the anatomic structure, but because of the higher soft-tissue contrast of MRI and no extra ionizing radiation, PET/MRI imaging seems to be preferred as the method of choice.

PET/MRI, though, has many technical difficulties to overcome, such as PM tubes, which cannot work properly in a magnetic field, and the inability to provide density information on the object for attenuation correction (Jung J. H. et al., 2016).

Hybrid PET/MRI offers potential benefits that help in the improvement of PET images. The combination of MR imaging – *anatomic imaging and highest resolution* – with PET – *metabolic imaging with high sensitivity* – has a high-clinical impact.

The advantage of MRI as synergistic modality in a hybrid system versus the CT is the ability of much better soft tissue contrast. High-resolution anatomical information from MRI can be used to enhance PET image quality and this is the desirable.

PET is a noninvasive metabolic imaging technique which traces the physiological and pathophysiological process at molecular level (Boss A. et al., 2010). However, PET does not offer any anatomical information, which is really the major drawback of it.

MRI, though, offers an immense 3D soft tissue contrast. This enables it to be a privileged technique as a first-line method of choice for the assessment of tumor.

By registration of high-resolution MR images, PET reconstruction improved with information derived from MRI. MRI with higher soft-tissue contrast in comparison with CT makes it easier to define the *regions of interest* in PET images that do not provide anatomic details (Jung J. H. et al., 2016).

Integrated *positron emission tomography (PET)/magnetic resonance imaging (MRI)*, which can provide complementary functional and anatomical information about a specific organ or a body system at the molecular level, has become a powerful imaging modality to understand the molecular biology details, disease mechanisms, and pharmacokinetics.

The system produces a better performance in terms of detection and technical possibilities. PET stability metrics demonstrated that PET quantitation was not affected during simultaneous MRI (Musafargani S. et al., 2018).

There were various technical challenges in integrating PET and MRI in a single system, with minimum interference between PET and MRI (Hu Z. et al., 2014). Multicenter and multivendor Quality Control (QC) and harmonization of quantitative PET/MR performance are feasible by using dedicated phantom protocols (EANM Boellaard et al., 2015).

Simultaneous PET/MRI will be best suited for clinical situations that are disease-specific, organ-specific, related to diseases of the children, or in those patients undergoing repeated imaging for whom cumulative radiation dose must be kept as low as reasonably achievable.

It appears logical that PET/MRI may be the multimodality of choice in view of less radiation exposure and all that is offered by MRI and PET for a comprehensive structural and metabolic assessment of pediatric patients. Hybrid PET/MRI may play a significant role in the care of children, particularly with cancer, who may need to undergo repeated diagnostic imaging sessions, and it is desirable to minimize their cumulative radiation exposure (Jadvar H. and Colletti P. M., 2014).

Figure 1.21 shows a fusion image of F-18 fluoroethyl-L-tyrosin registered PET/MRI of a grade II astrocytoma. Anatomical information on MRI allows for a region of interest definition.

Software registration with MRI is utilized in clinical applications and allowed more accurate localization of metastatic lesions, especially in the head and neck area.

However, software-based registration suffers from anatomical inaccuracies cause of different positioning of the patient in the two separate imaging devices. In addition, software-based registration becomes difficult in case that a radiopharmaceutical is highly specific for a certain tissue as the images of its distribution are, with

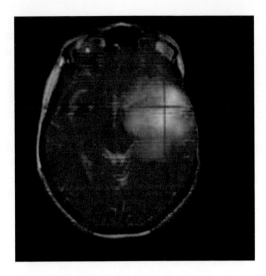

FIGURE 1.21   Image fusion following software registration. Software registered PET/MRI image of a grade II astrocytoma. (Extracted from Slomka P. J. and Baum R. P., 2009.)

regard to anatomical detail, poorer. These limitations are greatly reduced in hardware-based registration that should therefore offer a higher anatomical accuracy of image fusion.

1.2.5.7   Hybrid SPECT/MRI

The combination of SPECT and MRI raises great expectations for further advantage of multimodal imaging. The combination of SPECT and MRI is currently absent from the range of clinical multimodality systems, although work is in progress to produce the first pattern.

In medical imaging, SPECT can provide specific functional information while *magnetic resonance imaging (MRI)* can provide high spatial resolution anatomical information as well as complementary functional information.

A miniaturized dual-modality SPECT/MRI (MRSPECT) system was developed by Hamamura M. J. et al. (2010) and shown the feasibility of simultaneous SPECT and MRI data acquisition. In this MRSPECT system, a *cadmium-zinc-telluride (CZT)* detector was interfaced with a specialized radiofrequency (RF) coil and placed within a whole-body 4Tesla MRI system.

The metallic components of the SPECT hardware alter the B field and generate nonuniform reduction in the *signal-to-noise ratio (SNR)* of the MR images. The presence of a magnetic field creates a position shift and resolution loss in the nuclear projection data. Various techniques were proposed to compensate for these adverse effects. Overall, these results demonstrate that accurate, simultaneous SPECT and MRI data acquisition is feasible, justifying the further development of MRSPECT for either small-animal imaging or whole-body human systems by using appropriate components.

The radionuclide-based SPECT imaging techniques have relatively poor resolution but, like PET, are sensitive to picomolar tracer concentrations, while MRI gives high-resolution anatomical information but suffers from much lower sensitivity to concentration of contrast agent. The combination of both imaging techniques can offer synergistic advantages. With a SPECT/MRI system, simultaneous dynamic imaging of both structure and function would be feasible, even using multiple tracers simultaneously, providing valuable and accurate information with regard to disease staging and therapeutic outcome. However, all the works that are currently being executed to develop hybrid SPECT/MRI agents are applied in their preclinical stage of development.

Recently, there has been great interest in evolving nanoparticles with dual modality SPECT/MRI properties. Nanoparticles represent very attractive candidates for hybrid imaging due to their unique size and physical properties, allowing visualization of biological events at subcellular levels.

Functional process of nanoparticles increases the affinity of them to the biological target through a phenomenon known as "multivalency", allowing for targeted imaging of the disease. Furthermore, their large surface area permits multiple *marking* of imaging agents (Bouziotis P. and Fiorini C., 2014).

A major limitation in the development of simultaneous SPECT/MRI systems is the compatibility of detectors and electronics with MRI operation. In SPECT/MRI developments, arrays of pixelated CdTe and CdZnTe (CZT) γ-detectors are employed as they offer the advantage of a high-energy resolution. Furthermore, the simultaneous use of multiple radionuclides which emit γ-rays at different energies can be enabled.

These detectors are composed of a large number of pixels which require a large number of electronic readout channels, executed by the use of application-specific integrated circuits (ASICs). Such detectors have shown that a shift of the signal charge inside the detector, caused by Lorentz forces of the MRI, affects performance and requires corrections.

MRI compatible SPECT systems can also be based on scintillators, read out by *silicon photomultiplier (SiPM)* photo-detectors. This approach offers the advantage of using SiPM technology, which is intrinsically compatible with MRI; but the energy resolution offered by scintillation detectors is poorer than that achievable using solid-state CdTe and CZT detectors, which makes multiple radionuclide acquisition more challenging.

Preclinical systems have been developed, but there are several obstacles in developing a clinical model including the need for the system to be compact and stationary with MRI-compatible components.

There is still much to do to reach a stage of demonstrating robustness of SPECT/MRI and evaluating its clinical utility. Whether solid-state detectors or SiPM readout systems will become the design of choice remains to be seen. Extension of design ideas to permit whole-body acquisition may require a larger bore than typical of current MRI systems. Wide bore systems are clinically appealing to improve patients' comfort; so, the MRI system development may be dictated by clinical demands.

Clinical performance similar to that available on conventional SPECT systems should be possible with relatively compact detector/collimator combinations, but further innovation may be needed when the Field of View (FOV) is enlarged to cover the whole body (Hutton B. F. et al., 2018).

The dual radionuclide capability has a particular appeal. The development of appropriate technology remains challenging, but ultimately may lead to more general superior SPECT performance; although the clinical need for a simultaneous SPECT/MRI acquisition remains to be demonstrated.

1.2.5.7.1 MR-Based Attenuation Correction: PET/MRI versus SPECT/MRI

In hybrid PET/MRI systems, attenuation correction (AC) algorithms based in MRI data are used. As clinical SPECT/MRI systems are under dynamic growth, it is important to ascertain whether MR-based AC algorithms validated for PET can be applied to SPECT. Two imaging experiments (Marshall H. R. et al., 2011) were performed, one by an anthropomorphic chest phantom and one by two groups of canines to study and compare the PET/MRI and SPECT/MRI results (Figure 1.22).

The quality of the Nuclear Medicine reconstructions using MRI-based attenuation maps was compared between PET and SPECT. The sensitivity of these reconstructions to variations in the attenuation map was ascertained. It was found that the SPECT reconstructions were of higher fidelity than the PET reconstructions and they were less sensitive to changes to the MRI-based attenuation map. Thus, it is expected that MRI-based AC algorithms designed for PET/MRI to reveal better performance and quantification when used for SPECT/MRI.

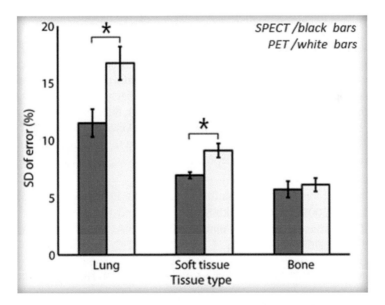

FIGURE 1.22 Results of the sensitivity analysis per tissue type; the black bars correspond to SPECT and the white bars to PET. The asterisks denote statistically significant differences. The error bars represent the standard deviation (SD) of the sample. (From Marshall H. R. et al., 2011.)

1.3 QUALITY ASSURANCE (QA) OF IMAGING SYSTEMS BASED ON INTERNATIONAL STANDARDS

Many guidelines of international organizations and professional bodies for Quality Assurance (QA) and Quality Control (QC) of the instrumentation in Nuclear Medicine are available. Guidelines in Nuclear Medicine continue to be developed for modern systems. World organizations generally coordinate the development of guidelines, but they recognize the need for national guidelines.

The purpose of the guidelines is to offer to Nuclear Medicine team a framework, to be helpful in daily practice.

QA program in Nuclear Medicine necessitates the use of QA standards, protocols, and phantoms that have been set by the international organizations. Many world organizations, as International Atomic Energy Agency (IAEA), have published relative guidelines; some of them include:

AAPM: American Association of Physicists in Medicine
DIN: German Institute of Standardization (Deutsches Institut fur Normung)
EANM: European Association of Nuclear Medicine
EFOMP: European Federation of Medical Physics
IAEA: International Atomic Energy Agency
ICRU: International Commission of Radiation Units and Measurements
IEC: International Electrotechnical Commission
NEMA: National Electrical Manufacturers Association
SNMMI: Society of Nuclear Medicine and Molecular Imaging; and many others
            offer publications, guidelines, standards, and reports

IAEA cooperates with the World Health Organization (WHO) and the Pan American Health Organization (PAHO) among others, in the area of radiation medicine and of radiological safety, security, and emergencies. The arrangement covers collaboration in quality and safety in radiation medicine (radiotherapy, diagnostic radiology, and Nuclear Medicine); radiological safety, security, and emergencies; information sharing; and medical physics. Joint activities focus particularly on Latin America and the Caribbean.

In IAEA Human Health series, guidelines relative to Nuclear Medicine Quality Assurance of instrumentation are deposited as:

  i. Guidelines concerning the Quality Assurance for PET and PET/CT Systems in IAEA (2009) Human Health Series, No. 1; and
 ii. PET/CT Atlas on Quality Control and Image Artefacts in IAEA (2014) Human Health Series No. 27.
     As well as,
iii. Quality Assurance for SPECT Systems, IAEA (2009) Human Health Series No. 6; and
 iv. SPECT/CT Atlas of Quality Control and Image Artefacts, IAEA (2019) Human Health Series No. 36, are addressed.
     • **EANM** networks with other associations, including EFOMP, IAEA, SNM, World Federation of Nuclear Medicine and Biology (WFNMB;

www.wfnmb.org), and WHO for Quality harmony of Nuclear Medicine in Europe. EANM has established guidelines for instrumentation QA taking into account the recent technological advances, the education and training requirements (EANM, 2010b).

- Acceptance Testing for Nuclear Medicine Instrumentation (EANM, 2010a).
- Routine Quality Control Recommendations for Nuclear Medicine Instrumentation (EANM, 2010b).
- Curriculum for education and training of Medical Physicists in Nuclear Medicine. EANM and EFOMP recommendations (EANM, 2013).

NEMA, National Electrical Manufacturers Association, (www.nema.org) is the "Association of Electrical Equipment and Medical Imaging Manufacturers" and has published many standards, as:

- NEMA NU1-2018. This NEMA Standard provides a uniform criterion for the measurement and reporting of γ-camera performance parameters for single and multiple crystal cameras and tomographic devices that image a section or reconstruction image volume, or both.
- NEMA NU2-2018 Performance Measurements of Positron Emission Tomographs (PET). This Standard provides a uniform and consistent method for measuring and reporting performance parameters of PET systems. NU2 makes it possible to objectively compare PET systems in purchase decision process; after delivery, tests are defined to ensure that a quality system is included.

MITA, "Medical Imaging and Technology Alliance" is a division of NEMA. The MITA is the collective voice of medical imaging equipment manufacturers, radio-pharmaceuticals, innovators, and product developers. Division members interact with health-care professionals to establish standards, reduce regulatory barriers and for consultation of the medical imaging industry.

DICOM, "Digital Imaging and Communications in Medicine" is the international standard to transmit, store, retrieve, print, process, and display medical imaging information. It makes medical imaging information interoperable.

DICOM makes medical imaging information interoperable, integrates image-acquisition devices, PACS, workstations from different manufacturers and is actively developed and maintained to meet the progressing technologies and needs of medical imaging, https://www.dicomstandard.org/.

In the *United States*, many guidelines are available by the Society of Nuclear Medicine and Molecular Imaging (**SNMMI**) and the American Society of Nuclear Cardiology (**ASNC**).

The ASNC at June 21, 2018, has published an update to its guidelines for SPECT, with a focus on advances in myocardial perfusion imaging (MPI). https://www.dicardiology.com/channel/spect-imaging/. The new guideline features updates on new hardware, collimators, and *cadmium zinc telluride (CZT)* detectors. This guideline earned the endorsement of the *Society of Nuclear Medicine and Molecular Imaging (SNMMI)*.

In *Australia*, courses or scientific meetings are prepared by the Australian/New Zealand Standards (**ANZSNM**) about imaging systems and radiation detection, physics, nuclear medicine technology, radiopharmaceutical sciences, radionuclide therapy, https://www.anzsnm.org.au.

In the *European Union (EU)*, the **European Commission** (EC), provides guidelines and reports, addresses to Member States, the needs of medical procedures applying ionizing radiation in today's growing use of higher-dose medical equipment as hybrid PET/CT or SPECT/CT imaging systems.

In 2009, European Commission by No-159 "guidelines on clinical audit for medical radiological practices (Diagnostic Radiology, Nuclear Medicine and Radiotherapy)" provides advice and detailed guidance to responsible professionals in Member States, on the implementation of part of the MED Directive (Council Directive 97/43/Euratom, 1997), https://ec.europa.eu/energy/sites/ener/files/documents/159.pdf.

No-175 guidelines of the EC, published in 2014, refer to radiation protection education: "Guidelines on radiation protection education and training of medical professionals in the European Union", https://ec.europa.eu/energy/sites/ener/files/documents/175.pdf.

ICRP, the *International Commission on Radiological Protection*, (www.icrp.org) has developed the System of Radiological Protection which is the basis of all standards, regulations, and practice of radiological protection worldwide.

- ICRP Publication 128, 2015 is referred to "Radiation Dose to Patients from radiopharmaceuticals: A Compendium of Current Information Related to Frequently Used Substances" (Ann. ICRP 44[2S]). The data presented in this report are intended for diagnostic Nuclear Medicine and not for therapeutic applications.
- ICRP Publication 140, 2019. Radiological protection in therapy with radiopharmaceuticals. Ann. ICRP 48(1).

   The recommendations and guidance of ICRP are all referred in the commission's journal, the *"Annals of the ICRP"*. Annals of the ICRP are the authoritative source of recommendations and guidance of the International Commission on Radiological Protection (ICRP).

The principal objective of **ICRU**, the International Commission of Radiation Units and measurements, is the development of internationally accepted recommendations for:

1. Quantities and units of radiation and radioactivity;
2. Procedures suitable for the measurement and application of these quantities in diagnostic radiology, radiation therapy, radiation biology, nuclear medicine, radiation protection, and industrial and environmental activities; and
3. Physical data needed in the application of these procedures, the use of which assures uniformity in reporting.

NCRP, the National Council on Radiation Protection and Measurements, has prepared Report No. 176 for "Radiation safety aspects of nanotechnology" in 2017.

This report is intended primarily for operational health physicists, radiation safety officers, and internal dosimetrists responsible to establish and use radiation safety programs involving radioactive nanomaterials. It describes the current knowledge relating to nanotechnology that is relevant to radiation safety programs.

It considers operational health physics practices that may be modified when nano-technology is involved and includes specific guidance for conducting internal dosimetry programs when nanomaterials are being handled.

AAPM, the American Association of Physicists in Medicine. Medical Physicists are concerned with three areas of activity: Clinical Service and Consultation, Research and Development, and Teaching. On the average, their time is distributed equally among these three areas as it is stated in the "Definition of a Qualified Medical Physicist Standard", 2019, https://www.aapm.org/medical_physicist/fields.asp#nuclear.

All guidelines are referred to *"procedures of good quality without needless details"*, that is *"good practice through minimum requirements"* and their target has to be:

- The skeleton of a local protocol
- Adapted in changes without loss of quality
- Easy to handle results in communication between different centers
- Good standards for training programs

The purpose is to provide Nuclear Medicine practitioners (physicists, technologists, physicians) with guidelines in creating QC protocols of all instruments of a Nuclear Medicine Department (counting devices, imaging systems) with a special focus on new, sophisticated, digital technologies.

In order to be the "skeleton" of a local protocol, guidelines would be adapted in changes without loss of quality.

Imaging guidelines must be established, in each country where Nuclear Medicine applications are used, taking under consideration the recommendations of the international standards as a guide. Reports of the assessment of equipment performance, typical patient dose, and the Image Quality compared to the standards should be checked locally in time intervals. A suggestion of the frequency that checks have to be done after installation is annually, monthly, at maintenance, following a service of detectors or electronics and daily tests.

Every Nuclear Medicine department ought to organize its own program according to its needs and workload. It must be efficient keeping the instrumentation in good condition preventing any deterioration. Logbooks are suggested to be deposited in every Nuclear Medicine department, as proper record keeping detection of gradual deterioration of performance over a long time interval. Analysis of the results for degradation and initiating the necessary corrective action is the way for *good practice*.

The QA program must also be harmonized with the world legislation against the dangers arising from Ionizing Radiation.

General reference for Quality Control (QC) procedures are covered by following summary schedule:

1. **Acceptance quality control:** To ensure that the performance of the system meets the technical and performance specifications quoted by the manufacturer.

2. **Reference quality control:** At the time of system installation and repeated annually regularly, or even after major failure of any part of the system, to give a new set of reference data.
3. **Routine quality control:** Should be carried out regularly, to ensure optimum performance of the system or detect any deterioration in a daily base.

1.4   THE QUALITY OF SCIENTIFIC AND CLINICAL OUTPUT

Quality of clinical output data can refer to the accuracy and precision that these data are obtained and the criteria applied to obtain them. The optimization of results extraction in Nuclear Medicine departments is dependent on the use of:

* optimally performing equipment; and
* successful clinical image processing.

Information Technology now plays a crucial role in Nuclear Imaging. During the last decade, Nuclear Medicine Technology has evolved not only to hardware systems developments, but also offers advanced informatics, quality, and quantification software.

Nuclear Medicine imaging has become increasingly complex, and management of image information requires common concepts, terminology, and measurement methodology. This is essential for the benefit of the patient to ensure maximum diagnostic information with minimum potential risk.

ICRU in 1996 prepared Report No. 54 on subject related to medical imaging; "Medical Imaging – Assessment of Image Quality".

A theoretical framework on image quality and evaluation of medical imaging systems, including Nuclear Medical Imaging, is provided now in *Current Activities of ICRU, 2020.*

Easy to use, fast, and precise data from imaging is a necessity. Access to more accurate data drives to improved management of Nuclear Medicine departments and customized patient treatment.

Software to enhance image quality or automate quantification and measurements is an assistance tool in Medical Physicists' hands, to monitor and possibly reduce tracer dose, offer analytics on procedures, imaging statistics, and reliable data of equipment performance. Dedicated software in high-level analytical information is a contributor to the Quality Management (QM) of the department (Fornell D., 2018).

1.4.1   DATA QUALITY

Data are of high quality if they correctly represent the real world to which they refer. Figure 1.23 refers to two examples of how to measure data or reflect data distribution.

Imaging system assessment depends on the task for which the system is intended, so there is:

* A need to understand not only *the physics of the equipment* but also the way that acquisition data affects results.

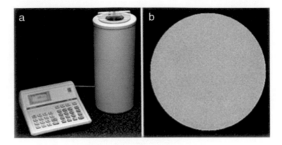

FIGURE 1.23 (a) Accurate measurements in a dose calibrator or (b) an homogeneous representation of a uniform object (homogeneous flood source of Co-57).

- Crucial to deep knowledge, of the clinical impact of equipment performance and acquisition methods.
- Essential to ensure that data is acquired correctly.

There is no point in optimizing the resolution of the camera if the user does not acquire clinical data in such a way as to maximize resolution (e.g., distance effects).

The ICRU Report No. 54, Medical Imaging, The Assessment of Image Quality, 1996, proposes a framework, based on statistical decision theory, within which imaging system performance may be measured, optimized, and compared. It is proposed that the analysis based on the acquired data could provide an assessment of the potential performance of the imaging system and may link the purely objective measures of device performance to the subjective assessment of image quality. The analysis of the acquired data has the advantage of allowing one to investigate the effect on performance by altering various parameters of the imaging system.

1.4.2 DATA RESOLUTION – NOISE DATA

Data resolution should match, at least, the imaging system resolution. When the data are digitized, the continuous coordinates become discrete coordinates. The spatial and temporal resolution will be lost unless the digitation is much finer than the system resolution. Imaging devices introduce blurring in the data of the image.

Data density is low and it is based on random events (disintegrations), so the image is subject to random noise. Random noise in the frequency domain gives a final image equal with real data plus noise data.

Image noise is the degree of variation of pixel values caused by the statistical nature of radioactive decay and detection processes. Even if we acquire an image of a uniform (flat) source on an ideal γ-camera with perfect uniformity and efficiency, the number of counts detected in all pixels of the image will not be the same (Figure 1.24) (Moore S., 1997).

The counting noise in Nuclear Medicine is Poisson noise, so that the pixel noise variance is equal to the mean number of counts expected in a given region of the image. Percentage image noise is expressed by the equation:

$$\% \; image \; noise = (\sigma/N) \times 100\% \qquad (1.1)$$

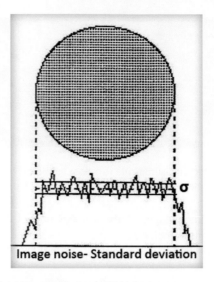

FIGURE 1.24 Image noise impression (standard deviation of noise = square-root of average pixel count N). (Courtesy of Moore S. C., 1997.)

where *standard deviation* σ is the square-root of the *variance*. N is the *average pixel count*.

Variance is the expectation of the squared deviation of a random variable from its mean. Informally, it measures how far a set of (random) numbers are spread out from their average value. The variance is the square σ^2 of the *standard deviation,* σ.

The noise variance is the mean number of counts expected on a certain part of the image. It is desirable to obtain as many counts as possible to reduce the percentage of image noise. Main limitations are:

- the selected radiopharmaceutical activity to be injected; and
- the image acquisition time.

Nuclear Medicine imaging systems distort and suppress real data more than do with noise. The deformation that the detector introduces in the "image" of an "object" is described by the *Point Spread Function (PSF)* and results into blurring. Blurring is caused by the relative suppression of the high-frequency components of the image in the frequency domain. To find the shape of an "object" must deconvolute the "blurring image" from the "image" of the object.

If *Resolution* could be better, *Noise* should be less than a problem. The effect of the resolution loss depends on the processing procedure that is performed.

- **Image processing** endeavors for the *"impossible"* target of a noise-free imaging with improved resolution but:

- **Noise improvement** requires some averaging procedures which degrade resolution and any attempt to improve resolution involves amplification of high-frequency components which make the image noisier.

Nuclear Medicine images illustrate the problem of image processing due to their relatively poor resolution and noise appearance. For example, enhancing the contribution from the high-frequency components of the image selectively, the contribution of noise into the image is enhanced more.

By image processing we try to correct image parameters as exposure, contrast, or low-count statistics, as well as we may also correct artefacts. However, Nuclear Medicine images, often, suffer from low contrast; that is further degraded by the noise introduced in the process of imaging (Lyra M. and Ploussi A., 2011). The image optimization tools are:

- **Filtering** which is aimed at restoring the image blurred because of the detector characteristics; and
- **Smoothing** (spatial and temporal) in order to reduce the effect of random noise.

1.4.3   ATTENUATION CORRECTION

The attenuation correction is dependent on a correct pixel size and attenuation factor, μ.

A cylindrical phantom, 20 cm in diameter, containing a homogeneous solution of Technetium-99m (Tc-99m) can be used to acquire a high-count dataset. Attenuation correction was applied using different linear attenuation coefficients. A profile was drawn across one of the transverse slices (IAEA, Quality Control Atlas for Scintillation camera Systems, 2003).

Figure 1.25 shows that, for the attenuation correction using $\mu = 0.11$ cm^{-1}, the profile through the slice is essentially flat, apart from statistical fluctuations indicating that the attenuation correction software is correct.

The attenuation correction determined by a homogeneous water-filled phantom is not correct when applied to a nonhomogeneous patient. So, attenuation correction in clinical imaging ought to be used carefully.

IAEA encourages and assists research development and practical use of atomic energy and its applications for peaceful purposes throughout the world. It brings together research institutions to collaborate on research projects of common interest, so-called Coordinated Research Projects (CRPs). An IAEA Co-ordinated Project, *"Development and Validation of an Internet-based clinical and technical study communication system for Nuclear Medicine"* https://www.iaea.org/projects/crp/e11013, was referred to software for imaging data communicating through the internet. The participating countries included: Argentina, Austria, China, Cuba, Greece, India, Thailand, United Kingdom of Great Britain and Northern Ireland, and United States of America. A part of it (*multimedia training package on planar γ-camera and QC practice*) is shown at http://www.medimaging.gr/cd/pages/intro.html.

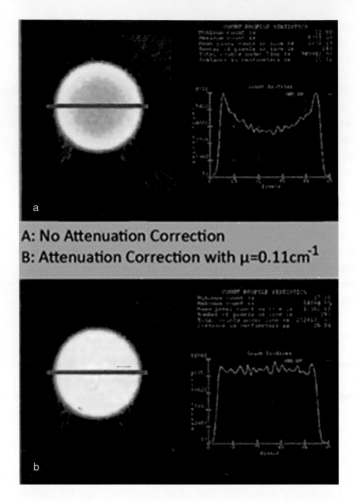

FIGURE 1.25 Profiles to check the accuracy of attenuation correction. (a) No attenuation correction. (b) Attenuation correction with $\mu = 0.11$ cm^{-2}. (Extracted of IAEA, Quality Control Atlas for Scintillation Camera Systems, 2003.)

1.5 ACCURACY AND PRECISION IMPROVEMENT IN QUANTIFICATION OF IMAGING DATA

Anyone who operates instruments will use terminology that refers to metrology. The definition of metrology in encyclopedias and branch dictionaries is a science of measurements covering all aspects concerning theory and practice.

Measuring instrument operators also have contact with metrology, the *scientific and technological metrology.*

One basic activity referring to this type of metrology is the calibration process. Operators are well aware of the importance of calibration results and of working in accordance with the norms of quality systems. Results obtained from test procedures have become more objective since knowledge on measurements and their errors have become more widespread (Costa Monteiro E. et al., 2015).

Accuracy (validity) and **precision** *(reliability, repeatability, or reproducibility)* are two important methodological issues in all fields of researches. When we use a single test in clinical care, for appropriate management of individuals and correct diagnosis in clinical care, the *accuracy* and *precision* of a single test are important.

Good diagnosis and proper management of the individuals in routine clinical care can be achieved using appropriate tests to assess accuracy and precision. Whenever a test or a measuring device is used as part of the data collection process, the accuracy and precision of that test or device are important.

Accuracy refers to the degree in which a test or a measuring device is truly measuring what is designated to measure; consequently accuracy of a measuring system is the degree of closeness of measurements of a quantity to that quantity's true value.

Precision of a measurement system, related to reproducibility and repeatability, is the degree to which repeated measurements under unchanged conditions show the same results; precision assesses the extent to which results agree when obtained by different approaches; that is, various procedures or study instruments or even, by the same approach at different times.

In repeated measurements of the same quantity, the set can be said to be *precise* if the values are close to each other, while the set can be said to be *accurate* if their average is close to the *true value* of the quantity being measured.

The two concepts are independent of each other, so a particular set of data can be said to be either accurate, or precise, or both, or neither. *JCGM 200:2008 International vocabulary of metrology – BIPM Basic and general concepts and associated terms (VIM).*

Accuracy and precision are being assessed by different statistical tests. The field of statistics, where the interpretation of measurements plays a central role, prefers to use the terms *bias* and *variability* instead of *accuracy* and *precision*:

- **Bias** *is the amount of inaccuracy*; and
- **Variability** *is the amount of imprecision.*

A measurement system can be accurate but not precise, precise but not accurate, neither, or both.

For example, if an experiment contains a systematic error, then increasing the sample size generally increases precision but does not improve accuracy. The result would be a steady yet inaccurate string of results. Eliminating the systematic error improves accuracy and does not change precision.

Counting precision (or counting reproducibility) is a measure of the stability of the overall electronics and cable connections; it is also a measure of the statistical error among repeated measurements.

A measurement system is considered **valid** *if it is both* **accurate** *and* **precise**. Ideally a measurement device is both accurate and precise, with measurements all close to and around the true value (Figure 1.26) (Sabour S. et al., 2017).

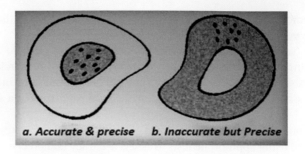

a. Accurate & precise   b. Inaccurate but Precise

FIGURE 1.26   Schematically a set of data that is (a) accurate and precise and (b) a set of data that is inaccurate but precise are shown.

1.5.1 Accuracy and Precision in Nuclear Medicine

The ability to reliable quantify activity in Nuclear Medicine has a number of important applications.

Dosimetry for targeted therapy treatment planning or for approval of new imaging agents requires accurate estimation of the activity in organs, tumors, or voxels at several imaging time points.

Another important application is the use of quantitative metrics derived from images, such as the standard uptake value (SUV) commonly used in PET, to diagnose and follow treatment of tumors. These measures require quantification of organ or tumor functioning and characteristics in Nuclear Medicine images. However, there are a number of physical or technical factors and patient individualities that limit the quantitative precision of Nuclear Medicine images (Frey E. C. et al., 2012).

Factors affecting reliability of image-based activity measurements are:

1. Physical factors affecting radioactivity quantification as: attenuation and scatter, imaging system resolution, partial volume effects, image noise; and
2. Anatomy, physiology, and function of the patient can also affect the image-based measurements.

The earliest recognition for the need of metrological classification of health technologies occurred in the area of radiation dosimetry, since the 1960. In 1976, the *International Atomic Energy Agency (IAEA) together with the World Health Organization (WHO)* established a network of Secondary Standard Dosimetry Laboratories (IAEA/WHO SSDL Network, 1976). The IAEA/WHO SSDL Network assists individual members in carrying out their functions involving the measurement of ionizing radiation, the creation of expertise in applied dosimetry to improve accuracy in radiation dosimetry. The Network conducts dose comparison to assure the unity of the Network standards and supports the operation of a quality program at every level of the measurement chain.

Collaboration with the Network not only helps SSDL members achieve confidence in their measurements but enables international cooperation of the whole measurement system.

1.6 THE NEED FOR MATHEMATICAL METHODS IN NUCLEAR MEDICINE IMAGE EVALUATION

The need of absolute quantification of radionuclide distribution for diagnostic and therapeutic purposes has emerged many efficient mathematical methods to make quantification a clinical reality.

The current state of the art of image quantification provides Medical Physicists and other related scientists with a strong enough background of tools and methods. It describes and analyses the physical effects that degrade image quality affecting the accuracy of quantification; moreover, it proposes methods to compensate for them in planar, SPECT, PET, and hybrid images.

Both hardware and software have made radical changes in the image quantification performance.

Hybrid equipment, by the use of SPECT, PET in connection of *Computerized Tomography (CT) or Magnetic Resonance Imaging (MRI)* systems explores functional and anatomical data to give individualized information for the patient.

The movement to digital imaging detectors for improvement of image quality was the main push of nuclear imaging systems to make advancements in recent years. Developments in the areas of software to improve image reconstruction quality, offer better clinical qualification and analytics data (Dougherty G., 2009).

The introduction of SPECT cameras using digital *cadmium zinc telluride (CZT)* detectors replaced in many systems the *photomultiplier tubes (PMTs)*, decreasing the size of SPECT systems which is crucial in some cases, as in heart imaging by SPECT. The advantage of CZT detectors is the upgraded image quality with direct digital transfer of photons into electrical signals to create images. The digital detectors are more sensitive than PMT-based cameras, so it enables the usage of much less amount of radiopharmaceutical comparing to examinations by conventional PMT-based cameras. The usual dose can be used and the imaging time can be reduced by 50%, known as half-time imaging.

Hybrid PET-CT and SPECT-CT systems use CT for attenuation correction on the nuclear images, while adding CT anatomical image overlaps, it improves the accuracy of diagnosis, e.g., by visualizing the coronary anatomy to better location where blockages are causing perfusion defects (Fornell D., 2018).

Reconstruction techniques, as statistical iterative reconstruction procedures, are enabled to obtain reliable quantitative information from planar, SPECT, PET, and hybrid image and have been advanced to allow for more reliable compensation for physical effects and imaging system restrictions.

Quantitative Imaging, by mathematical and physical ways, is a valuable tool for the Medical Imaging. Its application to clinical problems can be undertaken for establishment of better and more accurate diagnostic and therapeutic well validated protocols (IAEA, Human Health Report, No. 9, 2014). Advanced techniques of image processing and analysis applied to Nuclear Medicine images for diagnosis improve the acquired image qualitatively as well as offer quantitative information data useful in patients' therapy and carefulness. A number of software packages, mostly provided by vendors, are available for tomographic quantification and 3D rendering images of myocardial, brain, kidneys, and other organs.

Mathematical algorithms created by **MATLAB** or **ImageJ** or **IDL**, **CASToR**, **Python**, and other software are used for Nuclear Medicine images evaluation.

ImageJ is a Java-based image processing program developed at the US National Institutes of Health (NIH) and the Laboratory for Optical and Computational Instrumentation, University of Wisconsin. Routine acquisition, analysis, and processing plugins can be developed using ImageJ's built-in editor and a Java compiler. User-written plugins make it possible to solve many image processing and analysis problems; from 3D living cell imaging to medical image processing or data in comparison of multiple imaging systems. ImageJ is a common platform for teaching image processing (Burger W. and Burge M., 2007).

IDL, **I**nteractive **D**ata **L**anguage, is a programming language used for data analysis. It is prevalent in particular scientific areas and in Medical Imaging. IDL appeared, firstly, in 1977 and its last version is IDL 8.7.2 published in 2019. IDL is a reliable scientific programming language used to extract significant visualizations from compound numerical data. IDL is commonly used for interactive processing of large amounts of data as in image processing. The syntax includes many constructs from FORTRAN and some from C, https://en.wikipedia.org/wiki/IDL_(programming_language).

CASToR, **C**ustomizable and **A**dvanced **S**oftware for **T**omographic **R**econstruction. Its last version 2.1, published in 2019. CASToR is an open-source C++ parallel software platform for tomographic image reconstruction. Input PET, SPECT, and CT raw data and related metadata file formats are used by CASToR. For creators, specific advanced manuals about different parts of the platform are available on the CASToR website, http://www.castor-project.org/features (Merlin Th. et al., 2018).

STIR, **S**oftware for **T**omographic **I**mage **R**econstruction 2004–2019, is Open Source software for use in tomographic imaging. Its aim is to provide a Multi-Platform Object-Oriented framework for all data manipulations in tomographic imaging. The emphasis is on image reconstruction in emission tomography PET and SPECT (Thielemans K. and Tsoumpas Ch., 2012).

Python is an open-source interactive language, a general purpose programming language with many potential advantages for image handling and data presentation in Nuclear Medicine. Examples of each of the basic tasks that can be performed are: interacting with a graphical user interface, reading DICOM or Interfile images, assigning regions of interest, image processing functions, time-activity curve construction and display, and saving data and calculated results for later use. Python Software Foundation website is https://www.python.org/psf/.

MATLAB (Matrix-**Lab**oratory) is a programming formal language, which encompasses a set of instructions and produces numerous kinds of output; as programming language, it is used in computer programming to create and apply algorithms. It is written in C/C++. MATLAB is developed by MathWorks, an American privately held corporation that specializes in mathematical computing software. Its major products include **MATLAB** and **Simulink**, which support data analysis and simulation.

MATLAB is designed to operate primarily on whole matrices and arrays. A matrix is a 2D array, often used for linear algebra, which consists of both rows and columns. All MATLAB variables are multidimensional arrays, no matter what type of data. In a matrix, the two dimensions are represented by rows and columns. In the MATLAB

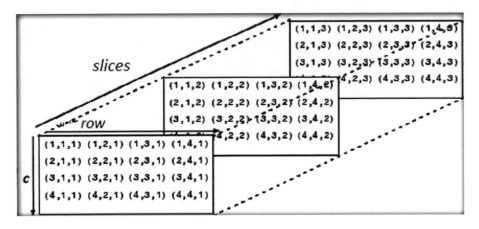

FIGURE 1.27   Multidimensional array/3D array uses three subscripts: Column and row just like a matrix, but the third dimension represents pages or sheets of elements (tomographic slices).

matrix, the rows and columns are created by using commas (,)/line-spaces ( ) and semicolons (;), respectively. The matrix is represented by square brackets "[ ]".

An array having more than two dimensions is a multidimensional array. In MATLAB for the generation of a multidimensional array, firstly a 2D array is created and then extended.

Each element is defined by two subscripts, the row index and the column index. Multidimensional arrays are an extension of 2D matrices and use additional subscripts for indexing. A 3D array uses three subscripts. The first two are just like a matrix, but the third dimension represents *pages* or *sheets* of elements (*tomographic slices*) (Figure 1.27).

MATLAB allows matrix manipulations, plotting of functions and data, execution of algorithms, and interfacing with programs written in other languages.

- **MATLAB Software Synergy in Nuclear Medicine**
  Ovtchinnikov E et al. in April 2020, published their work, in the 50th anniversary issue of Computer Physics Communications, about the **S**ynergistic **I**mage **R**econstruction **F**ramework (**SIRF**) software, intending to show that data acquired by PET/MR systems could obtain accuracy in the tomographic reconstruction via synergy of the two imaging techniques,
- SIRF provides user-friendly Python and MATLAB interfaces built on top of C++ libraries.
- SIRF also uses advanced PET and MR reconstruction software packages **S**oftware for **T**omographic **I**mage **R**econstruction (STIR) for PET; **Gadgetron** software for MR reconstruction, for image registration tool **NiftyReg** is used, which is open-source software for effective medical image registration. The SIRF C++ code provides a thin layer on top of these existing libraries. This C++ layer provides the basis for a simple and in-built Python and MATLAB interface, enabling users to quickly develop

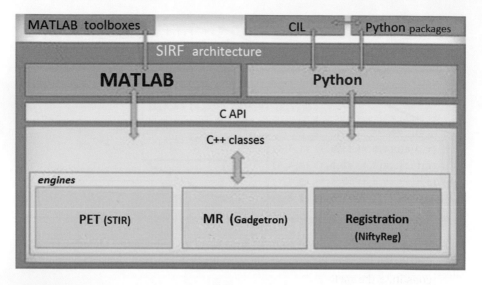

FIGURE 1.28 The architecture of SIRF: interfacing of underlying engines and most of the development is in C++. Interfacing **MATLAB** and **Python** with the underlying C++ code is done via a thin C API. CIL is a Python library.

their reconstruction algorithms using these scripting languages. MATLAB and Python are two languages and environments widely used by researchers, partly because of the excellent prototype abilities of these languages. MATLAB provides flexible, two-way integration with many programming languages, including Python.

A short description of the SIRF approach is presented in Figure 1.28.

- *ImageM: Graphical User Interface (GUI) for Image Processing with MATLAB*

ImageM is an interactive Graphical User Interface (GUI) for image processing. Analysis and visualization with MATLAB similar to **ImageJ,** the open-source software for image analysis is written in java.

ImageM integrates into a GUI several algorithms for interactive image processing and analysis. It allows to load images, apply filtering and/or segmentation, and run basic analyses. Interface is largely inspired from the open source software ImageJ. The interface is packaged as a MATLAB application. ImageM relies on MATLAB Image Processing Toolbox and on the GUI Layout Toolbox. Most algorithms are wrapped into a specific image class that allows to process in a unified way grayscale, binary, and color images, 2D or 3D, and that can be used independently.

ImageM integrates into a GUI several algorithms for interactive image processing and analysis. It allows to load images, apply filtering and/or segmentation, and run basic analyses. The interface is packaged as a MATLAB application (Legland D., 2020, **ImageM**). The development of ImageM started as a proof of concept for creating Graphical User Interface(GUI) dedicated to image processing.

1.6.1   MIRD Schema – MIRDOSE – OLINDA/EXM – RADAR Software

1.6.1.1   MIRD Schema

In 1948, Marinelli et al. published three articles where radionuclide dosimetry of that time was presented and was widely accepted. These publications marked the beginning of a new radiation dosimetry in Nuclear Medicine.

The equations for internal dosimetry can be formulated in general terms, independent of the characteristics of particular radiations. In 1968, Loevinger and Berman were participated into the newly formed **M**edical **I**nternal **R**adiation **D**ose (**MIRD**) Committee, and the original MIRD Schema was published as MIRD Pamphlet No. 1.

The MIRD Schema is a general approach for medical internal radiation dosimetry. Although the schema has mostly been used for organ dosimetry, it is also applicable to dosimetry at the sub-organ, voxel, multicellular, and cellular levels.

- In the MIRD Perspective 1999 commentary, the MIRD Committee reviews the history of the MIRD Schema, and presents the key equations that constitute the method. It clarifies misinterpretations on limitations of the MIRD Schema in its use for nonuniform distributions of radioactivity and spatial measures ranging from organ, sub-organ, multicellular, and cellular. It describes the importance of connecting MIRD absorbed dose calculations with observed biological effects; it sets the connection of the MIRD Schema and radiobiology (MIRD Perspective, 1999).

The MIRD Schema was formulated by the *Society of Nuclear Medicine and Molecular Imaging (SNMMI)* committee on **M**edical **I**nternal **R**adiation **D**ose to facilitate the calculation of radiation absorbed dose from distributed sources of radioactivity.

In the MIRD pamphlets following the first issue, new tools for radionuclide dosimetry applications were provided, including the dynamic bladder model, S values for small structures within the brain as sub-organ dosimetry, voxel S values for constructing 3D dose distributions, Dose-Volume Histograms (DVHs) and techniques for acquiring quantitative distribution and pharmacokinetic data (Zanzonico P. B., 2000; MIRD Bolch et al., 1999).

The concept of Dose-Volume Histograms (DVHs) can be used to display the nonuniformity in the absorbed dose distribution from radionuclide procedures. The mean absorbed dose in internal dosimetry may be a reduced representation of the absorbed dose to the tissue.

A *differential DVH* shows the fraction of the volume that has received a certain absorbed dose as a function of the absorbed dose. DVHs might be used to assist the correlation between absorbed dose and biological effect (Figure 1.29).

Internal dosimetry deals with the determination of the amount and the spatial and temporal distribution of radiation energy deposited in tissue by radionuclides within the body.

In Nuclear Medicine model, average organ dose evaluations are derived for risk estimation. But, specific patients deviate kinetically and anatomically from the model used and such dose estimates will be unreliable.

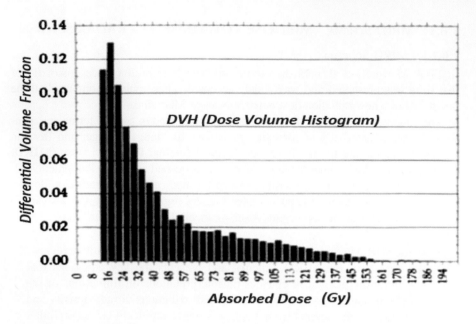

FIGURE 1.29 Cumulative Dose-Volume Histogram. The cumulative DVH graphs the volume (in the y-axis) that receives the corresponding absorbed dose in the x-axis. For example, here it is indicated which is the absorbed dose that 40% of the volume receives.

Increasing therapeutic application of internal radionuclides and need for greater accuracy, radiation dosimetry in Nuclear Medicine is turning from population- and organ-average to *patient- and patient/organ-specific dose* estimation.

In internal radionuclide therapy, a growing interest in voxel-level estimates of tissue-absorbed dose (MIRD Dewaraja et al. Pamphlet No. 23, 2012, Quantitative SPECT for Patient-specific 3-Dimensional Dosimetry in Internal Radionuclide Therapy) has been driven by the desire to report radiobiological quantities that account for the biologic consequences of both spatial and temporal nonuniformities in these dose estimates.

A synopsis of 3D SPECT methods and requirements for internal dosimetry at both regional and voxel levels are expressed in MIRD Pamphlet No. 23, which is the introduction to an upcoming series of MIRD pamphlets with detailed radionuclide-specific recommendations intended to provide best-practice SPECT quantification-based guidance for radionuclide dosimetry. SPECT imaging and reconstruction techniques for quantification in radionuclide therapy are not necessarily the same as those designed to optimize diagnostic imaging quality.

MIRD Pamphlet No. 23 presents a general overview of methods that are required for quantitative SPECT imaging.

The accuracy of absorbed dose calculations in personalized internal radionuclide therapy is directly related to the precise estimation of the activity obtained at each of the imaging time points. If personalized plans based on patient-specific dosimetry calculations are followed in targeted radionuclide therapies, the obtained result could be closer to the desirable. Though, these calculations require exact information of the biodistribution of the radioactive therapeutic agent in the patient body.

MIRD Pamphlet No. 26 follows the general information that was provided in MIRD-23. It focuses on Lu-177 (Lutetium) and its application in radiopharmaceutical therapy.

The medium-energy β-emissions of Lu-177 make it useful for many radionuclide therapy applications; more, its γ-emissions are suitable for quantitative imaging necessary for dosimetry calculations. MIRD Pamphlet No. 26 describes methods that are specifically recommended for use in Lu-177 quantitative studies.

Software tools of the MIRD Committee are:

- **Dynamic Bladder Software Tool** is an interactive software application that implements the dynamic bladder model described in MIRD Pamphlet No. 14, Revised (MIRD Thomas et al., 1999). The Dynamic Bladder Software is intended for use by the Nuclear Medicine research community for the purpose of evaluating radiation dose to the urinary bladder wall surface as a function of variable parameters.

  The software was designed as an educational and research tool providing investigators with the capability to compare the effect of varying protocols on bladder wall dose. It is not intended that the output of this software be used in the clinical management of patients.

- **MIRDcell V2.1** is an interactive software application described in MIRD Pamphlet No. 25, 2014 that performs radiation absorbed dose and response modeling for cells and numerous geometries of multicellular clusters.

  MIRD Committee has published 26 pamphlets, 20 reports, 4 commentaries, and 6 MIRD books relative to the action of radioisotope to live beings. The last, up to now, book published by MIRD Committee is the monograph:

- **MIRD Radiobiology and Dosimetry for Radiopharmaceutical Therapy with Alpha-Particle Emitters** reviews studies related to targeted α-particle-emitter therapy and provides guidance and recommendations for human dosimetry. Because of the nature of α-particle radiation, this new treatment modality is fundamentally different from existing therapies. This monograph provides guidance and recommendations for human dosimetry.

1.6.1.2 MIRDOSE

The main function of the MIRDOSE software is to provide estimates of the radiation dose per unit administered activity from user entered source organ residence times for a given radionuclide and one or more phantoms.

It uses libraries of radionuclide decay data and specific absorbed fractions to develop S-values for the source organs chosen by the user and the target organs desired. All source organ contributions to total dose may be viewed if desired. All model input and assumptions are given with the program output. The program can also provide tables of S-values for all source and target organs for a given phantom or phantoms.

Concerning the MIRDOSE software, written by Dr. Stabin G. M. (1996), in 8 years of use by the international community, no calculation bugs have been marked, although several inquiries regarding possible problems were existed.

The program was, and is, widely cited in the published literature as providing the basis for dose estimates for many radiopharmaceuticals in theoretical dose estimate tables. MIRDOSE 3.0 and 3.1 are predecessors of the OLINDA code developed later also by Dr. Stabin.

1.6.1.3   OLINDA (Organ Level INternal Dose Assessment)

The OLINDA code was written by Dr. Stabin. It is based on the MIRDOSE 3.0 and 3.1 codes developed earlier by Dr. Stabin too.

It is really an update on MIRDOSE 3.1; OLINDA employs essentially the same calculation algorithms as MIRDOSE, with some updates that include:

1. Addition of some new organ models (head/brain, prostate, peritoneal cavity).
2. Update of decay data (inclusion of >800 nuclides, including alpha emitters).
3. Inclusion of some alpha emitters, so alpha dose models for hollow organs and bone/marrow were applied.
4. Ability to fit retention and excretion data.
5. New organ phantoms permit users to modify organ masses and radiation quality factors (weighting factors).
6. The ability to modify organ masses to patient-specific values.

1.6.1.3.1   OLINDA/EXM (Organ Level INternal Dose Assessment/EXponential Modeling)

The OLINDA/EXM personal computer code performs dose calculations and kinetic modeling for radiopharmaceuticals.

- *OLINDA*, the **O**rgan **L**evel **IN**ternal **D**ose **A**ssessment code, is an update and replacement of the *MIRDOSE* personal computer software.
- *OLINDA* calculates radiation doses to different organs of the body from systemically administered radiopharmaceuticals and performs regression analysis on user-supplied biokinetic data to support such calculations for Nuclear Medicine drugs.
- *OLINDA* performs internal dose calculations, principally for radiopharmaceuticals, using the *RADAR* method of dose calculations and *RADAR* dose factors.
- The **EXM** module (**EX**ponential **M**odeling) fits kinetic data to sums of exponential functions.

1.6.1.4   RADAR (RAdiation Dose Assessment Resource)

RADAR, the **RA**diation **D**ose **A**ssessment **R**esource, is a working group that maintains resources for internal and external dose calculations, given on a website (www.doseinfo-radar.com) and in a number of open literature publications.

RADAR has updated the dosimetry world by replacing the old, *stylized* dosimetry phantoms of the 1980s and 1990s with image-based, *voxelized* models.

The new phantoms were used with various biokinetic models and the OLINDA/EXM2.0 software code to produce two collections in the RADAR Dosimetry

Compendia; one for adults and pediatric subjects, and another for women at different stages of pregnancy.

RADAR website dose factors are applied by OLINDA in a code that permits users to enter kinetic data for radiopharmaceuticals or fit them from time-activity data.

- *Overview of the RADAR System*
- *Internal Sources:* Kinetic Models – Radiopharmaceuticals, Radionuclide Decay Data, Specific Absorbed Fractions – Phantoms, Kinetic Models – Occupational Radionuclides.
- *External Sources:* External Source Configuration, Dose Conversion Factors, Radiation Dose Estimates and Distributions, Linking of Radiation Doses to Biological Effects.
- *RADAR Software*
  Software codes relative to dosimetry in Nuclear Medicine are included in RADAR website.

MIRDOSE AND OLINDA/EXM (NUCLEAR MEDICINE INTERNAL DOSE)

The Visual Monte Carlo Program (External dose Monte Carlo simulator)
- The VARSKIN code (Skin dose)
- Dynamic Bladder Model code (MIRD Pamphlet No. 14)
- Virtual Cell Radiobiology Software

1.6.2 COMPUTATIONAL PHANTOMS

Phantoms are physical or virtual representations of the human body to be used for the determination of absorbed dose to radiosensitive organs and tissues. In many situations, the most useful information is gained from computer simulations where a more or less realistic anthropomorphic distribution is generated and used to either estimate radiation dose or to model the imaging process for estimating reconstruction algorithms or other image processing techniques.

A computational anatomic phantom is a computerized model of human anatomy for use in radiation transport simulations of a medical imaging or a radiation therapy procedure.

Since the 1960s, the radiological science society has developed and applied these models for ionizing radiation dosimetry studies. Computational models have become increasingly accurate with respect to the internal structure of the human body.

In Nuclear Medicine, detailed 3D patient anatomy comes through the CT portion of SPECT/CT or PET/CT imaging systems. Turning from *stylized* phantoms based on simple quadratic equations to new *voxelized* phantoms based on *computed tomography (CT) and magnetic resonance (MR)* medical images of the human body was a significant change. The newest models are based on more advanced mathematics, such as Non-Uniform Rational B-Spline (NURBS) or polygon meshes, which allow for 4D phantoms where simulations can take concern not only of 3D space but of time as well.

Computational phantoms have been developed for a wide variety of humans, from children to adolescents and adults, male and female, as well as pregnant women. By this variety of phantoms, many kinds of simulations can be used in medical imaging procedures. Over the years, the results of these simulations have created several standards. *Phantoms' development was parallel with computing progress.*

Morphometric categories of the computational phantom are:

a. Reference/patient matching by age only
b. Patient-dependent/harmonized with patient nearest height or weight
c. Patient-shaped/corresponding to average height, weight, and body contour
d. Patient-specific/phantom uniquely alike patient morphometry, which is the target for real representation

Types of Computational Anatomic Phantoms are as follows.

1.6.2.1 Stylized (First-Generation) Computational Phantoms

Flexible but anatomically unrealistic

The first generation computational phantoms were developed to cover the need to better evaluate organ doses from internally deposited radioactive materials in patients and workers. Until the late 1950s, the International Commission on Radiological Protection (ICRP) still used very simple models. In those calculations, each organ of the body was assumed to be represented as a *sphere* with an *effective radius*. The radionuclide of interest was assumed to be located at the center of the sphere and the *effective absorbed energy* was calculated for each organ. Additionally, phantoms such as the Shepp–Logan Phantom (Shepp L. and Logan B. F., 1974) were used as models in the development and testing of image reconstruction algorithms, of a human head.

Researchers attempted to model individual organs of the body and finally the entire human body in a realistic manner. These efforts directed to stylized anthropomorphic phantoms that resemble the human anatomy.

Stylized computational phantom (Figure 1.30) is, generally, a mathematical representation of the human body which, when coupled with a Monte Carlo radiation transport computer code, can be used to track the radiation interactions and energy deposition in the body. By means of radiation transport calculations performed in a mathematical phantom, it is possible to register the energy deposited by the radiation in interactions with the atoms of the body tissues.

This computational phantom is structured by adjusting individual parameters of the mathematical equations, which describe the volume, position, and shape of individual organs. It has a long history of development through years 1960s to 1980s.

Examples of reference-stylized phantoms are those of the Oak Ridge National Laboratory (ORNL) series (Cristy M. and Eckerman K., 1987; Han E. et al., 2006).

In mathematical human phantoms, size and form of the body and its organs are described by mathematical expressions representing combinations and intersections of planes, circular and elliptical cylinders, spheres, cones, etc.

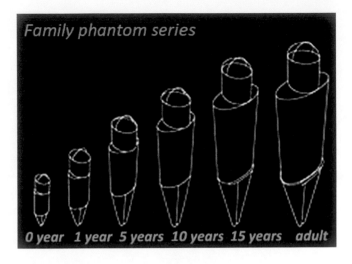

FIGURE 1.30 Family phantom series.

1.6.2.1.1 MIRD Phantom

The MIRD phantom (ICRP, 1975) was developed by Snyder and the group at Oak Ridge National Laboratory (ORNL) in the 1960s with 22 internal organs and more than 100 subregions. It is the first anthropomorphic phantom representing a (man/woman) hermaphrodite adult for internal dosimetry. Body height and weight as well as the organ masses of these MIRD-type phantoms are in accordance with the Reference Man data from 1975 (ICRP, 1975).

1.6.2.1.2 Phantoms Derived from MIRD

Based on MIRD phantom, many derivations of phantoms were developed for future decades. The major types of phantom include: stylized *Family* phantom series (Figures 1.30 and 1.31a) developed in the 1980s by (Cristy M. and Eckerman K., 1987) or the Male (Adam) and Female (Eva) Adult Mathematical Phantoms developed by the National Research Center for Environment and Health (GSF), Germany (Kramer R. et al., 1982).

External views of the age-specific phantom are presented; phantoms representing an adult and children at 1, 5, 10, 15 years, and 0 years of age. When used for an adult female, the 15-year phantom has breasts appropriate for a reference adult female, which are not shown (Cristy M. and Eckerman K., 1987).

Although many efforts were undertaken to extend its applications in radiation protection, radiation therapy, and medical imaging, stylized phantom in-built limitation cannot be overcome. The representation of internal organs in this type of mathematical phantom was basic, by presenting only the most general description of the position and geometry of each organ.

In the late 1980s the powerful computer and tomographic imaging technologies became available and a new era of *voxel* phantoms started (Figure 1.31b).

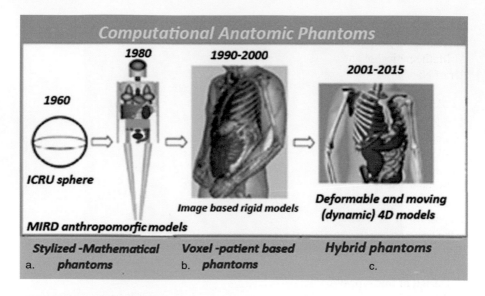

FIGURE 1.31 (a) Stylized mathematical phantoms are flexible but anatomically unrealistic. (b) Voxel phantoms are anatomically realistic but not very flexible. (c) The 21st century dynamic 4D models are realistic and flexible.

1.6.2.2 Voxel (Second-Generation) Phantoms

Anatomically realistic but not very flexible

The stylized phantoms provide only basic information. More accurate methods of simulating a human body were necessary to be set on. The real innovation occurred when *computed tomography (CT) and magnetic resonance imaging (MRI)* devices could generate highly accurate images of internal organs in digital format and in three dimensions. Researchers discovered that they could take that diagnostic data and transform it into a voxel (**vo**lumetric pi**xel**) format, essentially recreating the human body in digital form, 3D (Zaidi H. and Xu X. G., 2007).

Voxel phantoms include certain properties (Figure 1.31b). The raw data are obtained from CT scans, MRI imaging, or direct imaging through photography. The components of the body are identified and the density of each component is defined; the data are unified into a single 3D structure to be used for analysis.

The earliest work on voxelized phantoms occurred independently at about the same time by Dr. Pujol and Dr. Gibbs, 1982, of Vanderbilt University, and by Dr. Zankl and team, 1988, at the National Research Center for Environment and Health (GSF) in Germany.

Dr. Pujol and Dr. Gibbs' work started with X-ray images, for the reconstruction of a human phantom which was used for medical dose simulations (Pujol A. Jr and Gibbs, S. J., 1982).

Dr. M. Zankl and team have used CT imaging to create 12 phantoms, ranging from baby to adult human (Zankl M. et al., 1988).

The ICRP Reference Phantoms (ICRP, 2009), and the MAX and FAX phantoms (Kramer R. et al., 2006) are examples of reference phantoms in a voxelized format.

1.6.2.3   Hybrid (Third-Generation) Phantoms

Realistic and flexible

Mathematical phantoms of increasing complexity have been developed for a number of Nuclear Medicine applications.

1.6.2.3.1   Boundary Representation Phantom (BREP)

BREPs are computational human models that contain exterior and interior anatomical features of a human body using boundary representation method. In the Medical Physics these computational phantoms are mostly used for ionizing radiation dosimetry.

In the development of these human phantoms, the concept of a *deformable* phantom whose geometry can be accessibly transformed to fit particular physical organ shapes, volumes, or body postures is the principal interest (Figure 1.31c). Deformation is the transformation of a body from a *reference* configuration to an *in-progress* configuration and refers to the change in size or shape of an object (Truesdell C. and Noll W., 2004).

Design of this type of phantom is realized by NURBS method or polygonal mesh method, which are jointly called BREP methods. Compared to the voxel phantoms, BREP phantoms are better suited for geometry deformation and adjustment, because a larger set of computerized operations are available. A major advantage of BREP phantoms is their ability to convert into an existing reference phantom or into the anatomy of a real person, which makes individual-specific absorbed dose calculation possible (Na Y. H. et al., 2010).

1.6.2.3.2   NURBS-Based Phantom

Surfaces of an NURBS-based phantom are defined by NURBS equations which are formulated by a set of control points. This feature is useful in designing a time-dependent 4D human body modeling. An example is the **NURBS**-based **CA**rdiac-**T**orso NCAT phantoms, which are used to simulate cardiac and respiratory motions with more realistic modeling of the cardiac system.

1.6.2.3.2.1   NCAT/4D XCAT Phantoms   The 4D NURBS-based CArdiac-Torso (NCAT) phantom was developed to provide a realistic and flexible model of the human anatomy and cardiac and respiratory motions.

With NURBS as a basis, the Mathematical CArdiac-Torso (MCAT) phantom, a simple geometrically based model, progressed to the 4D eXtended CArdiac-Torso (XCAT) whole body, detailed models of the male and female (XCAT) anatomies. The extended 4D NCAT or XCAT phantom has found application in higher resolution imaging modalities (Segars W. P. and Tsui B. M. W., 2009). The 4D XCAT phantom has also been used in conjunction with accurate simulation models of the imaging process, e.g., through the use of Monte Carlo techniques. When these accurate models of the imaging process are applied to the XCAT, it is capable of generating pragmatic multimodality imaging data, as SPECT, PET, MRI, and CT, (Veress A. I. et al., 2011).

The XCAT phantoms are gaining a wide use in biomedical imaging research. They can provide a virtual patient base to quantitatively evaluate and improve imaging instrumentation, data acquisition, image reconstruction techniques, and processing methods which can lead to improved image quality and radiation dosimetry (Segars W. P. et al., 2018).

1.6.2.3.3   Polygonal Mesh-Based Phantom

The polygonal mesh specifies the shape of a polyhedral object in 3D space. The surfaces of the phantom are defined by a large amount of polygonal meshes. The polygonal mesh has three notable advantages in developing whole-body phantoms.

A. Mesh surfaces depicting human anatomy can be conveniently obtained from real patient images or commercial human anatomy mesh models.
B. The polygonal mesh-based phantom has considerable flexibility in adjusting and fine-tuning its geometry, allowing the simulation of very complex anatomies.
C. Many commercial computer aided design (CAD) software provide built-in functions able to rapidly convert polygonal mesh into NURBS (Na Y. H. et al., 2010).

1.6.2.4   Voxelization of Hybrid Phantoms

In dosimetry calculations, stylized phantoms that describe human anatomy through mathematical surface equations and voxel phantoms that use 3D voxel matrices to define body parts have been widely used.

Stylized phantoms are flexible in that changes to organ position and shape are possible, while voxel phantoms are typically fixed to a given patient anatomy; yet can be proportionally scaled to match individuals of larger or smaller stature but of equivalent organ anatomy.

To avoid the weaknesses of stylized and voxel phantoms, hybrid phantoms based on NURBS surfaces are used where anthropomorphic flexibility and anatomic reality are both well-kept.

The University of Florida (UF) series of hybrid phantoms preserve the nonuniform scalability of stylized phantoms while holding the anatomical pragmatism of patient-specific voxel phantoms with respect to organ shape, depth, and inter-organ distance (Kim C. H. et al., 2016, The reference phantoms: voxel vs polygon; Kim C. H. et al., 2018, New mesh-type phantoms and their dosimetric applications, including emergencies).

The voxelization of hybrid phantoms is, though, necessary as most radiation transport codes require a voxelized structure for particle tracking.

1.6.2.4.1   Hybrid-Voxel Phantom

The voxelization process is crucial for the resulting NURBS phantoms to be ported to radiation transport codes as currently there are no Monte Carlo codes available that can directly handle NURBS or polygon-mesh geometry for radiation transport.

A user-defined MATLAB routine (*Voxelizer*™) was written to fill the NURBS/PM structures with cubic voxels of any user-defined size. The resultant anatomic model is termed a *hybrid-voxel* phantom, where the adjective signifies that the consequent voxel structure was derived from NURBS/PM surfaces, and not directly from a segmented medical image. An essential difference between a voxel phantom and a *hybrid-voxel* phantom is that in the hybrid-NURBS/PM format, significant changes can be made to fit into different phantom categories.

In 2007, Lee C. and team used an upgraded in-house MATLAB code, *Voxelizer 4*, where speed-up algorithms and other user-friendly features were applied. All objects in the phantom were saved in one ASCII *Raw Triangles* file[1] and voxelized simultaneously. After voxelization, the skin identifier was assigned to the single outermost voxel layer of each phantom (Lee C. et al., 2007, 2008).

Alterations of the phantoms' outer body contours and reassignment of residual soft tissue densities allow someone to model body profiles and tissue structures that are unique to individual persons or to other weight percentile means in the population (Hurtado J. L. et al., 2012).

1.6.2.5   Reference Phantoms Adopted by the ICRP

- Examples of *reference-stylized* phantoms are those of the Oak Ridge National Laboratory (ORNL) series (Cristy M. and Eckerman K., 1987; Han E. et al., 2006).
- The ICRP Reference Phantoms (ICRP, 2009), and the MAX and FAX phantoms (Kramer R. et al., 2006) are examples of *reference-voxelized* phantoms.
- Two examples of *reference-hybrid* phantoms are those of the University of Florida (UF) series (Lee C. et al., 2010; Hurtado J. L. et al., 2012).

ICRP Publication 89 presents detailed information on age-and gender-related differences in the anatomical and physiological characteristics of reference individuals (ICRP, 2002). These reference values provide needed input to prospective dosimetry calculations for radiation protection purposes for both workers and members of the general public. The purpose of this report is to consolidate and unify in one publication, important new information on reference anatomical and physiological values that has become available since ICRP Publication 23 was published by the ICRP in 1975.

Moving from the past "Reference Man", the new report presents a series of reference values for both male and female subjects of six different ages. In selecting reference values, the Commission has used data on Western Europeans and North Americans because these populations have been well studied with respect to anatomy, body composition, and physiology. When appropriate, comparisons are made between the chosen reference values and data from several Asian populations.

The University of Florida, UF hybrid phantom representing the ICRP 89 reference 15-year-old female was voxelized at 3 mm isotropic resolution using the in-house MATLAB code *Voxelizer 4*. The resulting *hybrid-voxel* phantom was next imported to the radiation transport code MCNPX 2.5 (Pelowitz D. B., 2005).

Photon and electron emission spectra from Tc-99m were then abstracted from an updated monograph published by the MIRD Committee (Eckerman K. F. and Endo A., 2008).

---

[1] Raw Triangles (raw) Import/Export. This file format contains polygon mesh objects that consist of triangular polygon faces only. Raw import supports ASCII format only. ASCII is binary code used by electronic equipment to handle text using the English alphabet, numbers, and other common symbols. ASCII was developed in the 1960s and was based on earlier codes used by telegraph systems.

Two MC simulations were then run; one for photon emissions (X and γ rays) and one for electron emissions (β, internal conversion [IC], and Auger electrons) – each with 10^7 particle histories. Photon and electron components of the S value were then combined to yield total spectral S values (absorbed dose per nuclear transformation in the source tissue).

1.6.2.6 Statistical Phantoms

A computational framework is offered, based on statistical shape modeling, for construction of race-specific organ models for internal radionuclide dosimetry and other Nuclear Medicine applications. The proposed technique used to create the race specific statistical phantom maintains anatomic realism and provides the statistical parameters for application to radionuclide dosimetry (Mofrad F. B. et al., 2010).

Wang et al. in 2018 published their work in the construction of deformable torso phantoms that can be deformed to match the personal anatomy of Chinese adults (male and female). The phantoms were created based on a training set of trunk computed tomography (CT) images from normal Chinese adults. Major torso organs were segmented from the CT images, and the Statistical Shape Model (SSM) approach was used to learn the inter-subject anatomical variations. To match the personal anatomy, the phantoms were registered to individual body surface scans or medical images using the active shape model method. The created *Statistical Shape Model (SSM)* demonstrated anatomical variations in body height, fat quantity, respiratory status, organ geometry, male muscle size, and female breast size. The masses of the deformed phantom organs were consistent with Chinese people organ mass ranges. The performance of personal anatomy modeling was validated by registration of the phantoms to the body surface scan and CT images. These phantoms can serve as computational tools for personalized anatomy modeling for the Chinese population (Wang H. et al., 2018).

Computational phantoms and computer simulation are playing a significant role in the results of Nuclear Medicine research applications. As the computational models get more and more realistic, newer methods and applications including textured organs and dynamic perfusion are expected to expand the utility of the computational phantoms in virtual simulations.

Computer simulation has a great importance today and MATLAB is big tool which can help with Computer simulation. It starts with creation of a mathematical model and the obtained equations are solved by using an appropriate calculation method.

1.6.3   Monte Carlo (MC) Simulation in Nuclear Medicine

Simulation is the technical discipline which shows the behavior and reactions of any system on its model. There are many types of models; the main categories are real models and computer models.

Monte Carlo (MC) simulation methods are especially useful in studying a variety of problems difficult to calculate by experimental or analytical approaches. They are extensively applied to simulate Nuclear Medicine instrumentations such as *single photon emission computed tomography (SPECT) and positron emission tomography*

(PET) for assisting system design and optimizing imaging and processing protocols (Assié K. et al., 2004). Monte Carlo (MC) simulation packages are used as an important tool for the optimal design of detector systems or to improve image quality and acquisition protocols.

Monte Carlo method was first introduced by Snyder at Oak Ridge National Laboratory in order to assess the fraction of photon and electron energy emitted from radionuclides in source tissues, deposited in various target tissues. This was the concept of absorbed fraction defined within the Medical Internal Radiation Dose (MIRD) method of Internal Dosimetry stated later (Snyder W. S. et al., 1969).

The Monte Carlo method is a well-known technique for solving problems involving statistical processes and is very useful in Nuclear Medicine due to the stochastic nature of radiation and radioactivity; it provides unique possibilities to calculate and quantify essential parameters.

Monte Carlo principles and applications in radiation physics and Nuclear Medicine have been given by researchers some decades ago by Raeside D. E.(1976) or Turner J. E. et al. (1985).

There is a necessity of mathematical modeling for the evaluation of important parameters of photon imaging systems. A Monte Carlo program which simulates medical imaging nuclear detectors has been developed by Ljungberg M. and Strand S. E. (1989). Different materials can be chosen for the detector, various source shapes can be simulated. Photoelectric, incoherent, coherent interactions, and pair production are simulated, too. Simulation of different collimators permits studies of spatial resolution and sensitivity. Comparisons of the simulation's results with experimental data had shown worthy agreement. MC simulation has grown into quite flexible and accurate to permit for detailed investigation of a broad range of detector configurations (Staelens S. and Buvat I., 2009).

Monte Carlo simulations in emission tomography are not only used to study specific aspects of the imaging system response or to assess reconstruction and correction techniques, but is also applied to the patient personalized imaging, contributing to the SPECT and PET imaging datasets (Fahey F. H. et al., 2018).

Many *general purpose* MC codes as EGS, MCNP, GATE/Geant4, PENELOPE or *dedicated codes* as SIMIND, SimSET, SimSPECT, PETSIM have been developed aiming to provide accurate and fast results.

1.6.3.1 General Purpose Monte Carlo (MC) Codes

General purpose codes such as EGS4, EGS5/EGSnrc, Geant4, or MCNP/MCNPX/MCNP6 can be used in a variety of nuclear physics applications. They are useful in many areas of Medical Radiation Physics simulating particle tracks and energy depositions of available programs. They are widely used and validated and are covered of continuous support and update. Their limitations may be on specific applications as time management or simulation of low-energy photons.

EGS: the **E**lectron **G**amma **S**hower code is a general purpose package for the Monte Carlo simulation of the coupled transport of electrons and photons in an arbitrary geometry, for particles with energies from a few keV up to several hundreds of GeV.

In 1996, Monte Carlo simulation **E**lectron **G**amma **S**hower (**EGS4**) code and physical phantom measurements were used to compare accuracy and noise properties of two scatter correction techniques: the Triple-Energy Window (TEW), and the Transmission-Dependent Convolution Subtraction (TDCS) techniques (Narita Y. et al., 1996). Development of the original EGS code ended with version EGS4.

The code has been rewritten by two new physics groups:

- **EGSnrc**: Software tool to model radiation transport is maintained by the Ionizing Radiation Standards Group, Measurement Science and Standards, National Research Council of Canada. **EGSnrc** is distributed with a wide range of applications that utilize the radiation transport physics to calculate specific quantities. The documentation for **EGSnrc** is also available online.
- **EGS5**: Is maintained by the Japanese particle physics research facility known as KEK, *EGS at KEK Web Page, rc*www.kek.jp. These codes have been developed by numerous authors, and it is possible to calculate quantities such as absorbed dose, kerma, or particle fluency, with complex geometrical conditions.

MCNP: **M**onte **C**arlo **N**-**P**article Transport is a general purpose, continuous-energy, generalized geometry, time-dependent Monte Carlo radiation transport code designed to track many particle types over broad ranges of energies. It is developed by Los Alamos National Laboratory. Areas of application include radiation protection and dosimetry, radiation shielding, radiography, medical physics, nuclear safety, detector design and analysis.

MCNP-X: **M**onte **C**arlo **N**-**P**article e**X**tended, was also developed at Los Alamos National Laboratory and is capable of simulating particle interactions of 34 different types of particles and 2000+ heavy ions at nearly all energies, including those simulated by MCNP; MCNP5 and MCNP6 are further development of MCNP-X.

FLUKA is developed using the FORTRAN language. FLUKA has many applications in particle physics, high energy experimental physics and engineering, shielding, detector design, dosimetry, medical physics, radiobiology.

Geant4 (for **GE**ometry **AN**d **T**racking) is a platform for the simulation of the passage of particles through matter using Monte Carlo methods. It is the successor of the GEANT series of software toolkits, in C++ programming language. Its development, maintenance, and user support are taken care by the international Geant4 Collaboration. Application areas include high energy physics and nuclear experiments, medical applications, and particle accelerator studies. The software is used by a number of research projects around the world. The Geant4 software and source code is freely available from the project website.

GATE is an application based on the extensively validated GEANT4 MC code. GATE has been widely used for imaging application but also articles have reported its application and trustworthiness in internal dosimetry and in individualized patient-specific, image-based dosimetry.

Using the simulation dedicated code GATE, the γ-camera, Philips SPECT/CT Precedence 16 system, was modeled. The spatial resolution, system sensitivity, and scatter fraction of various collimators were examined according to NEMA specifications following a clinical data acquisition protocol of 120 projections on a 128×128

pixel matrix. A simulation of the clinical projection pixel size of 4.6 × 4.6 mm was performed (Georgantzoglou A. et al., 2009).

The absorbed dose for Tc-99m-DMSA using GATE MC simulation based on data from hybrid planar/SPECT imaging in patients with genitourinary abnormalities was evaluated. An image-based GATE method for patient-specific 3D absorbed dose calculation was applied (Bagheri M. et al., 2018).

1.6.3.2   Specific Monte Carlo (MC) Codes

The similarity of physical and geometrical characteristics of the emission imaging systems, as SPECT and PET, suggested specific codes to be developed, dedicated to emission tomography formation simulations, in general.

Several SPECT and PET dedicated Monte Carlo software packages were developed to be used in optimization of imaging system design, development of correction methods for improved image quantitation, valuation of image reconstruction algorithms or pharmacokinetic modeling.

A number of Monte Carlo specific packages state problems dedicated to emission computed either SPECT (SimSET, SIMIND, SimSPECT) or PET (PETSIM); other similar packages are suitable for applications in dosimetry.

The main advantage of those packages is the relatively simple geometry construction, the fast execution of simulation code and commonly the performance advantages since they are optimized for specific requests. SIMIND is an example:

SIMIND: The **Sim**ulation of **I**maging **N**uclear **D**etectors is a Monte Carlo simulation code that calculates basic detector parameters such as absolute sensitivity and intrinsic efficiency, the energy pulse-height distribution and an image describing the position of the photon interaction events (Toossi M. B. et al., 2010). Prof. Michael Ljungberg developed SIMIND in Lund University, Sweden. The SIMIND is based in FORTRAN-90 and describes a standard clinical SPECT camera. It can be adjusted for almost any calculation in SPECT imaging procedure.

Several works have been concerned for SIMIND to determine different components of the detected radiation.

Bremsstrahlung imaging, for example, makes difficult to choose collimator and energy window; hence, the optimal clarity of the image is not always achievable.

The SIMIND program was chosen to evaluate resolution in order to select the appropriate collimator and energy window settings for bremsstrahlung imaging. The SIMIND was used to generate the Y-90 bremsstrahlung SPECT projection of a point source. In Y-90 imaging, image quality is very dependent on the selection of energy window and collimator design.

By SIMIND, accurate valuation of the geometric, penetration, and scatter contribution inside the photo-peak window can be obtained (Asmi H., et al., 2020). Excellent correspondence is found between experimental measurement and simulation results for Point Spread Functions (PSFs).

1.6.3.3   Validation of Monte Carlo (MC) Codes in Nuclear Medicine

One of the most important issues related to the use of a Monte Carlo code is how the code has been validated. The problem of validation is firmly connected with the problem of accuracy: only the results of thorough validation studies can warrant the accuracy of a code (Buvat I. and Castilioni I., 2002).

The output of a simulation is generally a large file, where the data of each event, from its growth up to its detection, are recorded. This file can be used in order to produce or reconstruct images, separate scattered from non-scattered events or determine the components of the system where interactions occurred.

The validation of the simulation has to be carried out first. The accuracy of a code is determined by its ability to generate data identical to those that are acquired on real imaging systems.

Simulation prediction of the physical response of the SPECT or PET system is usually checked by comparing the simulated values with the empirical values of physical parameters that can be experimentally measured.

A simulator is then considered to be validated if it accurately reproduces the response of the experimental system. The parameters of interest that are used mostly are the spatial resolution, scatter fractions, sensitivity, or count rates.

A study for comparison of the dosimetric calculations, obtained by Monte Carlo (MCNP-X code) simulation and the MIRDOSE model, in therapeutic schemes of skeleton metastatic lesions, with Rhenium-186 (Sn) – HEDP and Samarium-153 – EDTMP was performed by (Andreou M. et al., 2011). A good agreement between the results derived from the two pathways (the patient specific and the mathematical, MCNP-X code) has been showed, with a deviation of less than 9% for both radiopharmaceuticals.

For estimation of absorbed doses in the lesions and in critical organs after administration of therapeutic dose of In-111 octreotide (Argyrou M. et al., 2014), a Monte Carlo simulation (MCNP code) was employed. The overall standard deviation was about 11% for planar data calculations compared to those obtained by Monte Carlo simulation.

1.6.3.4 Quantification Accuracy in Monte Carlo (MC) Simulation

The optimization of dosimetry results by Monte Carlo simulation is an effective and easy method for dose estimation taking into account anatomical features. An essential point when Monte Carlo simulations are used for evaluating the accuracy of quantification methods is to make sure that the characteristics of the data analyzed by that method are realistic. To study the relevance of statistical reconstruction methods, for example, care should be taken that the statistical properties of the simulated data are identical to those of experimental data.

When considering scatter correction methods relying on energy information collected over a wide spectral range, the energy spectra of the simulated events should be identical to those that would be physically acquired. When assessing quantification methods relating to a given isotope, the code should have been firstly validated for this specific isotope (Loudos G., 2007).

Patient-specific dosimetry calculations help the physician to optimize the planning of the treatment, avoid side effects to healthy tissue and allocate administered activity dose to treatment results.

The In-111 Octreotide therapy is an example:

- Scintigraphic data of patients diagnosed for neuroendocrine tumors in liver and had received therapeutic doses of In-111 Octreotide were used in order

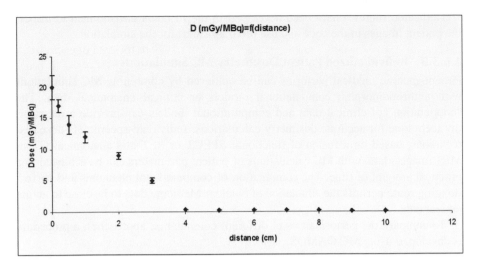

FIGURE 1.32 Monte Carlo estimated absorbed dose (mGy/MBq) in the tumor by the MCNPX code (Monte Carlo N – particle Extended Code). (From Argyrou M., 2018.)

to estimate the absorbed doses from radiopharmaceutical volume distributions and absorbed doses in the lesions and in critical organs, employing the MIRD Schema, as well as the Monte Carlo simulation (Argyrou M. et al., 2018). For the MC (MCNP-X) simulation, two patients, one male and one female of similar weights (70 kg), diagnosed of neuroendocrine tumors located in liver, were considered to be as representative cases, to be used in the simulation process. The Monte Carlo geometries, used in Monte Carlo simulation input files, were prepared in order to approach the patient's anatomy. Then, a theoretical estimation of the accuracy of the results can be achieved with an MCNP-X simulation (Figure 1.32).

Monte Carlo simulation gives the opportunity to receive fast and accurate patient-specific dosimetric calculations.

1.6.3.5 Use of Voxel Models/Monte Carlo Simulations

The use of voxel models in MC radiation transport codes has contributed to the improvement of dose calculations. 3D voxel dosimetry is the absorbed dose calculation at the voxel level with the scope to obtain the average calculation at the organ level (Argyrou M., 2019).

FLUKA MC Code used to perform patient-specific 3D dosimetry through direct MC simulation on PET/CT and SPECT/CT images. The absorbed dose maps were compared at the voxel level by the use of FLUKA in patient-specific, image-based dosimetry (Botta F., 2013).

Voxel dosimetry advantage is the availability of a 3D-absorbed dose matrix which allows a managing in terms of absorbed dose-volume restrictions. It also considers the absorbed dose inhomogeneity from the dosimetric and radiobiological approach.

Dedicated routines have been developed in the FLUKA environment to import the patient images in the code and use them as a start for the simulation.

1.6.3.6   Individualized Patient Dosimetry/MC Simulations

Patient-specific medical performs can be achieved by combining MC simulations with anthropomorphic computational models or clinical anatomical data. The "cooperation" of clinical data and computational models can provide a trustworthy technique for accurate dosimetry calculations. Individual-specific 3D dose distributions, based on a fusion of functional SPECT or PET and anatomical CT or MRI images data, with MC calculations of patient parameters can be achieved in a practical amount of time. The combination of computational phantoms and correct imaging route permits the simulation of Nuclear Medicine data to be close to actual patient data.

To compare the performances of different computation approaches, a procedure is developed using MC GAMOS.

3D internal dosimetry is increasingly used in planning **T**rans-**A**rterial **R**adio-**E**mbolization (TARE) of Hepatocellular Carcinoma (HCC). GAMOS is a **GE**ANT4-based **A**rchitecture for **M**edicine-**O**riented **S**imulation. 3D absorbed dose maps, dose profiles, and Dose-Volume Histograms (DVHs) were produced for liver through MC simulations and a convolution software.

The absorbed dose distribution in presence of relevant tissue inhomogeneities can be measured accurately by means of MC GAMOS simulation (Auditore L. et al., 2019).

Consequently, *an accurate model of the interactions associated with the radiations emitted by the radiopharmaceuticals is crucial:*

- For photon interactions, this may involve Thomson and Compton scattering, the photoelectric effect and pair production with the respective probabilities of interaction depending on the energies of the emitted photons and the material with which these radiations interact.
- For the charged particles, it is necessary to model the exchange of energies between these particles and the surrounding media leading to excitation and ionization of atoms close of the path of the charged particle.
- In the case of positron interactions, the chance of the follow-on 511-keV annihilation photons must be considered.
- The nature of the objects within which the radiation is emitted and detected it is necessary to be taken under consideration.

1.6.4   MATLAB Computing Software in Nuclear Medicine

Computer simulation is of a great importance today, and MATLAB is a big tool which offers assistance to this. It starts with creation of a mathematical model and the obtained equations are solved by using an appropriate calculation method.

Image processing and analysis applied to Nuclear Medicine images for diagnosis improve the acquired image qualitatively as well as offer quantitative information data useful in patients' therapy and care.

MATLAB and Image Processing Toolbox enable both quantitative analysis and visualization of Nuclear Medicine images acquired as planar or angle projected images to reconstruct tomographic SPECT, PET slices, and 3D volume surface rendering images.

- Image Processing Toolbox performs image processing, visualization, and analysis. It provides a comprehensive set of reference-standard algorithms, workflow applications (apps) and algorithm development. The Toolbox supports processing of 2D, 3D, randomly large images and automation of common image processing workflows.
- Segmentation of image data can be done interactively, perform image enhancement, noise reduction, geometric transformations, and image registration using deep learning and traditional image processing techniques.
- Comparison of image registration techniques can be performed.
- Visualization functions and apps permits to explore images, 3D volumes, and videos; adjust contrast; create histograms; and manipulate *regions of interest (ROIs)*.
- Acceleration of algorithms by running them on multicore processors and **G**raphics **P**rocessing **U**nits (GPUs) can be obtained.
- Many toolbox functions support C/C++ code generation for desktop prototyping and embedded vision system deployment.

1.6.4.1   MATLAB (The MathWorks, Inc.) in Developing Nuclear Medicine Algorithms

A great number of research and articles published present useful algorithms created on the base of MATLAB and Image Processing Toolbox. Some of these are described in the text that follows.

ImSim — A nuclear imaging teaching simulator using MATLAB has been developed by (Pointon B., 2018) for education purposes in Nuclear Medicine imaging physics by MATLAB.

Nuclear medicine imaging physics involves many mathematically sophisticated concepts that are often difficult for technologist-students to learn and apply. Students have to understand planar projection imaging, sinograms, tomographic reconstruction, digital filtering, image quality and much more. Neither the background nor the time to master these mathematical formalisms exists. These concepts are used in selecting acquisition and processing parameters for optimal clinical imaging.

ImSim simulator instantly generates and displays medical images (SPECT, PET, and CT) based on user-defined acquisition parameters. Image data may be analyzed, processed, and reconstructed.

MATLAB scripts can be compiled by ImSim to create distributable applications. MATLAB has also greatly increased functionality allowing for more complex simulation capability.

ImSim simulates planar and SPECT imaging in real or computing time.

- The user chooses whether to include the effects of attenuation, image noise, and/or depth-dependent blurring.

- Can select from a wide range of digital phantoms, from simple geometric shapes, standard planar, and SPECT phantoms to a digital anthropomorphic phantom for simulating different types of clinical imaging.

ImSim allows users to choose different acquisition parameters including camera and collimator type, detector distance and orientation, acquisition matrix, imaging time/counts, SPECT dual detector orientation, number of gantry stops, and much more. Reconstruction choices include reconstruction method (FBP, OSEM) and parameters (filtering parameters, number of iterations, and subsets). Image analysis may be done to view profiles, calculate contrast, noise, uniformity, and more. Simulation of detector calibrations (tuning, linearity, energy, and flood corrections) and of PET and CT imaging are at various stages of development (Pointon B., 2018).

ImSim may be used for classroom demos and for independent and group homework tasks. Through ImSim, students gain the necessary comprehensions to relate imaging parameters to the resulting image quality.

- *CALRADDOSE, Software Using the MATLAB as a Tool for Internal Dosimetry* The calculation process of this software proceeds from collecting time-activity data from image data followed by residence time calculation and absorbed dose calculation using MIRD method. The CALRADDOSE is a user-friendly, GUI-based software for internal dosimetry. It provides fast and accurate results, useful for a routine work.

  The development process of software "CALRADDOSE" covered the design of GUIs, data structure, and calculation algorithms of MATLAB.

The Graphic User Interface Development Environment (GUIDE) tool of MATLAB was used to create software GUIs.

The workflow of the MATLAB-based software includes "Image Processing UIs", "Residence Time Calculation UI", and "Dose Calculation UI" (Figure 1.33).

1. The calculation process in this software was divided into three major steps: Time-activity data collection,
2. Residence time calculation, and
3. Dose Calculation.

Software evaluation was estimated on the accuracy of residence time calculation and organ dose calculation by comparing with those from OLINDA/EXM. The selected source organs were liver, spleen, and total body. These time-activity data were used to calculate residence time for all source organs and then organ doses with the CALRADDOSE and OLINDA/EXM (Chaichana A. and Tocharoenchai C., 2016).

CALRADDOSE is user-friendly, GUI-MATLAB-based software. It can perform all steps of internal dosimetry within single environment to reduce calculation time and diminish the possibility of error. CALRADDOSE also provides fast and accurate results which are useful for dose calculations in a Nuclear Medicine department.

Spectalyzer is a MATLAB-based fully automated 3D basal ganglia activity measurement software in dopamine transporter Nuclear Medicine imaging. The

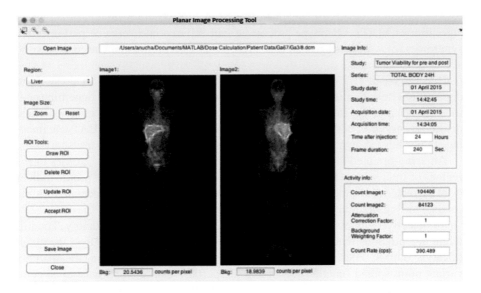

FIGURE 1.33 A simple screenshot of the "Planar Image Processing UI" includes two planar images with liver ROIs.

software is written in MATLAB using its functions of Image Processing Toolbox and can be easily run in an ordinary computer with Windows software. Functional imaging of the dopamine transporter (DAT) with 123-I-labeled radioligands defines integrity of the dopaminergic system. The primary supposition is that the basal ganglia represent the maximal tracer uptake in the DAT-scan Nuclear Medicine examination.

The overall activity distribution on the whole brain is expected to be approximated by a mixture of two multivariate Gaussian distributions. The capture range of a local maximum is defined by the set of points in space which will lead to the local maximum. The results showed that these assumptions are fulfilled by all the investigated test datasets including three anthropomorphic striatal phantoms (Mirzaei S. et al., 2010).

Spectalyzer is highly reproducible and observer-independent MATLAB software for 3D *region of interest* determination in DATscan. This software using MATLAB codes could be an alternative approach for semiquantitative analysis of DAT imaging.

- *Volume Quantification of 123I-DATscan Imaging by MATLAB*
- *Grading of Parkinsonism and Essential Tremor Dopamine Transporter (DAT)*

By creating a MATLAB algorithm, the ability to detect loss of striatal dopamine transporters has been shown. The dopamine transporter (DAT) is a transmembrane protein that is responsible for the reuptake of dopamine (DA) from the synaptic split and for the termination of dopaminergic transmission.

The differential diagnosis of Parkinson Disease (PD) to Essential Tremor (ET) is challenging. Patients who present only with a tremor may have Parkinson Disease (PD) or may have Essential Tremor (ET), a completely different disorder.

An evaluation whether mathematical approach of striatum imaging data by an algorithm of MATLAB image processing can differentiate between parkinsonian syndromes, of various stages, and Essential Tremor (ET), thus increased diagnostic accuracy is obtained.

I123-DATscan (I123-Ioflupane) is a cocaine analog that binds to presynaptic dopamine. The dopamine transporter (DAT) expressed exclusively in dopamine neurons. Imaging of DATs, situated in the membrane of dopaminergic neurons, could detect *degeneration* of the dopaminergic nigrostriatal pathway. I123-DATscan imaging studies have shown the ability to detect loss of striatal dopamine transporters.

An array-oriented language and a graphical display technique by profiles of the regions of interest data is intergraded by MATLAB use. Quantitative profile indices were extracted and ratios of specific to nonspecific binding were calculated for the caudate nucleus and putamen imaging of all subjects (PD and ET).

To objectively assess striatal DAT binding, quantification is mandatory. The different degradations that are involved in the reconstruction process affect the quantification output. Thus, image-degrading effects such as attenuation, the spatially variant point spread function scatter and the partial volume effect have to be compensated to achieve an accurate quantification. Moreover, in the I-123 decay scheme, there are a few high-energy photons that have a non-minor impact to the final image; this high-energy contamination influence has to be balanced.

Background intensity, on the image, plays a crucial role both in visual and quantitative diagnostic evaluation. Background intensity isocontours by MATLAB algorithm defines the threshold of the striatum specifically in each case. For the slices's delineation a histogram technique for the lower threshold was used. A threshold for each patient was defined as the 60% of the maximum striatal count density (Figure 1.34).

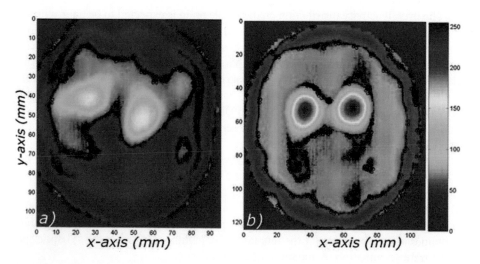

FIGURE 1.34 Isocontouring on central coronal slice. (a) PD subject with high eccentricity difference between the left and right subregion of the striatum. (b) Low eccentricity difference of the two subregions in an ET subject. (Adapted from Lyra M. et al., 2011b.)

Furthermore,

- Analysis of the image data by creating a DICOM format image file. The DICOM images saved in an uncompressed bitmap format in order to be processed by Interactive Data Language (IDL) tools and MATLAB program.
- Ellipse ROIs are created around the striatum while rectangular ROIs include caudate and putamen uptake data.
- Mesh Intensity plotting by MATLAB algorithm of the striatum image gives a semiquantitative index using parameters of area, intensity and distribution. Loss in uniformity-Mesh plotting distribution is used for grades extraction: ratio of intensities and areas.
- ISOCONTOUR Intensity Distributions are specific for each case and they are important factors included in categorization (Figure 1.35).

Measurement properties of the SPECT images regions are calculated by an algorithm in MATLAB.

This algorithm calculates shape measurements (eccentricity, major and minor axis length, orientation, area, equivalent diameter) and pixel measurements (min, max, mean, and integral intensity). A specific ratio of both ganglia parameters gives a classification indication and contributes in diagnosis decision. The pixel size is made up from a combination of field of view, matrix size, and zoom factor and it is 3.2 mmm.

- *A method of determination of thyroid gland's volume* by the use of SPECT γ-camera data and MATLAB assistance in order to calculate thyroid's mass a quantity crucial in Nuclear Medicine therapy procedures (Lyra M. et al., 2008, Thyroid volume determination by single photon tomography and 3D processing for activity dose estimation, IEEE International Workshop on Imaging Systems and Techniques).

  The method consists of the determination of the borders and area of the gland from a set of transaxial tomographic images by use of a specific for each case threshold value and the length of the gland by coronal slices.

  Patients with hyperthyroidism were referred to Nuclear Medicine imaging department for evaluation for I-131 therapy. Each patient received Tc-99m pertechnetate intravenously; 15 minutes later tomographic images of the resulting radioactive distribution were obtained using a GE SPECT, STARCAM 4000 γ-camera. Tomographic images at various levels were obtained with each slice 1 pixel thick. That is, the size of the elementary voxel is one pixel in the x- and y-axis and 1 pixel, too, in the z direction. Pixel size was 6.4 mm in a 64×64 matrix.

  After correction for the contribution of background events, the pixel values in a 2D transverse or coronal section represent the radioisotope concentration within that section.

  Analysis of the slices' data producing interfile format images, transferring these images in a PC and saved in an uncompressed bitmap format was processed by IDL tools and MATLAB program algorithm.

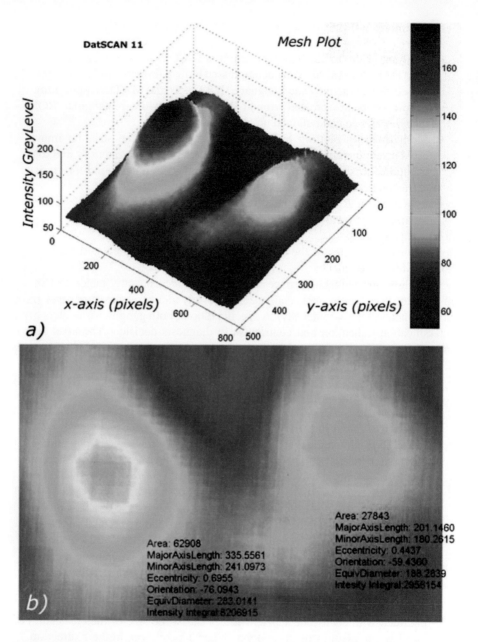

FIGURE 1.35 (a) Intensity of dopamine receptors area is plotted in mesh plot. Processing by the MATLAB algorithm gives the intensity distribution. (b) Useful parameters of each subregion of striatum are calculated for classification. Magnified striatum images were obtained for processing, segmentation, and data collection by MATLAB. Major axis length, areas, and integral intensity differences are calculated for the ET and PD subjects. (Adapted from Lyra M. et al., 2011b.)

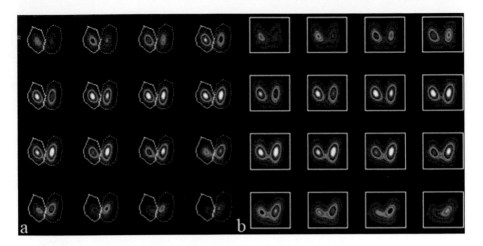

FIGURE 1.36 Sequential tomographic slices of a thyroid gland produced by MATLAB. (a) Automatic irregular ROIs around the lobes. (b) Rectangular ROIs cover the whole thyroid gland.

The accuracy of the volume estimations depends on the correct delineation of the borders of the thyroid tissue. The precise borders of the thyroid tissue are sometimes hard to define. This difficulty is due to various factors including fluctuations in the background and an unavoidable smoothing of the thyroid boundary during reconstruction.

Although an edge determines the organ pixels, a threshold to every single case was adapted, as its value is dependent upon size of the imaged organ and the contrast. The thyroid lobes were enclosed with the help of *region of interest (ROI)* MATLAB tool and a threshold was selected to identify the thyroid boundaries. The threshold selected for each slice of a tomographic set of data was specific for it.

Ellipse ROIs are created around each thyroid lobe, for separate data concentration of its lobe (Figure 1.36a) by using the images.roi.Ellipse function. Rectangular ROIs (the drawrectangle function to create a Rectangle ROI) include both lobes uptake data including first-level threshold to indicate the background limits (Figure 1.36b).

MATLAB language was used in computing and creating the matrices of the maps of the counted voxels of all the tomographic slices of the organ. Their total is multiplied by the elementary voxel size in mm to obtain the final organ volume (Lyra M. et al., 2008).

Data were extracted and *area, eccentricity, major and minor axis length, orientation, equivalent diameter,* and *integral intensity* of both left and right lobes were calculated. MATLAB profiles were created to extract area data in three dimensions. Elliptical ROIs cutting off the background are automatically designed and matrices of the maps of the counted voxels of the slices of the organ give the intensity numbers in x-axis and y-axis (Figure 1.37a and b).

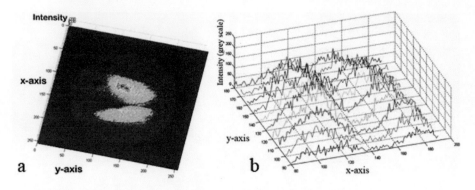

FIGURE 1.37 Mesh plot of a thyroid scintigraphic image. (a) Thyroid image projection. (b) Isointensity curves derived from map-matrix.

- *Thyroid Mesh Plots*
  3D plots are useful to present data having more than two variables. The command plot3 (x,y,z) in MATLAB help to create three-dimensional (3D) plots. The general form of the command is:

$$\text{plot3}(x, y, z, \text{"line specifiers"})$$

  where, x, y, and z are vectors of equal size.
  Mesh plots are 3D plots. The general form of the mesh command is: mesh(x,y,z).
  They plot a graph over the functions in the form of $z = f(x,y)$, where x and y are the independent variables and z is the dependent variable.
  In thyroid mesh plot x,y are the horizontal and in depth axis dimensions and z expresses intensity (Number of pixels in the organ).
  MATLAB label commands: ylabel('text as string') put the label on the y-axis of the plot, xlabel('text') put the label on the x-axis of the plot, command axis[xmin,xmax,ymin,ymax] adds an axis to the plot and title('text') command puts the title on the plot (Figure 1.37).
- ***Thyr-Vol** MATLAB Algorithm/Thyroid 3D SPECT Volume Quantification*
  A MATLAB algorithm which integrates 3D visualization is used for volume estimation of the thyroid gland. This is a technique of determination of the thyroid gland's volume using the SPECT data in order to calculate the *absorbed patient dose* from the radioactive substances (I-131) injected to the patient during thyroid treatment.
  Thyroid SPECT examinations were completed by a γ-camera GE SPECT Star Cam4000. SPECT acquisition was performed for angles −90 to +90 degrees, in 32 projections. Images of thyroid gland were reconstructed by the GE Volumetrix software in the GE Xeleris-2 processing system using Ramp and Hanning filters. DICOM data was extracted for each patient and an algorithm that integrates 3D visualization has been used for image processing analysis by MATLAB R2012b.

```
SPECT=dicomread('F:\Thyroid\MATLAB1\TOM.g1001_DS');
%%Apeikinoisi tomis #9
I=SPECT(:,:,1,36);
imagesc(I);
C=contour(I);
clabel(C);

%%3D reconstruction
mask=logical(zeros(128,128,77));
for k=1:77
    for i=1:128
        for j=1:128
            if (SPECT(i,j,k)>5);
                mask(i,j,k)=1;
            end
        end
    end
end
 isosurface(mask,0.9)


%%Number of pixels
mask=logical(zeros(128,128,77));
numberOfPixels=0;
for k=1:77
    for i=1:128
        for j=1:128
            if (SPECT(i,j,k)>5);
                mask(i,j,k)=1;
                numberOfPixels=numberOfPixels+1;
            end
        end
    end
end
```

FIGURE 1.38 Code of thyroid isosurface contours.

Thyroid transaxial slices in DICOM format were obtained and MATLAB Thyr-Vol algorithm was created. Ellipse ROIs are shaped around each thyroid lobe while rectangular ROIs including both lobes uptake data were formed.

Matrices of the maps of the counted voxels of the slices of the organ were created. Their total is multiplied by the elementary voxel size in mm to obtain the final organ volume. The appropriate threshold value was identified by creating intensity isocontours (Figure 1.39).

The Thyr-Vol MATLAB algorithm, detects the segmented regions automatically and the outer contour can be specified. Therefore, the thyroid volume was evaluated and reconstructed as 3D image based on the threshold value. By this technique, numerical errors are eliminated because the images are filtered and reconstructed (Synefia S. et al., 2014).

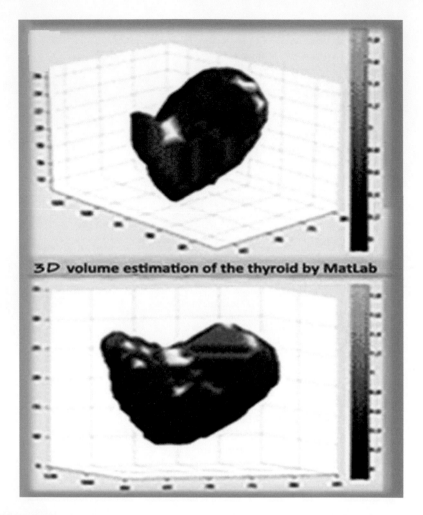

FIGURE 1.39   3D volume images of thyroid gland by Thyr-Vol MATLAB algorithm. (a) Iso-Intesity contours of Thyroid gland Image (b) 3D volume images of Thyroid gland by *Thyr-Vol MATLAB algorithm*. (Courtesy of Synefia S., MSc University of Athens, 2013.)

Thyroid gland volume estimation in 3D images will give confidence in accurate absorbed patient treatment dose calculation. This method determines the thyroid gland volume by using SPECT γ-camera data. The data are being processed by a mathematical algorithm Thyr-Vol that was created using MATLAB language, in order to calculate thyroid volume and mass. Volume and mass calculations are very useful in Nuclear Medicine Therapy.

- *Segmentation of Alzheimer's Disease in PET scan datasets using MATLAB* is an article-example of using the MATLAB in PET imaging to segment image datasets (Meena A. and Raja K., 2013).

The PET scan images require expertise in the segmentation where clustering plays an important role in the automation process. Clustering is commonly known as unsupervised learning process of n-dimensional datasets that are clustered into k groups (k<n) so as to maximize the inter cluster similarity and to minimize the intra-cluster similarity. It is proposed to apply the commonly used K-Means and Fuzzy C-Means (FCM) clustering algorithms.

This is realized using MATLAB and tested with sample PET scan image, collected from Alzheimer's disease Neuro imaging Initiative (ADNI). The motive of automatic medical image segmentation is to describe the image content based on its features.

This algorithm is used to automate process of segmentation on the datasets classified by its type, size, and number of clusters. The basic concept of clustering is described and the algorithm of K-Means and Fuzzy C-Means clustering is presented. The application method and the results of these MATLAB algorithms are shown.

PET scan is a biomedical nuclear imaging technique that provides a solution for abnormal cells. The segmented image provides the clear picture about the affected portions. The basic K-Means algorithm and the Fuzzy C-Means, both algorithms are executed in MATLAB. The cluster distribution is based on less distance and high-membership value in FCM.

MATLAB is widely used for employing algorithms in numerical environment. The Image Processing Toolbox has a stable, well supported set of software tools for wide range of digital image processing and segmentation. The major applications are intensity transformation, image restoration, registration, image data compression, morphological image processing, regions and boundary representation, and description.

- *MATLAB Coding to Determine the Distribution of a Collimated Ionizing Radiation Beam*

    In the case of the production of radionuclides used in Nuclear Medicine, knowing the geometric parameters of the beam involved is very important. In the nuclear physics-related experiments involving particles acceleration for obtaining radionuclides used in Nuclear Medicine, characterization from the point of view of the dimensions and distributions of the involved energetic beams must be done.

    A novel method to determine the distribution and the dimension of an ionizing radiation collimated beam is described. A high-quality picture of the glass sample placed in the beam where the distribution must be determined is taken, after the beam is off. The image is then analyzed by making use of a MATLAB developed code.

A MATLAB code, following the logical structure that can be seen in Figure 1.40, was developed.

By using the MATLAB code developed for this experiment, and reading the information stored in the taken picture, a conversion of it to grayscale was performed.

The code was used to obtain a gray levels histogram (Figure 1.41) of the resulted image. The histogram is very important for obtaining the binarized images (0/1).

READ THE RGB IMAGE
SHOW THE RGB IMAGE
CONVERT IMAGE TO GREY SCALE
SHOW GREY SCALE IMAGE
PLOT GREY SCALE IMAGE HISTOGRAM
BINARIZE THE GREY IMAGE (0/1)
SHOW THE BINARIZED IMAGE
PLOT BINARIZED IMAGE HISTOGRAM
COUNT THE TOTAL NUMBER OF PIXELS
COUNT THE NUMBER OF BLACK (1) AND WHITE (0) PIXELS
COMPUTE THE RATIO BETWEEN THE BLACK PIXELS AND ALL PIXELS
CONVERT PIXELS NUMBER TO SIZE

FIGURE 1.40   The logical structure of the developed MATLAB code. (Modified from Ioan M. R., 2016.)

Using the gray levels histogram, the image was further binarized for different values of the background level. A correlation between the absorbed dose and the color tone of the exposed glass materials can be done. By using the developed MATLAB code, the image can be binarized for different values of the gray tone, resulting images of the spot having certain requested characteristics.

The geometrical analysis part of the method was preceded by a calibration in terms of pixels to lengths conversion and then tested on a circular graphite plate having the diameter of 12.2 mm.

The diameter value obtained with the MATLAB code proposed method was 12.4 mm. The two results are in a good agreement, with a relative difference of 1.6%, demonstrating the good capabilities of the proposed technique by MATLAB code.

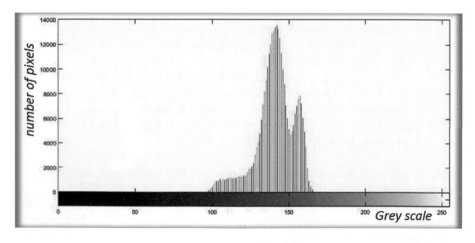

FIGURE 1.41   This is the histogram of the grayscale image. (From Ioan M. R., 2016.)

In **Image Processing Toolbox™** (www.mathworks.com/help/images) a broad number of MATLAB functions, standard algorithms, examples, and workflow apps is provided for image processing, analysis, visualization, and algorithm development; they are tools valuable in Nuclear Medicine imaging.

Image segmentation, image enhancement, noise reduction, geometric transformations, and image registration can be performed, using deep learning and traditional image processing techniques. The toolbox supports processing of 2D, 3D, and arbitrarily large images.

Image Processing Toolbox apps give the opportunity for automation of common image processing workflows. Interactively, contrast can be adjusted, histograms be created or manipulate *regions of interest (ROIs)* and more.

Few hints of what in the Image Processing Toolbox pages are included, follow:

- *A function example from MathWorks website*, for the determination of full width at half maximum (FWHM) of a peak if the input data has two columns: Column 1 = x, Column 2 = y (Figure 1.42) (Ewusi-Annan E., 2020).
- ***Create Image Histogram:*** An example published in Mathworks.com for the creation of an image histogram, https://www.mathworks.com/help/images/ref/imhist.html.

```
                                                          Editor - fwhm.m
1        function[Fullw]=fwhm(data)
2        % This function determines   full width at half
3        % maximum of a peak if the input data  has two columns:
4        % Column 1 = x
5        % Column 2 = y
6        %Coded by Ebo Ewusi-Annan
7        %University of Florida
8        %August 2012
9  -     x = data(:,1);
10 -     y= data(:,2);
11 -     maxy = max(y);
12 -     f = y==maxy;
13 -     cp = x(f);% ignore Matlabs suggestion to fix!!!
14 -     yl= y./maxy;
15 -     ydatawr(:,1) = yl;
16 -     ydatawr(:,2) = x;
17 -     newFit1=find(x>= cp);
18 -     newFit2=find(x < cp);
19 -     ydatawr2 = ydatawr(min(newFit1):max(newFit1),:);
20 -     ydatawr3 = ydatawr(min(newFit2):max(newFit2),:);
21 -     sp1 = spline(ydatawr2(:,1),ydatawr2(:,2),0.5);
22 -     sp2 = spline(ydatawr3(:,1),ydatawr3(:,2),0.5);
23 -     Fullw = sp1-sp2;
24 -     end
25
```

FIGURE 1.42 MATLAB code of FWHM (a MATLAB suggestion at line 13: If "x" is an index variable, performance can be improved using logical indexing instead of FIND, has been ignored).

This example shows how to create a histogram for an image using the *imhist* function. An image histogram is a chart that shows the distribution of intensities in an indexed or grayscale image. The *imhist* function creates a histogram plot by defining *n* equally spaced bins, each representing a range of data values, and then calculating the number of pixels within each range. The information in a histogram is used to select an appropriate enhancement operation. For example, if an image histogram shows that the range of intensity values is small, you can use an intensity adjustment function to spread the values across a wider range.

An image is read into the workspace and is displayed.

```
I = imread('tumor.png');
imshow(I)
figure;
imhist(I);
```

The *imhist* function displays the histogram. In this example of an image of nonhomogeneous tissue, *imhist* creates a histogram with 64 bins.

Another useful information, written in Mathworks.com, shows the way to convert a grayscale image into a binary image.

- **Imbinarize**, Binarize 2D grayscale image or 3D volume by thresholding.

In image analysis techniques many times it is necessary to convert grayscale images into binary images.

The general Imbinarize function used, e.g., for edge detection, is the following:

```
[BW] = edge (image, 'method', threshold)
```

where *[BW]* is the new binary image produced, *image* is the initial one; "method" refers to the method of edge detection and "threshold" to the threshold applied.

BW = imbinarize(I) creates a binary image from 2D or 3D grayscale image I by replacing all values above a globally determined threshold with 1s and setting all other values to 0s. *Imbinarize* uses Otsu's method, which chooses the threshold value to minimize the intraclass variance of the thresholded black and white pixels. imbinarize uses a 256-bin image histogram to compute threshold.

```
Read grayscale image into the workspace.
I = imread('liver.jpg');
Convert the image into a binary image.
BW = imbinarize(I);
Display the original image next to the binary version.
figure
imshowpair(I,BW,'montage')
```

- *Deconvblind* function, *Deblurring Images Using the Blind Deconvolution Algorithm* (Mathworks.com).

This example shows how to use the MATLAB blind **deconvolution** algorithm (by *deconvblind* function) to deblur **images** when you have no information about the blurring or the noise.

The blind deconvolution algorithm can be used effectively when no information about the distortion (blurring and noise) is known. The algorithm restores the image and the point-spread function (PSF) simultaneously. Optical system characteristics can be used as input parameters that could help to improve the quality of the image restoration. PSF limits can be passed in through a user-specified function.

Read a grayscale image into the workspace. The *deconvblind* function can handle arrays of any dimension.

Although blind **deconvolution** algorithm does not require information about the blurring or noise, it enables you to deblur **images** that have complicated distortions such as nonuniform image quality.

By the *deconvblind* function, reduction of the effect of noise on the restoration, account for nonuniform image quality (e.g., bad pixels) or handle Nuclear Medicine system read-out noise.

- *(PVcorr2D.m)* Partial volume correction method using ***reverse diffusion interpolation***

  To correct for the Partial Volume Effect – especially useful in Nuclear Medicine to optimize the image qualitative and quantitative – the Interpolation with Reverse Diffusion can be used (Salvado O., 2020). This function implements a reverse diffusion scheme to interpolate images to correct for partial volume effect. There is a script *(PVcorr2D.m)* that asks to select multiple DICOM files and then runs the main function. Any file format could be read, upon a small modification *(imread* or *dicomread).*

```
function [I,Flow,H] = PVcorr2D(Ir,ratio,option);
% PVcorr2D: partial vollume correction with reverse difusion
% [I,Flow,H] = PVcorr2D(Ir,ratio,option);
% Ir: image to be interpolated
% ratio: 2 for x2 interpolation
% option: structure with parameters
```

Image processing and analysis applied to Nuclear Medicine images improve the acquired image qualitatively and offer quantitative information data useful in patients' therapy and care. MATLAB and Image Processing Toolbox enable both quantitative analysis and visualization of Nuclear Medicine images acquired as planar or angle projected images to reconstruct tomographic (SPECT, PET) slices and 3D volume surface rendering images.

Specific potentials of MATLAB which are also applicable in Nuclear Medicine imaging are referred in the next chapters of the book.

A list of documents of the Help Center, in Image Processing Toolbox™ will be an analytical assistance about.

https://www.mathworks.com/help/images/

Getting Started
Learn the Basics of Image Processing Toolbox
Import, Export, and Conversion
Image data import and export, conversion of image types and classes
Display and Exploration
Interactive tools for image display and exploration
Geometric Transformation and Image Registration
Scale, rotate, perform other N-D transformations, and align images using intensity correlation, feature
  matching, or control point mapping
Image Filtering and Enhancement
Contrast adjustment, morphological filtering, deblurring, ROI-based processing
Image Segmentation and Analysis
Region analysis, texture analysis, pixel and image statistics
Deep Learning for Image Processing
Perform image processing tasks, such as removing image noise and creating high-resolution images
  from low-resolutions images, using convolutional neural networks (requires Deep Learning Toolbox™)
3D Volumetric Image Processing
Filter, segment, and perform other image processing operations on 3D volumetric data
Code Generation
Generate C code and MEX functions for toolbox functions
GPU Computing
Run image processing code on a graphics processing unit (GPU)

REFERENCES

AAPM; Judy PF, Balter S, Bassano D, McCullough EC, Payne JT, Rothenberg L 1977, Report No. 001 – Phantoms for Performance Evaluation and Quality Assurance of CT Scanners. Diagnostic Radiology Committee Task Force on CT Scanner Phantoms. ISBN: 978-1-888340-04-4.

AAPM 2019a, Definition of a Qualified Medical Physicist Standard, https://www.aapm.org/medical_physicist/fields.asp#nuclear

AAPM; Mawlawi OR, Jordan DW, Halama JR, Schmidtlein CR, Wooten WW 2019b, Report No. 126 – PET/CT Acceptance Testing and Quality Assurance. The Report of AAPM Task Group 126 – published: October 2019, ISBN: 978-1-936366-73-6, https://doi.org/10.37206/193

AAPM; Halama JR et al. 2019c, Report No. 177 – Acceptance Testing and Annual Physics Survey Recommendations for Gamma Camera, SPECT, and SPECT/CT Systems. Task Group No. 177 (TG177), ISBN: 978-1-936366-68-2, https://doi.org/10.37206/184

Andreou M, Lagopati N, Lyra M 2011, Re-186 and Sm-153 dosimetry based on scintigraphic imaging data in skeletal metastasis palliative treatment and Monte Carlo simulation, J Phys Conf Ser, 317: 012013, DOI:10.1088/1742-6596/317/1/012013.

ANZSNM, Australian/New Zealand Standards, https://www.anzsnm.org.au

Argyrou M, Andreou M, Lagopati N et al. 2014, A comparison of patient specific dosimetric calculations obtained by planar images and Monte Carlo simulation in I-111octreotide therapy, Physica Medica, 30 (Suppl. 1): e29, DOI:10.1016/j.ejmp.2014.07.094 2014.

Argyrou M, Andreou M, Lagopati N et al. 2018, Patient specific dosimetric calculations obtained by planar images and Monte Carlo simulation in 111In octreotide therapy, Case Rep Imag Surg, 1 (3): 1–5, ISSN: 2516-8266, DOI:10.15761/CRIS.1000118.

Argyrou M, Lyra M 2019, Monte Carlo simulation in radionuclide therapy dosimetry, Biomed J Sci Tech Res, 15 (1): 2019.

Asmi H, Bentayeb F, Bouzekraoui Y et al. 2020, Energy window optimization for Y-90 Bremsstrahlung SPECT imaging: a Monte Carlo simulation study, Iranian J Med Phy, DOI:10.22038/ijmp.2020.44763.1691.

ASNC, American Society of Nuclear Cardiology, SPECT, https://www.dicardiology.com/channel/spect-imaging/

Assié K, Breton V, Buvat I et al. 2004, Monte Carlo simulation in PET and SPECT instrumentation using GATE, Nucl Instrum Methods Phys Res A: Accelerators, Spectrometers, Detectors and Associated Equipment, 527 (1–2): 180–189.

Auditore L, Amato E, Italiano A et al. 2019, Internal dosimetry for TARE therapies by means of GAMOS Monte Carlo simulations, Physica Medica, 64 (245–251): 1120–1797, Elsevier Ltd. https://doi.org/10.1016/j.ejmp.2019.07.024

Bagheri M, Parach AA, Razavi-Ratki SK, Nafisi-Moghadam R, Jelodari MA 2018, Patient specific dosimetry for paediatric imaging of 99mTc-dimercaptosuccinic acid with GATE Monte Carlo code, Radiat Prot Dosim, 178 (2): 213–222, DOI:10.1093/rpd/ncx101.

Bailey DL, Willowson KP 2013, An evidence-based review of quantitative SPECT imaging and potential clinical applications, J Nucl Med, 54 (1): 83–89, DOI:10.2967/jnumed.112.111476.

Bolch W, Lee C, Wayson M, Johnson P 2010, Hybrid computational phantoms for medical dose reconstruction: review, Radiat Environ Bioph, 49 (2): 155–168.

Boss A, Bisdas S, Kolb A et al. 2010, Hybrid PET/MRI of intracranial masses: initial experiences and comparison to PET/CT, J Nucl Med, 51 (8): 1198–1205, DOI: 10.2967/jnumed.110.074773.

Botta F, Mairani A, Hobbs RF et al. 2013, Use of the FLUKA Monte Carlo code for 3D patient specific dosimetry on PET-CT and SPECT-CT images, Phys Med Biol, 58 (22): 8099–8120, DOI:10.1088/0031-9155/58/22/8099.

Bouziotis P, Fiorini C 2014, SPECT/MRI: dreams or reality?, Clin Transl Imaging, 2: 571–573, DOI:10.1007/s40336-014-0095-6.

Burger W, Burge M 2007, Digital Image Processing: An Algorithmic Approach Using Java. Springer, Berlin Heidelberg, ISBN 1-84628-379-5.

Buvat I, Castilioni I 2002, Monte Carlo simulations in SPET and PET, QJ Nucl Med, 46: 48–61.

Chaichana A, Tocharoenchai C 2016, Development of a software tool for an internal dosimetry using MIRD method, J Phys Conf Ser, 694: 012056.

Cook GJR, Goh V 2019, What can artificial intelligence teaches us about the molecular mechanisms underlying disease?, EJNMMI, 46: 2715–2721, https://doi.org/10.1007/s00259-019-04370-z

Corrigendum to ICRP Publication 133, Jan 30, 2018 Correction Volume: 46 (3–4): 487, https://doi.org/10.1177/0146645317741391

Costa Monteiro E, Leon LF 2015, Metrological reliability of medical devices, J Phys Conf Ser, 588: 012032.

Cristy M, Eckerman K 1987, Specific Absorbed Fractions of Energy at Various Ages from Internal Photon Sources, Report ORNL/TM 8381.

Data Spectrum Corporation, http://www.spect.com/products.html

Data Spectrum Corporation, NEMA IEC Body Phantom Set™, Model PET/IEC-BODY/P, http://www.spect.com/pub/NEMA_IEC_Body_Phantom_Set.pdf

Demir M, Toklu T, Abuqbeitah M, Çetin H, Sezgin H S, Yeyin N, Sönmezoğlu K 2018, Evaluation of PET scanner performance in PET/MR and PET/CT systems: NEMA Tests, (MIRT) Mol Imaging Radionucl Ther, 27 (1): 10–18.

DICOM, Digital Imaging and Communications in Medicine, https://www.dicomstandard.org/

Dictionary and Handbook of Nuclear Medicine and Clinical Imaging). Experts in Medical & Radiologic Imaging Medical Dictionary, 1990, ISBN: 9781315892276, 1315892278, eText ISBN: 9781351088275, 1351088270 (https://www.xmri.com/resource-center/dictionary.html?term=60002).

Dougherty G 2009, Digital Image Processing for Medical Applications, Cambridge University Press, Cambridge, UK, ISBN 978-0-521-86085-7.

EANM; Carrio I et al. 2006, Best Practice in Nuclear Medicine, Part 1, https://www.eanm.org/publications/technologists-guide/best-practice-nuclear-medicine-part-1/

EANM; Cuocolo A et al. 2007, Best Practice in Nuclear Medicine Part 2, Technologist's Guide, https://www.eanm.org/publications/technologists-guide/best-practice-nuclear-medicine-part-2/

EANM 2007, Radiopharmacy Committee, cGRPP-guidelines, version2, Guidelines on Current Good Radiopharmacy Practice (cGRPP) in the Preparation of Radiopharmaceuticals.

EANM 2010a, Acceptance Testing for Nuclear Medicine Instrumentation, https://www.eanm.org/publications/guidelines/physics/

EANM 2010b, Routine quality control recommendations for nuclear medicine instrumentation, Eur J Nucl Med Mol Imaging, 37: 662–671, DOI:10.1007/s00259-009-1347-y, https://www.eanm.org/publications/guidelines/physics/.

EANM 2013, Curriculum for education and training of Medical Physicists in Nuclear Medicine, EANM and EFOMP Recommendations, https://www.eanm.org/publications/guidelines/physics/

EANM; Boellaard R et al. 2015, FDG PET/CT: EANM procedure guidelines for tumor imaging: version 2.0, Eur J Nucl Med Mol Imaging, 42: 328–354.

EANM/EFOMP; Del Guerra A et al. 2013, Curriculum for education and training of medical physicists in nuclear medicine, Physica Medica (EJMP), 29 (2): 139–162, DOI:10.1016/j.ejmp.2012.06.004, Recommendations from the EANM Physics Committee, the EANM Dosimetry Committee and EFOMP.

Ewusi-Annan E 2020, Calculate full width at half maximum (FWHM) of a peak. MATLAB Central File Exchange, https://www.mathworks.com/matlabcentral/fileexchange/37990-calculate-full-width-at-half-maximum-fwhm-of-a-peak

Eckerman KF, Endo A 2008, MIRD: Radionuclide Data and Decay Schemes, The Society of Nuclear Medicine, Reston, VA.

EGSnrc: software tool to model radiation transport, Canada, Government of Canada, National Research Council, EGS5: EGS at KEK Web Page, rcwww.kek.jp

Euratom 2014, Official Journal of the European Union OJ L13, 17.1.2014, basic safety standards for protection against the dangers arising from exposure to ionizing radiation, and repealing Directives 89/618/Euratom, 90/641/Euratom, 96/29/Euratom, 97/43/Euratom and 2003/122/Euratom, http://data.europa.eu/eli/dir/2013/59/oj

European Commission 2009, No-159, Guidelines on clinical audit for medical radiological practices, https://ec.europa.eu/energy/sites/ener/files/documents/159.pdf

European Commission 2014, No-175, Guidelines on radiation protection education and training of medical professionals in the European Union, https://ec.europa.eu/energy/sites/ener/files/documents/175.pdf

Fahey FH, Grogg K, El Fakhri G 2018, Use of Monte Carlo techniques in nuclear medicine, J Am Coll Radiol, 15 (3PtA): 446–448, DOI:10.1016/j.jacr.2017.09.045.

Fisher HLJ, Snyder WS 1966, Variation of Dose Delivered by 137Cs as a Function of Body Size from Infancy to Adulthood, ORNL-4007, Oak Ridge, TN: Oak Ridge National Laboratory, p. 221.

Fisher HLJ, Snyder WS 1967, Distribution of Dose Delivered in the Body Size from a Source of Gamma Rays Distributed Uniformly in an Organ, ORNL-4168, Oak Ridge, TN: Oak Ridge National Laboratory, p. 245.

Fornell D 2018, ITN Editor, Nuclear Imaging Moves towards Digital Detector Technology, June 15, Vendors offering options for direct digital imaging and advances in software.

Frey EC, Humm JL, Ljungberg M 2012, Accuracy and precision of radioactivity quantification in nuclear medicine images, Semin Nucl Med, 42 (3): 208–218, DOI:10.1053/j.semnuclmed.2011.11.003.

Gao G, Kim H, Kim BS et al. 2019, Tissue-engineering of vascular grafts containing endothelium and smooth-muscle using triple-coaxial cell printing, Appl Phys Rev, 6: 041402, https://doi.org/10.1063/1.5099306

Georgantzoglou A, Alzimami K, Livieratos L, Spyrou NM 2009, Monte Carlo simulation of a gamma-camera with general purpose and high resolution collimators for 99mTc and 131I data acquisition using GATE, Nucl Med Commun, 30 (5): 371, DOI:10.1097/MNM.0b013e32832b3490.

Gnesin S, Kieffer C, Zeimpekis K et al. 2020, Phantom-based image quality assessment of clinical 18F-FDG protocols in digital PET/CT and comparison to conventional PMT based PET/CT, EJNMMI Physics, https://doi.org/10.1186/s40658-019-0269-4

Hamamura MJ, Ha S, Roeck WW et al. 2010, Development of an MR-compatible SPECT system (MRSPECT) for simultaneous data acquisition, Phys Med Biol, 55: 1563.

Han E, Bolch W, Eckerman K 2006, Revisions to the ORNL series of adult and pediatric computational phantoms for use with the MIRD schema, Health Phys, 90: 337–356.

Hu Z, Yang W, Liu H et al. 2014, From PET/CT to PET/MRI: advances in instrumentation and clinical applications, Mol Pharm, 11 (11): 3798–3809, DOI:10.1021/mp500321h.

Hurtado JL, Lee C, Lodwick D, Geode T, Williams JL, Bolch WE 2012, Hybrid computational phantoms representing the reference adult male and adult female: construction and applications to retrospective dosimetry, Health Phys, 102 (3): 292–304, DOI:10.1097/HP.0b013e318235163f.

Hutton BF, Occhipinti M, Kuehne A et al. 2018, Development of clinical simultaneous SPECT/MRI, Br J Radiol, 90: 20160690.

IAEA 2003, Quality Control Atlas for Scintillation camera Systems, Vienna, ISBN 92-0-101303-5.

IAEA 2006, Nuclear Medicine Resources Manual, ISBN 92–0–107504–9, https://www-pub.iaea.org/MTCD/Publications/PDF/Pub1198_web.pdf

IAEA 2008, TECDOC 1597, Clinical Applications of SPECT CT: New Hybrid Nuclear Medicine Imaging System, ISBN 978-92-0-107108-8, ISSN 1011–4289.

IAEA 2014, Human Health Report No. 9, Quantitative Nuclear Medicine Imaging: Concepts, Requirements and Methods, Vienna, http://www.iaea.org/Publications/index.html

IAEA 2015a, Quality Management Audits in Nuclear Medicine Practices, IAEA Human Health Series No. 33, Vienna, 2nd edition, https://www.iaea.org/publications/search/type/human-health-series/year/2015

IAEA 2015b, TECDOC No. 1782, Good Practice for Introducing Radiopharmaceuticals for Clinical Use, ¦ 978-92-0-111215-6

IAEA 2018a, Quality Control in the Production of Radiopharmaceuticals, TECDOC-1856, IAEA, Vienna, https://www-pub.iaea.org/MTCD/Publications/PDF/TE-1856web.pdf

IAEA 2018b, Medical Physics Staffing Needs in Diagnostic Imaging and Radionuclide Therapy: An Activity Based Approach. Human Health Reports No. 15, Vienna, https://www.iaea.org/publications/12208/medical-physics-staffing-needs-in-diagnostic-imaging-and-radionuclide-therapy-an-activity-based-approach

IAEA Bulletin 2019, Cancer Control, Vol. 60-3, https://www.iaea.org/publications/magazines/bulletin/60-3

IAEA CRP 1072, 1998–2005, Development and Validation of an Internet Based Clinical and Technical Study Communication System for Nuclear Medicine, Coordinated Research Project, E11013, https://www.iaea.org/projects/crp/e11013

IAEA/WHO SSDL Network, 1997, Network of Secondary Standards Dosimetry Laboratories, 37: 1011–2669, https://www-pub.iaea.org/MTCD/Publications/PDF/Newsletters/SSDL-NL-37.pdf

ICRP; Snyder WS et al. 1975, Report of the Task Group on Reference Man, ICRP Publication 23, Oxford: Pergamon Press.

ICRP; Valentin J 2002, Basic anatomical and physiological data for use in radiological protection: reference values: annals of the ICRP, 32 (3–4): 1–277, Elsevier, Publication 89, https://doi.org/10.1016/S0146-6453 (03)00002-2

ICRP; Menzel HG, Clement C, DeLuca P 2009, Publication 110, Realistic reference phantoms: an ICRP/ICRU joint effort. A report of adult reference computational phantoms, Ann ICRP, 39 (2): 1–164.

ICRP; Petoussi-Henss N et al. 2014, Publication 116, The first ICRP/ICRU application of the male and female adult reference computational phantoms, Phys Med Biol, 59 (18): 5209, DOI:10.1088/0031-9155/59/18/5209.

ICRP 2015, Publication 128, Annals ICRP, 44 (2S).

ICRP; Bolch WE, et al. 2016, Publication 133, The ICRP computational framework for internal dose assessment for reference adults: specific absorbed fractions, Annals of ICRP, 45 (2): 1–74, http://journals.sagepub.com/doi/full/10.1177/0146645316661077ICRP; Zankl M et al. 2018, Computational phantoms and further developments, ICRP/ICRU Annals, 47 (3–4): 35–44.

ICRP 2019, Publication 140, Annals ICRP, 48 (1).

ICRU 1996, Medical Imaging – Report 54, The Assessment of Image Quality.

ICRU 2020, International Commission on Radiation Units & Measurements, Current Activities of ICRU, https://www.icru.org/

IDL, Interactive Data Language, https://en.wikipedia.org/wiki/IDL_(programming_language)

Ioan MR 2016, A novel method involving MATLAB coding to determine the distribution of a collimated ionizing radiation beam, JINST, 11: P08005, DOI:10.1088/1748-0221/11/08/P08005.

Israel O, Pellet O, Biassoni L et al. 2019, Two decades of SPECT/CT – the coming of age of a technology: an updated review of literature evidence, EJNMM Imaging, 46: 1990–2012, https://doi.org/10.1007/s00259-019-04404-6

ISTR 2019, The International Symposium on Trends in Radiopharmaceuticals, 28 October–1 November 2019, Vienna, Austria.

Jadvar H, Colletti PM 2014, Competitive advantage of PET/MRI, Eur J Radiol, 83 (1): 84–94.

JCGM 200:2008 International vocabulary of metrology – BIPM Basic and general concepts and associated terms (VIM).

Jha AK, Mithun S, Puranik AD, Purandare NC, Shah S, Agrawal A, Rangarajan V 2019, Performance characteristic evaluation of a bismuth germanate-based high-sensitivity 5-ring discovery image quality positron emission tomography/computed tomography system as per National Electrical Manufacturers Association NU 2-2012, World J Nucl Med, 18 (4): 351–360.

Jiang W, Chalich Y, Deen MJ 2019, Sensors for positron emission tomography applications, Sensors, 19 (22): 5019, https://doi.org/10.3390/s19225019

Jung J H, Choi Y, Im KC 2016, PET/MRI: technical challenges and recent advances, Nucl Med Mol Imaging, 50 (1): 3–12.

Kim CH, Yeom YS, Nguyen TT et al. 2016, The reference phantoms: voxel vs polygon, ICRP Annals, 45 (1_suppl): 188–201, https://doi.org/10.1177/0146645315626036

Kim CH, Yeom YS, Nguyen TT et al. 2018, New mesh-type phantoms and their dosimetric applications, including emergencies, Ann ICRP, 47 (3–4): 45–62.

Kramer R, Khoury HJ, Vieira JW, Lima VJ 2006, MAX06 and FAX06: update of two adult human phantoms for radiation protection dosimetry, Phys Med Biology, 51: 3331–3346.

Kramer R, Zankl M, Williams G, Drexler G 1982, The Calculation of Dose from External Photon Exposures using Reference Human Phantoms and Monte Carlo Methods: Part I, The Male (Adam) and Female (Eva) Adult Mathematical Phantoms, GSF Report S-885, GSF-National Research for Environment and Health, Neuherberg, Germany.

Lee C, Lodwick D, Hasenauer D, Williams JL, Bolch WE 2007, Hybrid computational phantoms of the male and female newborn patient: NURBS-based whole-body models, Phys Med Biol, 52: 3309–3333.

Lee C, Lodwick D, Hurtado J, Pafundi D, Williams JL, Bolch WE 2010, The UF family of reference hybrid phantoms for computational radiation dosimetry, Phys Med Biol, 55 (2): 339–363, DOI:10.1088/0031-9155/55/2/002.

Lee C, Lodwick D, Williams JL, Bolch WE 2008, Hybrid computational phantoms of the 15-year male and female adolescent: applications to CT organ dosimetry for patients of variable morphometry, Med Phys, 35: 2366–2382, DOI:10.1118/1.2912178, PMID: 18649470.Legland D 2020, ImageM, MATLAB Central File Exchange, https://www.mathworks.com/matlabcentral/fileexchange/45847-imagem

Ljungberg M, Celler A, Mark W, Konijnenberg MW, Eckerman KF, Dewaraja YK and Sjogreen-Gleisner K, in collaboration with the SNMMI MIRD Committee: Bolch WE, Bertrand Brill A, Fahey F., Fisher DR, Hobbs R, Howell RW, Meredith RF, Sgouros G, and Zanzonico Pat, and the EANM Dosimetry Committee: Bacher K, Chiesa C, Flux G, Lassmann M, Strigari L, Walrand S.

Ljungberg M, Strand SE 1989, A Monte Carlo program for the simulation of scintillation camera characteristics, Comput Meth Prog Bio, 29 (4): 257–272, https://doi.org/10.1016/0169-2607(89)90111-9

Loudos G 2007, Monte Carlo simulations in Nuclear Medicine, AIP Conf Proc, 958, 147, https://doi.org/10.1063/1.2825768

Lyra M, Ploussi A 2011, Filtering in SPECT image reconstruction, Int J Biomed Imaging, 2011, Article No.10, Hindawi Publishing Corp. NY.

Lyra M, Ploussi A, Georgantzoglou A 2011a, MATLAB as a Tool in Nuclear Medicine Image Processing, In Clara M. Ionescu (Ed) MATLAB – A Ubiquitous Tool for the Practical Engineer, IntechOpen, DOI:10.5772/19999. https://www.intechopen.com/books/matlab-a-ubiquitous-tool-for-the-practical-engineer/matlab-as-a-tool-in-nuclear-medicine-image-processing

Lyra M, Striligas J, Gavrilelli M, Chatzijiannis Ch, Skouroliakou K 2008, Thyroid volume determination by single photon tomography and 3D processing for activity dose estimation, IEEE International Workshop on Imaging Systems and Techniques – IST 2008, Chania, Greece, September 10–12, 2008.

Lyra M, Striligas J, Gavrilleli M, Lagopati N 2011b, Volume Quantification of 123I-DaTSCAN Imaging by MATLAB for the Differentiation and Grading of Parkinsonism and Essential Tremor, January 2011, DOI:10.1109/CSSR.2010.5773757.·Source: IEEE Explore, 2010 International Conference on Science and Social Research (CSSR).

Marshall HR, Stodilka RZ, Theberge J et al. 2011, A comparison of MR-based attenuation correction in PET versus SPECT, Phys Med Biol, 56 (2011): 4613–4629, DOI:10.1088/0031-9155/56/14/024.

Meena A, Raja K 2013, Segmentation of Alzheimer's disease in PET scan datasets using MATLAB, Source arXiv, DOI:10.18000/ijisac.50121.

Merlin Th, Stute S, Benoit D et al. 2018, CASToR: a generic data organization and processing code framework for multi-modal and multi-dimensional tomographic reconstruction, Phys Med Biol, 63 (18), http://www.castor-project.org/features

MIRD; Thomas SR, Stabin MG, Chen C-T, Samaratunga RC 1999, Pamphlet No. 14 Revised, Dynamic Bladder Software Tool.

MIRD; Bolch EW et al. 1999, Pamphlet No. 17, The dosimetry of non-uniform activity distributions radionuclide S values at the voxel level, J Nucl Med, 40: 11S–36S.

MIRD Perspective; Howell RW et al. 1999, In collaboration with the MIRD Committee of SNM, J Nucl Med, 40: 38–10S.

MIRD; Dewaraja YK et al. 2012, Pamphlet No. 23, Quantitative SPECT for patient-specific 3-dimensional dosimetry in internal radionuclide therapy, J Nucl Med, 53: 1310–1325, DOI:10.2967/jnumed.111.100123.

MIRD; Vaziri B, Wu H, Dhawan AP, Du P, Howell RW 2014, Pamphlet No. 25, in collaboration with the SNMMI MIRD Committee, MIRDcell V2.0 software tool for dosimetric analysis of biologic response of multicellular populations, J Nucl Med, 55 (9): 1557–1564.

MIRD 2016, Pamphlet No. 26, Joint EANM/MIRD guidelines for quantitative 177-Lu SPECT applied for dosimetry of radiopharmaceutical therapy, J Nucl Med, 57: 151–162, DOI:10.2967/jnumed.115.159012.MIRD Monograph: MIRD Radiobiology and Dosimetry for Radiopharmaceutical Therapy with Alpha-Particle Emitters, Editor: G. Sgouros, PhD.

Mirzaei S, Zakavi R, Rodrigues M, Schwarzgruber Th 2010, Fully automated 3D basal ganglia activity measurement in dopamine transporter scintigraphy (Spectalyzer), Ann Nucl Med, 24: 295–300, DOI:10.1007/s12149-010-0353-2.

MITA, Medical Imaging and Technology Alliance, https://www.nema.org/Products/Pages/Medical-Imaging.aspx

Mofrad FB, Zoroofi RA, Tehrani-Fard AA, Akhlaghpoor S, Hori M, Chen Y-W, Sato Y 2010, Statistical construction of a Japanese male liver phantom for internal radionuclide dosimetry, Radiat Prot Dosim, 140 (2): 140–148, DOI:10.1093/rpd/ncq164.

Moore SC 1997, BrighamRAD, Physical Characteristics of Nuclear Medicine Images scmoore@bwh.harvardedu.

Musafargani S, Ghosh KK, Mishra S, Mahalakshmi P, Padmanabhan P, Gulyás B 2018, PET/MRI: a frontier in era of complementary hybrid imaging, Eur J Hybrid Imaging, 2 (1): 12, DOI:10.1186/s41824-018-0030-6.

Na YH, Zhang B, Zhang J, Caracappa PF, Xu G 2010, Deformable adult human phantoms for radiation protection dosimetry: anthropometric data representing size distributions of adult worker populations and software algorithms, Phys Med Biol, 55: 3789.

Narita Y, Eberl S, Iida H, Hutton BF, Braun M, Nakamura T, Bautovich, G 1996, Monte Carlo and experimental evaluation of accuracy and noise properties of two scatter correction methods for SPECT, Phys Med Biol, 41: 2481.

NCRP; Hoover M et al. 2017, Report No. 176 – Radiation Safety Aspects of Nanotechnology – osti.gov, DOI:10.2172/1351238.

NEMA NU1-2018, Performance Measurements of Gamma Cameras, NEMA Standards, Document ID: 100113, https://www.nema.org/Standards/Pages/Performance-Measurements-of-Gamma-cameras.aspx

NEMA NU2-2018 Performance Measurements of Positron Emission Tomographs, Performance Measurements of Positron Emission Tomographs (PET).

Oddstig J, Svegborn LS, Almquist H et al. 2019, Comparison of conventional and Si-photomultiplier-based PET systems for image quality and diagnostic performance, BMC Med Imaging, https://doi.org/10.1186/s12880-019-0377-6

Ovtchinnikov E, Brown R, Kolbitsch C et al. 2020, SIRF: Synergistic Image Reconstruction Framework, CPC 50th anniversary article, Comput Phys Commun, 249: 107087.

Pelowitz DB 2005, MCNPX user's manual version 2.5.0., Los Alamos, NM: Los Alamos National Laboratory.

Pointon B 2018, ImSim A nuclear imaging teaching simulator using compiled MATLAB, Nucl Med, 59 (supplement 1): 1189.Pujol A Jr, Gibbs SJ 1982, A Monte Carlo method for patient dosimetry from dental x-ray, Dentomaxillofac Radiol, 11, p. 25, birpublications.org, Published Online: 23-1-2015, https://doi.org/10.1259/dmfr.1982.0003

Python Software Foundation 2001–2020, https://www.python.org/psf/

RADAR, the Radiation Dose Assessment Resource, www.doseinfo-radar.com

Radiology Key 2016, Digital Image Processing in Nuclear Medicine, Physics in Nuclear Medicine, Chapter 20.

Raeside DE 1976, Monte Carlo principles and applications, Phys Med Biol, 21: 181–197.

Sabour S, Abbasnezhad O, Mozaffarian S, Kangavari HN 2017, Accuracy and precision in medical researches; common mistakes and misinterpretations, WJRR, 4 (4): 58–60, ISSN:2455-956.

Salvado O 2020, Interpolation with Reverse Diffusion, MATLAB Central, https://www.mathworks.com/matlabcentral/fileexchange/14120-interpolation-with-reverse-diffusion

Segars WP, Tsui BMW 2009, MCAT to XCAT: The evolution of 4-D computerized phantoms for imaging research: computer models that take account of body movements promise to provide evaluation and improvement of medical imaging devices and technology, Proc IEEE Inst Electr Electron Eng, 97 (12): 1954–1968, DOI:10.1109/JPROC.2009.2022417.

Segars WP, Sturgeon G, Mendonca S, Grimes J, Tsui BMW 2010, 4D XCAT phantom for multimodality imaging research, Med Phys, 37 (9): 4902–4915, DOI:10.1118/1.3480985, PMC2941518.

Segars WP, Tsui BMW, Cai J, Yin F, Fung GSK, Samei E 2018, Application of the 4-D XCAT phantoms in biomedical imaging and beyond, in IEEE T Med Imaging, 37 (3): 680–692.

Shepp L, Logan BF 1974, The Fourier reconstruction of a head section, IEEE T Nucl Sci, 21 (3). DOI:10.1109/TNS.1974.6499235.

Slomka PJ, Baum RP 2009, Multimodality image registration with software: state-of-the-art, Eur J Nucl Med Mol Imaging, 36 (Suppl 1): S44–S55, DOI:10.1007/s00259-008-0941-8.

Snyder WS, Ford MR, Warner GG, Fisher HL 1969, Estimates of absorbed fractions for monoenergetic photon sources uniformly distributed in various organs of a heterogeneous phantom, New York, J Nucl Med, 3: 7–52.

Stabin GM 1996, MIRDOSE personal computer software for internal dose assessment in nuclear medicine, J Nucl Med, 37: 538–646.

Stabin GM, Spark RB, Crowe E 2005, OLINDA/EXM: the second-generation personal computer software for internal dose assessment in nuclear medicine, J Nucl Med, 46 (6): 1023–1027, Oak Ridge Institute for Science and Education, Oak Ridge, Tennessee.

Staelens S, Buvat I 2009, Monte Carlo simulations in nuclear medicine imaging, In: Advances in Biomedical Engineering, pp. 177–209, https://doi.org/10.1016/B978-0-444-53075-2.00005-8

Synefia S, Gavrilelli M, Valassi A et al. 2014, 3D SPECT thyroid volume quantification by Thyr-Vol algorithm in MATLAB, Physica Medica, 30 (Supplement 1): e115–e116, DOI:10.1016/j.ejmp.2014.07.330.

Thielemans K, Tsoumpas Ch 2012, STIR: software for tomographic image reconstruction release 2, IOP publishing Physics in Medicine and Biology, Phys Med Biol, 57: 867–883, DOI:10.1088/0031-9155/57/4/867.

Toossi MB, Islamian JP, Momennezhad M et al. 2010, SIMIND Monte Carlo simulation of a single photon emission CT, J Med Phys/Assoc Med Phys India, 35: 42.

Truesdell C, Noll W 2004, The non-linear field theories of mechanics, In: Antman SS (Eds) The Non-Linear Field Theories of Mechanics, Springer, Berlin, Heidelberg ISBN 978-3-642-05701-4, https://doi.org/10.1007/978-3-662-10388-3_1.

Turner JE, Wright HA, Hamm RH 1985, A Monte Carlo primer for health physicists, Health Phys, 48: 717–733.

Universal Medical, CIRS Model 610 AAPM CT Performance phantom, https://www.universalmedicalinc.com/aapm-ct-performance-phantom.html

Veress AI, Segars WP, Tsui BMW, Gullberg GT 2011, Incorporation of a left ventricle finite element model defining infarction into the XCAT imaging phantom, IEEE T Med Imaging, 30 (4): 915–927.

Wang H, Sun X, Wu T et al. 2018, Deformable torso phantoms of Chinese adults for personalized anatomy modelling, J Anatomy, 233: 121–134, DOI:10.1111/joa.12815.

Zaidi H, Xu XG 2007, Computational anthropomorphic models of the human anatomy: the path to realistic Monte Carlo modeling in radiological sciences, Ann Rev Biomed Eng, 9: 471–500.

Zankl M, Veit R, Williams G et al. 1988, The construction of computer tomographic phantoms and their application in radiology and radiation protection, Radiat Environ Bioph, 27: 153–164, https://doi.org/10.1007/BF01214605

Zanzonico BP 2000, Internal radionuclide radiation dosimetry: a review of basic concepts and recent developments, J Nucl Med, 41: 297–308.

2 Image Formation in Nuclear Medicine

Nefeli Lagopati

CONTENTS

2.1 INTRODUCTION AND GENERAL INFORMATION

Analog and digital images are the two main types of images that exist. All medical images recorded on film or presented on various display devices, like computer monitors, are offered in an analog form. In an analog image, various levels of brightness (or film density) and colors are shown (Cromey D. W., 2013). A digital image can be considered as a discrete representation of data possessing both spatial (layout) and intensity (color) information. The great advantage of digital images is that they can be processed, in many ways, by computer systems (van der Stelt P. F., 2008).

Image formation in nuclear medicine includes the processes of imaging acquisition and processing. It is the method for producing images by detecting radiation from different parts of the body, after a radioactive tracer is administered intravenously to the patient (Govaert G. A. M. et al., 2016). Actually, when gamma rays are absorbed inside the sodium iodide crystal, a visible light flash known as a scintillation

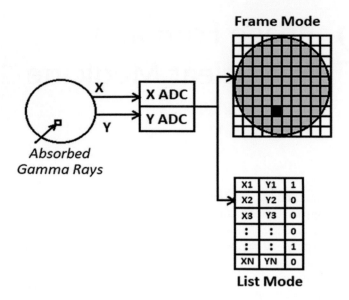

FIGURE 2.1 Computer acquisition modes. Digitalized signals from the γ-camera can be stored in list mode or in frame mode.

is created, sampled by an array of photomultiplier tubes, creating a set of electronic pulses, namely x and y position signals, which correspond to the location of the scintillation on the crystal, as well as an energy signal which is proportional to the energy absorbed in the interaction. If this energy signal falls within a pre-selected energy window, a pair of analog-to-digital converters (ADCs) can digitize these x and y signals (Madsen M. T., 1994).

The information, related to the position of the event (scintillation), can usually be stored in two ways, either as a sequential list of position entries (list mode), or by incrementing a matrix element (pixel) corresponding to the location of the position, known as frame or matrix mode (Figure 2.1). In both cases, images can be acquired in a zoom mode.

Thus, the images are digitally generated on a computerized system and transferred to a nuclear medicine physician, who interprets the images to make a diagnosis, as a part of his clinical routine (Faber T. L. et al., 1994).

2.1.1 BINARY REPRESENTATION

Computers, virtually, are based on the handling of information which is coded in the form of binary numbers. Binary number can have only one of two values, 0 or 1 and these numbers are referred to as binary digits or bits. A sequence of bits in computer language represents a piece of information and is equal to a word in physical language (Stokes J., 2007). When this sequence contains 8 bits, then a byte is formed, which is commonly used as the basic unit for expressing amounts of binary-coded information. Binary coding of image information is obligatory to store images in a computer. Most imaging devices used in medicine however generate information which can assume a continuous range of values between preset limits (analog form) and it is necessary to convert this analog information into the discrete form required for binary coding, via ADCs (Block F. E., 1987). Since, there are still many displays and photographic devices, in medical imaging, designed for handling images in analog format, it is important to reconvert the discrete, binary data into analog information, using Digital-to-Analogue Converters (DACs).

2.2 PIXEL

2.2.1 HISTORY AND ETYMOLOGY

The word "pixel" is a portmanteau of the shortening of the words "picture" and "element". It was first seen in the *Wireless World* magazine in 1927, in an article related to the new technology of television (Dinsdale A., 1927; Lyon R. F., 2006).

Although pixel was already being used in engineering circles a couple of years earlier, it did not appear in print until 1965, in a paper presented at SPIE (Society of Photo-Optical Instrumentation Engineers), by an engineer called Fred Billingsley (Billingsley F. C., 1965). The advent of Computed Tomography (CT) contributed to the introduction of this terminology. The term "pel" had also been used previously as another synonym for pixel, but it was never used in medical imaging.

2.2.2 DEFINITION

Pixel or pel or dot of illumination represents the smallest, constituent element in a digital image (Foley J. D. et al., 1982; Graf R. F., 1999). Thousands or even millions of individual pixels together make up an image. Pixel contains a numerical value which is the basic unit of information within the image at a given spatial resolution and quantization level. It is indexed, using geometric coordinates, as an (x, y) or column row (c, r), location from the origin of the image (Solomon C. et al., 2011) and it is represented by a dot or square on a computer monitor display screen. Pixels usually contain the color or intensity response of the image, as a small point-sample of colored light from the whole scene. All images do not always contain strictly visual information. Depending on the display monitor, the quantity, size, and color combination of pixels varies and is measured in terms of the display resolution. Each pixel has a unique logical address and size (Singh V. et al., 2004). The pixel resolution spread determines the quality of display. This means that more pixels per inch of monitor screen yields better image results (Rakhshan V., 2014). Particularly, if pixel is considered as a sample of an original image, thus, the more samples are created, the more accurate representation of the image is obtained. Hence, if the concentration of pixels is great the quality of image is better, because the image is clear.

2.3 IMAGE DIGITIZATION

Digitization is the process used to convert analog data into digital form that is saved in the computer memory (Block F. E., 1987). It can be achieved by two steps: sampling and quantization. Sampling is the process of dividing the image in pixels and it gives the resolution of a digital image. It is necessary to choose the appropriate resolution to preserve all the information which is crucial to be captured (Burger W. et al., 2009). The spatial sampling is the conversion of the continuous signal into its discrete representation and it depends on the geometry of the sensor elements of the acquisition device. Temporal sampling is carried out by measuring at regular intervals the amount of light incident on each individual sensor element.

Quantization is the procedure of assigning a numerical value to every pixel. The value represents color (brightness) of the pixel. For instance, an 8-bit quantization allows $2^8 = 256$ numerical values for each pixel and $2^{12} = 4096$ numerical values per pixel. A floating – point scale is usually used in professional applications such as medical imaging. Conversion is carried out by ADCs, embedded directly in the sensor electronics or is performed by spatial interface hardware (Pavlin M. et al., 2012).

2.3.1 IMAGE STORAGE

A digital or digitized image can be stored in the computer as an array or matrix of count values, as it was previously mentioned, and it is displayed by assigning a gray or color scale that depends on the number of counts in each element (Gurcan M. N. et al., 2009). These arrays are typically square matrices that have dimensions that range from 32×32 up to 1024×1024. Most nuclear medicine images have dimensions of either 64×64 or 128×128 or 256×256. Each matrix element (pixel) is a location in computer memory. From a mathematical point of view, a 64×64 matrix has 4096 pixels, while a 128×128 matrix is four times larger (16,384 pixels), and a 256×256 matrix is 16 times larger (65,536 pixels) (Karellas A. et al., 1992).

The number of counts which can be stored in a pixel depends on the number of bits which are allocated. It is most convenient to assign either 8 bits (1 byte) or 16 bits (2 bytes) to each pixel, due to the design and the mechanism of a computerized system (Zhang X., 2004). But there is some information that it must be taken into account, such as that a 1-byte pixel requires half the storage space of a 2-byte pixel but has a limited dynamic range. The maximum number of counts that can be stored in a 1-byte pixel is 255 ($2^8 - 1$) and in a 2-byte pixel is 65,535 ($2^{16} - 1$) (Tanley S. W. M. et al., 2013).

2.4 BITS – GRAYSCALE – COLOR SCALE

Each pixel can exist in only one color at a time. Although pixels are so small, they often blend together to form new colors (Cromey D. W., 2013). The number of distinct colors that can be represented by a pixel depends on the number of bits per pixel (bpp). In a 1-bpp image, each pixel can be either on or off. Each additional bit can double the number of available colors. Therefore a 2-bpp image can have four colors. Table 2.1 includes the main types of bpp images and the relevant number of colors (Dragoi I.-C. et al., 2016). In some systems that use the RGB color model, there are 2^{16}, or 65,636, possible levels for each primary color. When $R = G = B$ in this system, the image is known as 16-bit grayscale because the decimal number 65,536 is equivalent to the 16-digit binary number 1111111111111111 (Rossner M. et al., 2004).

TABLE 2.1

The Most Frequently Abandoned Bits per Pixel (bpp) Images with the Corresponding Color Matching

bpp	Number of Colors
1	$2^1 = 2$ (monochrome)
2	$2^2 = 4$
3	$2^3 = 8$
8	$2^8 = 256$
16	$2^{16} = 65,536$ (high color)
24	$2^{24} = 16,777,216$ (true color)
...	...

For color depths of 15 or more bpp, the depth is typically the total number of the bits allocated to each of the red, green, and blue components. High color image normally has 5 bits for red and blue each, and 6 bits for green, since the human eye is more sensitive to errors in green than in the other two primary colors. If the application involves transparency, then the 16 bits may be divided into 5 bits each of red, green, and blue, with one bit left for transparency (Thompson L. et al., 2013). As expected, a 16-bit digital grayscale image consumes far more memory than the same image, with the same physical dimensions, rendered in 8-bit digital grayscale. A 24-bit depth allows 8 bits per component (Larobina M. et al., 2014). A 32-bit or 64-bit depth is available on some systems. This means that each 24-bit pixel has an extra 8 bits to describe its opacity (for purposes of combining with another image).

Bit resolution is the parameter that defines the number of possible intensity values that a pixel may have and relates to the quantization of the image information. Thus, a binary image has just two colors; a grayscale image commonly has 256 different gray levels, whilst for a color image this number depends in the color range in use. The bit resolution is usually quoted as the number of binary bits required for storage at given quantization level (Hockley N. S. et al., 2012). The range of values a pixel may take is often referred to as the dynamic range of an image.

Images in grayscale have pixels that range from black to white, going through a number of intensity steps in between. Most images allow 256 different intensity levels and this number arises from the 256 different values that a byte can take (Halazonetis D. J., 2005). Grayscale is a range of shades of gray without apparent color. Digital radiography images are displayed in grayscale. The dose of radiation that hits a pixel is assigned a bit which is then expressed in the image as a shade of gray. The human eye can discern more than 720 shades of gray, approximately 9–10 bits (Gómez-Polo C. et al., 2014).

The darkest shade corresponds to black, which means that there is total absence of transmitted or reflected light. The lightest shade matches to white, where there is total transmission or reflection of light at all visible wavelengths (Sikri V. K., 2016). All the rest of intermediate shades of gray are represented by equal brightness levels of the three primary colors (red, green, and blue [RGB]) for transmitted light, or

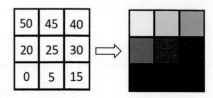

FIGURE 2.2 The matrix size defines the number of pixels in an image. The variation in counts shows a variation in the grayscale.

equal amounts of the three primary pigments (cyan, magenta, and yellow [CMY]) for reflected light (Qin T. et al., 2018).

The matrix size defines the number of pixels in an image. For instance, a 3×3 matrix contains 9 pixels. This matrix shows a count value in each cell. It decreases in descending order with the lower right containing 15 counts and upper left at 50 counts (Figure 2.2). This variation in counts shows a variation in the grayscale where increasing shades of gray relates to decease in the number of counts.

Particularly, the RGB scale is calibrated so that when all three red/green/blue numbers of a pixel are equal, the pixel is a shade of gray (Table 2.2), because when all the numbers are equal, it is drained any bias toward red, green, or blue. Thus, for every pixel in a red-green-blue (RGB) grayscale image, $R = G = B$ (Prasad S. et al., 2017). In fact, the original pure black/white colors fit this all-equal pattern too, just using the values 0 and 255. The lightness of the gray is directly proportional to the number representing the brightness levels of the primary colors. Therefore, black is represented by $R = G = B = 0$ or $R = G = B = 00000000$, and white is represented by $R = G = B = 255$ or $R = G = B = 11111111$. Because there are 8 bits in the binary representation of the gray level, this imaging method is called 8-bit grayscale. This system works because the displays and other systems using RGB are calibrated so that the all-equal cases are on the grayscale (Smith H. J. et al., 1992).

In the case of reflected light, the levels of cyan (C), magenta (M), and yellow (Y) for each pixel are represented as a percentage from 0 to 100. For each pixel in a cyan-magenta-yellow (CMY) grayscale image, all three primary pigments are present in equal amounts ($C = M = Y$) (Kanan C. et al., 2012). The lightness of the gray is inversely proportional to the number representing the amounts of each pigment. White is thus represented by $C = M = Y = 0$, and black is represented by $C = M = Y = 100$.

TABLE 2.2

Examples of Gray Colors, in Red-Green-Blue (RGB)

Red	Green	Blue	Color
50	50	50	Dark gray
120	120	120	Medium gray
200	200	200	Light gray
0	0	0	Pure black
255	255	255	Pure white

A color is typically represented by three (red, green, and blue – RGB scale) or four (cyan, magenta, yellow, and black) component intensities (McGavin D., 2005).

In analog practice, grayscale imaging is sometimes called "black and white", but technically this is a misnomer. In true black and white, also known as halftone, the only possible shades are pure black and pure white. The illusion of gray shading in a halftone image is obtained by rendering the image as a grid of black dots on a white background (or vice versa), with the sizes of the individual dots determining the apparent lightness of the gray in their vicinity (Chalazonitis A. N. et al., 2003). The halftone technique is commonly used for printing photographs in newspapers and not in medical imaging.

In some cases, three other parameters are used to define grayscale. These are hue, saturation, and value. Each of these three parameters can be interpreted as follows: Hue (H) is the dominant wavelength of the color (red, blue, etc.). Saturation (S) is the purity of color, in the sense of the amount of white light mixed with it. Value (V) is the brightness of the color. It is also known as luminance. In a grayscale image, the hue (apparent color shade) and saturation (apparent color intensity) of each pixel is equal to 0 (Teixidó M. et al., 2012). The lightness (apparent brightness) is the only parameter of a pixel that can vary. Values can range from a minimum of 0 (black) to 100 (white).

Clinical radiological images, such as CT and magnetic resonance (MR) images, have a grayscale photometric interpretation (Matsuda S. et al., 2017). Nuclear medicine images, such as positron emission tomography (PET) and single photon emission tomography (SPECT), are typically displayed with a color map (Jiemy W. F. et al., 2018). Ultrasound images are typically stored employing the RGB system. Also, color is used to encode blood flow direction (and velocity) in Doppler ultrasound. Additional functional information onto a grayscale anatomical image is presented in colored overlays, as in the case of fMRI activation sites. Moreover, PET/CT and PET/MRI allow simultaneous display of functional and anatomical images (Njemanze P. C. et al., 2017).

2.5   OTHER PICTURE PARAMETERS

2.5.1   Spatial Resolution

Spatial resolution is a term that refers to the number of pixels utilized in construction of a digital image (Oborska-Kumaszyńska D. et al., 2010). Practically, spatial resolution is considered as the smallest change in the position of the pixels that can be recognized. In a picture of low-resolution, the image looks blocky, or "pixelated" with detectable individual pixels. Spatial resolution is commonly quoted as C × R (640 × 480, 800 × 600), where C and R are the column and row dimensions of the image, respectively (Solomon C. et al., 2011).

2.5.1.1   Pixel Pitch

Pixel pitch describes the distance between pixels in an image. As the distance between pixels gets smaller, the resolution of the image gets more accurate. In digital radiography, usually, the optimum pixel pitch is between 100 and 150 μm

(Indrajit I. K. et al., 2009). In the case of nuclear medicine, the optimum pixel size is given by Equation 2.1,

$$PS_{opt} \approx \frac{FWHM}{3} \tag{2.1}$$

where PS_{opt} is the optimal pixel size, and FWHM is the Full-Width Half-Maximum (FWHM[1]) which is used in order to estimate the energy resolution[2] of γ-camera systems (currently the SPECT system resolution including both the detector resolution and the selected collimator resolution).

2.5.2 CONTRAST RESOLUTION

Contrast resolution is the ability to distinguish the differences between shades. It is quite subjective parameter and it is difficult to be defined because each person has unique ability to realize separate shades of colors. A good contrast ratio allows to better detect differences in the shades of gray of a particular pixel (Averbukh A. N. et al., 2005). The usual diagnostic monitors offer contrast ratios of 10,000:1. If the background is dark enough and the absolute white of a pixel is very bright, then the radiographic image is optically clear and easy to be observed and studied (Hillman E. M. C. et al., 2011).

2.5.3 TEMPORAL RESOLUTION

Temporal resolution for a continuous capture system is the number of images captured in a given time period. It is commonly quoted in frames per second (fps), where each individual image is referred to as a video frame. Fps values between 25 and 30 fps are considered as suitable for most visual surveillance (Ribeiro B. et al., 2013).

2.6 VOXEL

Voxel is the short of the words volume and element, so it is a volumetric pixel. A voxel is a unit of graphical information that defines a point in three-dimensional space (Zhang Q. et al., 2011). Since a pixel defines a point in two-dimensional space

---

[1] FWHM is a parameter which is used to characterize the width of a peak on a graph. In nuclear medicine, the FWHM is utilized in order to estimate the energy resolution of γ-camera systems. FWHM is an important measurement to assess the efficiency of the scintillation counting equipment in a nuclear medicine department and it should typically be less than 10% (Zanzonico P., 2008).

[2] Energy resolution is defined as the ability of a detector to accurately determine the energy of the incoming radiation. It is assumed that there is no flawless system, thus there is no system being capable of determining precisely what energy photon struck the crystal (Vandenberghe S. et al., 2016). Every system can only determine within a range of values, what energy radiation it is detecting. For this reason, the energy resolution is expressed as a percent of the energy of the incoming photons. The energy resolution is important parameter in evaluating the performance of a γ-camera, since this parameter allows a camera to differentiate between primary photons and Compton scattered photons (Ashoor M. et al., 2015). The equation for determining the percent energy resolution for a particular radionuclide is: %Energy resolution = FWHM × 100/photo peak.

with its x and y coordinates, a third z coordinate is needed. Each of the coordinates is defined in terms of its position, color, and density. Voxel can be considered as a cube, where any point on an outer side is expressed with an x, y coordinate (Ye H. et al., 2011). The third, z coordinate, defines its density and its color. Voxels are frequently used in the visualization and analysis of medical and scientific data (Figure 2.3).

A voxel represents a single sample, or data point, on a three-dimensional grid. This point can represent a single parameter or multiple parameters, and, consequently, the value of a voxel may represent various properties (Hutton C. et al., 2008). In CT scans, the values are Hounsfield units, giving the opacity of material to X-rays. Different types of value are acquired from MRI or ultrasound or γ-camera.

A voxel represents only a single point on this grid and not a volume. Thus, the space between each voxel is not represented in a voxel-based dataset (Loewe K. et al., 2014). Depending on the type of data and the intended use for the dataset, this missing information may be reconstructed via interpolation and other procedures.

While voxels provide the benefit of precision and depth of reality, they are typically large datasets and difficult to manage (Gillies R. J. et al., 2016). However, through efficient compression and manipulation these large data files can allow interactive visualization on common computers.

2.7 MEDICAL IMAGE FILE FORMATS IN NUCLEAR MEDICINE

A standardized way to store the information describing a medical image in a computer file is provided by the appropriate file formats of the medical images (Varma D. R., 2012). A medical image dataset consists of different information obtained by the projection of an anatomical volume (planar images), or by thin slices through a volume (two-dimensional imaging – tomography), or by a volume (three-dimensional imaging) or by multiple acquisition of the same tomographic image over time, to produce a dynamic series of acquisitions (four-dimensional imaging) (Kohli M. D. et al., 2017). The file format choice is crucial to describe in which way these data are organized inside the image file. Moreover, file format describes how the pixel data should be interpreted by software for the correct loading and visualization (Varma D. R., 2012).

Generally speaking, a medical image is the representation of the internal structure or function of an anatomic region in the form of an array of pixels or voxels (Gerber A. J. et al., 2008). It results from a sampling or reconstruction process that maps numerical values to positions of the space.

Pixel data can be stored as integers or floating-point numbers, utilizing the minimum number of bytes required to represent the values, according to the data type (Cromey D. W., 2013). Nuclear medicine modalities, PET and SPECT as well as radiological images like CT and MR store 16 bits for each pixel as integers (Cherry S. R., 2009). Images, generated by tomographic imaging modalities, are sent to a Picture Archiving and Communication System (PACS) or a reading station.

Medical image file formats can be divided in two categories. The first includes formats intended to standardize the images generated by diagnostic modalities (Koutelakis G. V. et al., 2012). Digital Imaging and Communications in Medicine (DICOM) is one of the most common formats of this category. The second includes

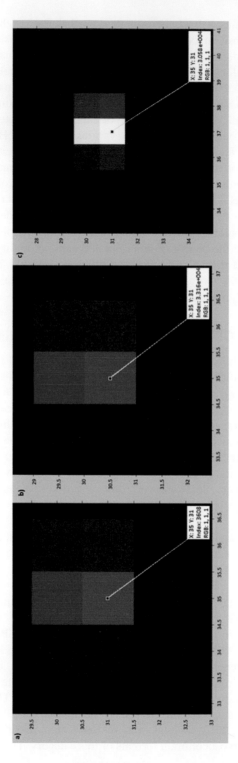

FIGURE 2.3 (a) A matrix with voxels that represent counts, (b) a matrix with voxels that represent cumulative activity, (c) A matrix with voxels that represent dose. Scintigraphy images are three-dimensional matrices and can be converted into dosimetric data. (From Vamvakas I. et al., 2015.)

formats which are created to facilitate and strengthen post-processing analysis (Yvernault B. C. et al., 2014). Analyze, Neuroimaging Informatics Technology Initiative (NIfTI) and Medical Imaging NetCDF (MINC) are the most representative ones.

Medical image files are typically stored in two possible configurations. The first configuration requires the files to contain both the metadata and image data, with metadata stored at the beginning of the file (Savaris A. et al., 2014). DICOM, MINC, and NIfTI files use this format. The second configuration stores the metadata in one file and the image data in another. The Analyze file format uses this two-files paradigm (Neu S. C. et al., 2012).

2.7.1 METADATA

The information of an image can be described through metadata. Metadata is typically stored at the beginning of the file as a header (Kohli M. D. et al., 2017). It contains at least the dimensions of the image matrix, the spatial resolution, the pixel depth, and the photometric interpretation. Metadata allows a software application to recognize and correctly open an image in a supported file format (Tan L. K., 2006).

In medical images, metadata have a wider role. Images from different diagnostic modalities typically include information about the origin and the way of the image development (Korenblum D. et al., 2011). A nuclear medicine image will have information about the radiopharmaceutical injected and the weight of the patient (Volkan-Salanci B. et al., 2012). These data allow the appropriate software like to on-the-fly convert pixel values in standardized uptake values (SUV) without the need to really write SUV into the file (Kumar Pandey A. et al., 2012).

Metadata are a powerful tool to exploit image-related information for clinical and research purposes. Additionally, metadata allow organizing and retrieving into archives images and associated data (Gutman D. A. et al., 2014).

2.7.2 DIGITAL IMAGING AND COMMUNICATIONS IN MEDICINE (DICOM)

The DICOM standard was established by the American College of Radiology and the National Electric Manufacturers Association, in 1993 (Mendelson D. S. et al., 2013). However, the real introduction of the DICOM standard into imaging departments takes place at late 1990s. Nowadays, the DICOM standard is the reference axis of every medical imaging department. DICOM is not only a file format but also a network communication protocol (Haak D. et al., 2016).

In a DICOM file format, the innovation, which is notable, is that pixel data cannot be separated from the description of the medical procedure. Metadata and pixel data are merged in a unique file (Varma D. R., 2012). The DICOM header, in addition to the information about the image matrix, contains the most complete description of the entire procedure used to generate the image ever conceived regarding to acquisition protocol and scanning parameters. Personal data and physical characteristics of each patient, such as name, gender, age, weight, and height are contained in the header. From a practical point of view, the header allows the image to be self-descriptive (Larobina M. et al., 2014).

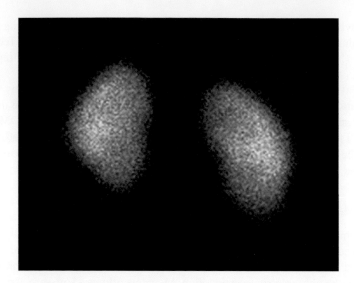

FIGURE 2.4 Posterior image of kidneys in a DICOM format. (From Lyra M. et al., 2011.)

With regard to the pixel data, DICOM can only store pixel values as integers. It supports compressed image data via a mechanism which allows a non-DICOM-formatted document to be encapsulated in a DICOM file (Clunie D. A. et al., 2016). An encapsulated DICOM file includes metadata related to the native document plus the metadata necessary to create the DICOM shell.

The DICOM interfaces are available for connection of any combination of the following types of digital imaging devices (Larobina M. et al., 2014): (a) image acquisition equipment (Nuclear Medicine Cameras, CT, MRI, Computed Radiography, Ultrasonography), (b) image archives, (c) image processing devices and image display workstations, and (d) hard-copy output devices (photographic transparency film and paper printers). DICOM is a comprehensive specification of information content, encoding, structure, and communications protocols for interchange of diagnostic and therapeutic images and image-related information (Figure 2.4) (Fatehi M. et al., 2015).

2.7.2.1 Anonymous DICOM

As already mentioned, the metadata elements include identifiable information about the patient, the study, and the institution. Sharing such sensitive data demands protection to ensure data safety and maintain patient privacy (Hollis K. F., 2016). Thus, for ethical and, of course, legal reasons, researchers are required to anonymize patient data before they can be made available to the public. Several studies have discussed techniques and methods for anonymizing DICOM image sets and generic DICOM files (Aryanto K. Y. et al., 2012).

There are numerous tools, anonymizers, and software that have been built to perform DICOM data de-identification, in order to fulfill the requirements of patient data protection, but two methods are usually used for de-identification of patient-related information in a DICOM header (Aryanto K. Y. et al., 2015). Anonymization is the first one, which removes information carried by header elements or replaces the information with random data, since the remaining information cannot be used to reveal the patient identity at all (Tamersoy A. et al., 2012). Pseudonymization is the second one, which is implemented by replacing the most identifying fields within a data record, utilizing artificial identifiers that could be used by authorized personnel to track down the real personal data of the patient. This method is commonly used in clinical analysis, processing, and research (Neubauer T. et al., 2012). It is chosen because good clinical practice requires this type of data manipulation. If additional findings are encountered that are crucial for the well-being of the patient, then it is important the real identity of the patient to be tracked in order to offer this information to him or her (Kornhaber R. et al., 2016).

2.8 IMAGE FORMATION AND PROCESSING BY MATLAB USE

The medical images are usually complex with various types, according to medical application and the clinical purpose. Nuclear medicine images need to show characteristic information about the physiological properties of the anatomic structures (European Society of Radiology (ESR), 2015). Appropriate image formation and image processing provide the high quality which is obligatory, by improving the quality of the acquired image and extracting quantitative information efficiently.

MATLAB (Matrix Laboratory) is a high-performance interactive software package for scientific and engineering computation developed by MathWorks. It allows simulation, matrix computation, implementation of algorithms, functions and data plotting, signal and image processing by the useful Image Processing Toolbox (Sobie E. A., 2011). Quantitative analysis and visualization of nuclear medical images of several modalities, such as SPECT, PET, or a hybrid system SPECT/CT is enabled by this toolbox (Ciarmiello A. et al., 2014). Furthermore, it offers the possibility to restore noisy or degraded images, analyze shapes and textures, and register two images. Most toolbox functions are written in open MATLAB language (Gou S. et al., 2013). In this way, they offer the opportunity to the user for fast and controllable algorithm development and custom functions creation. Its powerful interface and user-friendliness make it a tool of choice in many disciplines, including medical image processing (Ranganath K. A. et al., 2008).

Some scientists use MATLAB as a complementary tool, in parallel with another programming language, such as C++, when high requirements of a project need very fast processing speed (Lee L.-F. et al., 2016). MathWorks has provided a compiler to translate m-files (MATLAB programs) in C/C++ and FORTRAN. With proper programming practices, such as vectorization, pre-allocation, and specialization, applications in MATLAB can run fast enough even in large datasets and high-resolution image sets (Axelrod V., 2014). Fast implementations can also be achieved by bilinear interpolation, watershed segmentation, and volume rendering.

The Image Tool simply loads the image for further analysis with the command *imtool* (Ross B. D., 2016). It offers a package of widely used functions, already installed, in order to use it as simple image processing software, to regulate many important parameters, such as contrast, cropping, zooming in and out, color mapping, plotting, and many more (Wade M. J. et al., 2017).

MATLAB also supports DICOM files and it is very useful that it allows reading and writing metadata and image data of a DICOM file. The *dicomread* command provides the ability to open and read the image data from the DICOM file (Jodogne S., 2018). The *dicominfo* command allows metadata reading from a DICOM file. Then full details of the DICOM image appear in the command window. Toolbox image display functions *imshow* or *imtool* can be used to view the image data imported from a DICOM file (Gutman D. A. et al., 2014). The *dicomwrite* function is employed to modify or write image data or metadata to a file in DICOM format (Varma D. R., 2012).

Thus, MATLAB and Image Processing Toolbox enable formation, visualization, and quantitative analysis of Nuclear Medicine images acquired and is a very useful and user-friendly software.

REFERENCES

Aryanto K.Y., Broekema A., Oudkerk M., van Ooijen P.M. (2012) Implementation of an ano-
nymisation tool for clinical trials using a clinical trial processor integrated with an
existing trial patient data information system, Eur Radiol, 22: 144–151.
Aryanto K.Y.E., Oudkerk M., van Ooijen P.M.A. (2015) Free DICOM de-identification tools
in clinical research: functioning and safety of patient privacy, Eur Radiol, 25 (12):
3685–3695.
Ashoor M., Asgari A., Khorshidi A., Rezaei A. (2015) Evaluation of Compton attenuation and
photoelectric absorption coefficients by convolution of scattering and primary functions and counts ratio on energy spectra, Indian J Nucl Med, 30 (3): 239–247.
Averbukh A.N., Channin D.S., Homhual P. (2005) Comparison of human observer per-
formance of contrast-detail detection across multiple liquid crystal displays, J Digit
Imaging, 18 (1): 66–77.
Axelrod V. (2014) Minimizing bugs in cognitive neuroscience programming, Front Psychol,
5: 1435.
Billingsley F.C. (1965) Digital video processing at JPL, in Turner E.B. (Ed), Electronic
Imaging Techniques I, Proceedings of SPIE, 3: 1–19.
Block F.E. Jr. (1987) Analog and digital computer theory, Int J Clin Monit Comput, 4 (1):
47–51.
Burger W., Burge M.J. (2009) Principles of Digital Image Processing: Fundamental
Techniques, Undergraduate Topics in Computer Science. Springer, London.
Chalazonitis A.N., Koumarianos D., Tzovara J., Chronopoulos P., (2003) How to optimize
radiological images captured from digital cameras, using the Adobe Photoshop 6.0
program, J Digit Imaging, 16 (2): 216–229.
Cherry S.R. (2009) Multimodality imaging: beyond PET/CT and SPECT/CT, Semin Nucl
Med, 39 (5): 348–353.
Ciarmiello A., Giovannini E., Meniconi M., Cuccurullo V., Gaeta M.C. (2014) Hybrid
SPECT/CT imaging in neurology, Curr Radiopharm, 7 (1): 5–11.
Clunie D.A., Dennison D.K., Cram D., Persons K.R., Bronkalla M.D., Primo H. (2016)
Technical challenges of enterprise imaging: HIMSS-SIIM collaborative white paper, J
Digit Imaging, 29 (5): 583–614.
Cromey D. W. (2013) Digital images are data: and should be treated as such, Methods Mol
Biol, 931: 1–27.
Dinsdale A. (1927) Television Demonstration in America: A Successful Public Demonstration
of Television between Washington and New York, Wireless World and Radio Review,
20: 680–686.

Dragoi I.-C., Stanciu S.G., Hristu R., Coanda H.-G., Tranca D.E., Popescu M., Coltuc D. (2016) Embedding complementary imaging data in laser scanning microscopy micrographs by reversible watermarking, Biomed Opt Express, 7 (4): 1127–1137.

European Society of Radiology (ESR) (2015) Medical imaging in personalised medicine: a white paper of the research committee of the European Society of Radiology (ESR), Insights Imaging, 6 (2): 141–155.

Faber T.L., Folks R.D. (1994) Computer processing methods for nuclear medicine images, J Nucl Med Technol, 22: 145–162.

Fatehi M., Safdari R., Ghazisaeidi M., Jebraeily M., Habibi-koolaee M. (2015) Data standards in tele-radiology, Acta Inform Med, 23 (3): 165–168.

Foley J.D., Van Dam A. (1982) Fundamentals of Interactive Computer Graphics, Addison-Wesley, Reading, MA.

Gerber A.J., Peterson B.S. (2008) What is an image?, J Am Acad Child Adolesc Psychiatry, 47 (3): 245–248.

Gillies R.J., Kinahan P.E., Hricak H. (2016) Radiomics: images are more than pictures, they are data, Radiology, 278 (2): 563–577.

Gómez-Polo C., Gómez-Polo M., Celemin-Viñuela A., Martínez Vázquez De Parga J.A. (2014) Differences between the human eye and the spectrophotometer in the shade matching of tooth colour, J Dent, 42 (6): 742–745.

Gou S., Wang Y., Wang Z., Peng Y., Zhang X., Jiao L., Wu J., Wang G. (2013) CT image sequence restoration based on sparse and low-rank decomposition, PLOS ONE, 8 (9): e72696.

Govaert G.A.M., Glaudemans A.W.J.M. (2016) Nuclear medicine imaging of posttraumatic osteomyelitis, Eur J Trauma Emerg Surg, 42: 397–410.

Graf R.F. (1999) Modern Dictionary of Electronics, Newnes, Oxford, 569.

Gurcan M.N., Boucheron L., Can A., Madabhushi A., Rajpoot N., Yener B., (2009) Histopathological image analysis: a review, IEEE Rev Biomed Eng, 2: 147–171.

Gutman D.A., Dunn W.D. Jr., Cobb J., Stoner R.M., Kalpathy-Cramer J., Erickson B. (2014) Web based tools for visualizing imaging data and development of XNATView, a zero footprint image viewer, Front Neuroinform, 8: 53.

Haak D., Page C.-E., Deserno T.M. (2016) A survey of DICOM viewer software to integrate clinical research and medical imaging, J Digit Imaging, 29 (2): 206–215.

Halazonetis D.J. (2005) What do 8-bit and 12-bit grayscale mean and which should I use when scanning?, AJO-DO, 127 (3): 387–388.

Hillman E.M.C., Amoozegar C.B., Wang T., McCaslin A.F.H., Bouchard M.B., Mansfield J., Levenson R.M. (2011) In vivo optical imaging and dynamic contrast methods for biomedical research, Philos Trans A Math Phys Eng Sci, 369 (1955): 4620–4643.

Hockley N.S., Bahlmann F., Fulton B. (2012) Analog-to-digital conversion to accommodate the dynamics of live music in hearing instruments, Trends Amplif, 16 (3): 146–158.

Hollis K.F. (2016) To share or not to share: ethical acquisition and use of medical data, AMIA Jt Summits Transl Sci Proc, 2016: 420–427.

Hutton C., De Vita E., Ashburner J., Deichmann R., Turnerd R. (2008) Voxel-based cortical thickness measurements in MRI, Neuroimage, 40 (4): 1701–1710.

Indrajit I.K., Verma B.S. (2009) Monitor displays in radiology: part 2, Indian J Radiol Imaging, 19 (2): 94–98.

Jiemy W.F., Heeringa P., Kamps J.A.A.M., van der Laken C.J., Slart R.H.J.A., Brouwer E. (2018) Positron emission tomography (PET) and single photon emission computed tomography (SPECT) imaging of macrophages in large vessel vasculitis: current status and future prospects, Autoimmun Rev, 17 (7): 715–726.

Jodogne S. (2018) The orthanc ecosystem for medical imaging, J Digit Imaging, 31 (3): 341–352.

Kanan C., Cottrell G.W. (2012) Color-to-grayscale: does the method matter in image recognition?, PLOS ONE, 7 (1): e29740.

Karellas A., Harris L.J., Liu H., Davis M.A., D'Orsi C.J. (1992) Charge-coupled device detector: performance considerations and potential for small-field mammographic imaging applications, Med Phys, 19 (4): 1015–1023.

Kohli M.D., Summers R. M., Raymond Geis J. (2017) Medical image data and datasets in the era of machine learning—whitepaper from the 2016 C-MIMI meeting dataset session, J Digit Imaging, 30 (4): 392–399.

Korenblum D., Rubin D., Napel S., Rodriguez C., Beaulieu C. (2011) Managing biomedical image metadata for search and retrieval of similar images, J Digit Imaging, 24 (4): 739–748.

Kornhaber R., Walsh K., Duff J., Walker K. (2016) Enhancing adult therapeutic interpersonal relationships in the acute health care setting: an integrative review, J Multidiscip Healthc, 9: 537–546.

Koutelakis G.V., Anastassopoulos G.K., Lymberopoulos D.K. (2012) Application of multiprotocol medical imaging communications and an extended DICOM WADO service in a teleradiology architecture, Int J Telemed Appl, 2012: 11 pages.

Kumar Pandey A., Sharma P., Pandey M., Aswathi K.K., Malhotra A., Kumar R. (2012) Spreadsheet program for estimating recovery coefficient to get partial volume corrected standardized uptake value in clinical positron emission tomography-computed tomography studies, Indian J Nucl Med, 27 (2): 89–94.

Larobina M., Murino L. (2014) Medical image file formats, J Digit Imaging, 27 (2): 200–206.

Lee L.-F., Umberger B.R. (2016) Generating optimal control simulations of musculoskeletal movement using OpenSim and MATLAB, PeerJ, 4: e1638.

Loewe K., Grueschow M., Stoppel C.M., Kruse R., Borgelt C. (2014) Fast construction of voxel-level functional connectivity graphs, BMC Neurosci, 15: 78.

Lyon R.F. (2006) A Brief History of 'Pixel', Digital Photography II, IS&T/SPIE Symposium on Electronic Imaging, 15–19 January 2006, San Jose, California, USA.

Lyra M., Ploussi A., Georgantzoglou A. (2011) Chapter 23. MATLAB as a tool in nuclear medicine image processing. MATLAB – A Ubiquitous Tool for the Practical Engineer. InTech Editors, Clara M Ionescu, 477–500.

Madsen M.T. (1994) Computer acquisition of nuclear medicine images, J Nucl Med Technol, 22: 3–11.

Matsuda S., Yoshimura H., Yoshida H., Ryoke T., Yoshida T., Aikawa N., Sano K. (2017) Usefulness of computed tomography image processing by OsiriX software in detecting wooden and bamboo foreign bodies, Biomed Res Int, 2017: 3104018.

McGavin D. (2005) Color figures in BJ: RGB versus CMYK, Biophys J, 88 (2): 761–762.

Mendelson D.S., Rubin D.L. (2013) Imaging informatics: essential tools for the delivery of imaging services, Acad Radiol, 20 (10): 1195–1212.

Neu S.C., Crawford K.L., Toga A.W. (2012) Practical management of heterogeneous neuroimaging metadata by global neuroimaging data repositories, Front Neuroinform, 6: 8.

Neubauer T., Heurix J. (2012) A methodology for the pseudonymization of medical data, Int J Med Inform, 80 (3): 190–204.

Njemanze P.C., Kranz M., Amend M., Hauser J., Wehrl H., Brust P. (2017) Gender differences in cerebral metabolism for color processing in mice: a PET/MRI study, PLOS ONE, 12 (7): e0179919.

Oborska-Kumaszyńska D., Wiśniewska-Kubka S. (2010) Analog and digital systems of imaging in roentgenodiagnostics, Pol J Radiol, 75 (2): 73–81.

Pavlin M., Novak F. (2012) A wireless interface for replacing the cables in bridge-sensor applications, Sensors (Basel), 12 (8): 10014–10033.

Prasad S., Kumar Pal A. (2017) An RGB colour image steganography scheme using overlapping block-based pixel-value differencing, R Soc Open Sci, 4 (4): 161066.

Qin T., Han J., Geng Y., Ju L., Sheng L., Zhang S.X. (2018) A multiaddressable dyad with switchable cyan/magenta/yellow colors for full-color rewritable paper, Chemistry, 24 (48): 12539–12545.

Rakhshan V. (2014) Image resolution in the digital era: notion and clinical implications, J Dent (Shiraz), 15 (4): 153–155.

Ranganath K.A., Tucker Smith C., Nosek B.A. (2008) Distinguishing automatic and controlled components of attitudes from direct and indirect measurement methods, J Exp Soc Psychol, 44 (2): 386–396.

Ribeiro B., Perra N., Baronchelli A. (2013) Quantifying the effect of temporal resolution on time-varying networks, Sci Rep, 3: 3006.

Ross B.D. (2016) Demonstration of an inline publication image viewer: the future of radiological publishing, Tomography, 2 (1): 1–2.

Rossner M., O'Donnell R. (2004) The JCB will let your data shine in RGB, J Cell Biol, 164 (1): 11–13.

Savaris A., Härder T., von Wangenheim A. (2014) DCMDSM: a DICOM decomposed storage model, J Am Med Inform Assoc, 21 (5): 917–924.

Sikri V.K. (2016) Color: implications in dentistry, J Conserv Dent, 13 (4): 249–255.

Singh V., Marinescu D.C., Bakerb T.S. (2004) Image segmentation for automatic particle identification in electron micrographs based on hidden Markov random field models and expectation maximization, J Struct Biol, 145 (0): 123–141.

Smith H.J., Bakke S.J., Smevik B., Hald J.K., Moen G., Rudenhed B., Abildgaard A. (1992) Comparison of 12-bit and 8-bit gray scale resolution in MR imaging of the CNS. An ROC analysis, Acta Radiol, 33 (6): 505–511.

Sobic E.A. (2011) An Introduction to MATLAB, Sci Signal, 4 (191): tr7

Solomon C., Breckon T. (2011) Fundamentals of Digital Image Processing: A Practical Approach with Examples in MATLAB. Wiley-Blackwell, 49–50.

Stokes J. (2007) Inside the machine: an illustrated introduction to microprocessors and computer architecture, No Starch Press, 66.

Tamersoy A., Loukides G., Ercan Nergiz M., Saygin Y., Malin B. (2012) Anonymization of longitudinal electronic medical records, IEEE Trans Inf Technol Biomed, 16 (3): 413–423.

Tan L.K. (2006) Image file formats, Biomed Imaging Interv J, 2 (1): e6.

Tanley S.W.M., Schreurs A.M.M., Helliwell J.R., Kroon-Batenburgb L.M.J. (2013) Experience with exchange and archiving of raw data: comparison of data from two diffractometers and four software packages on a series of lysozyme crystals, J Appl Crystallogr, 46 (Pt 1): 108–119.

Teixidó M., Font D., Pallejà T., Tresanchez M., Nogués M., Palacín J. (2012) Definition of linear color models in the RGB vector color space to detect red peaches in orchard images taken under natural illumination, Sensors (Basel), 12 (6): 7701–7718.

Thompson L., Galvão Dias H., Passos Ribeiro Campos T. (2013) Dosimetry in brain tumor phantom at 15 MV 3D conformal radiation therapy, Radiat Oncol, 8: 168.

Vamvakas I., Lyra M. (2015) Voxel based internal dosimetry during radionuclide therapy, Hell J Nucl Med, 18 (Suppl 1): 76–80.

van der Stelt P.F. (2008) Better imaging: the advantages of digital radiography, J Am Dent Assoc, 139 (Suppl): 7S–13S.

Vandenberghe S., Mikhaylova E., D'Hoe E., Mollet P., Karp J.S. (2016) Recent developments in time-of-flight PET, EJNMMI Phys, 3: 3.

Varma D.R. (2012) Managing DICOM images: tips and tricks for the radiologist, Indian J Radiol Imaging, 22 (1): 4–13.

Volkan-Salanci B., Şahin F., Babekoğlu V. Uğur Ö. (2012) Experience with nuclear medicine information system, Mol Imaging Radionucl Ther, 21 (3): 97–102.

Wade M.J., Oakley J., Harbisher S., Parker N.G., Dolfing J. (2017) MI-Sim: a MATLAB package for the numerical analysis of microbial ecological interactions, PLOS ONE, 12 (3): e0173249.

Ye H., Greer T., Li L. (2011) From pixel to voxel: a deeper view of biological tissue by 3D mass spectral imaging, Bioanalysis, 3 (3): 313–332.

Yvernault B.C., Theobald C.D. Jr., Smith J.C., Villalta V., Zald D.H., Landman B.A. (2014) Validating DICOM transcoding with an open multi-format resource, Neuroinformatics, 12 (4): 615–617.

Zanzonico P. (2008) Routine quality control of clinical nuclear medicine instrumentation: a brief review, J Nucl Med, 49 (7): 1114–1131.

Zhang Q., Eagleson R., Peters T.M. (2011) Volume visualization: a technical overview with a focus on medical applications, J Digit Imaging, 24 (4): 640–664.

Zhang X. (2004) Estimation of saturated pixel values in digital color imaging, J Opt Soc Am A Opt Image Sci Vis, 21 (12): 2301–2310.

3 Nuclear Medicine Imaging Essentials

Nefeli Lagopati

CONTENTS

3.1 GENERAL INFORMATION

It is well known that nuclear medicine is based on the assessment of the distribution of a radionuclide in a target-organ after in vivo administration of a radiopharmaceutical, in order to detect differences between the normal and the abnormal tissues (Fernandes A. R. et al., 2017). γ-Camera performs this assessment, permitting the acquisition of both static and dynamic images (Sciammarella M. et al., 2019).

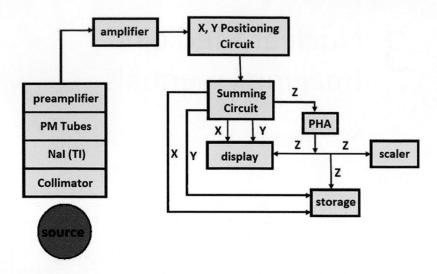

FIGURE 3.1   Schematic representation of γ-camera.

3.1.1 γ-CAMERA

The γ-camera consists of several components of critical importance, such as, detector, collimator, PM tubes, preamplifier, amplifier, pulsed-height analyzer (PHA), X-, Y-positioning circuit, and a display device (Figure 3.1) (Polemi A. M. et al., 2016).

The detector, amplifiers, and PM tubes are housed in the detector head, which is mounted on a stand (Saha G. B., 2013). It can be moved in order to align with the field of view (FOV) on the patient. The X-, Y-positioning circuits, PHA, and some recording devices are placed on a console (Saha G. B., 2013). The operation of the γ-camera is performed by a computer, built in it, equipped with the appropriate software (Peterson T. E. and Furenlid L. R., 2011). High voltage, window, and photopeaks are set by the operator (Saha G. B., 2013).

In brief, γ-rays from a source interact with the NaI (Tl) detector (Akkurt I. et al., 2014). Thus, light photons are emitted, striking the photocathode of PM tubes, and leading to a pulse generation. This pulse is then amplified by an amplifier and sorted out by a PHA. At the final step of this process, the pulse is positioned by an X-, Y-positioning circuit and stored in the computer, corresponding to the location of the γ-ray interaction in the detector (Peterson T. E. and Furenlid L. R., 2011).

3.1.2 DATA ACQUISITION

Data acquisition and processing of the data are carried out by the computer. The X- and Y-signals which are obtained in scintigraphy are digitized by ADCs and stored in frame or list mode. Magnification or zooming can be applied, whereby the pixel size is decreased by a zoom factor (Saha G. B., 2013).

The frame mode is the most common practice and widely used in static, dynamic, gated, and single photon emission computed tomography (SPECT) studies (Slomka P. et al., 2016). In this mode, a position (X, Y) in the detector corresponds to a pixel

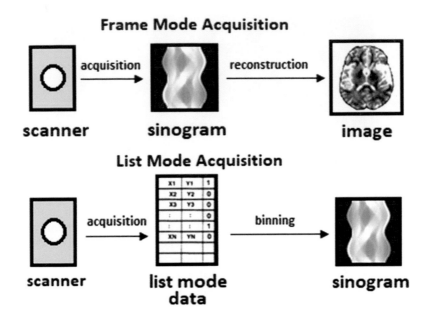

FIGURE 3.2 Frame and list mode acquisition.

position in the matrix (Jungmann J. H. et al., 2012). Digitized signals are stored in the corresponding positions (pixel) of the matrix of choice (Figure 3.2). The specification of the size and depth of the matrix and the number of frames per study is obligatory (Cromey D. W., 2013). The definition of the time of collection of data per frame, or the number of total counts to be collected, is of crucial importance, because data acquisition continues until the preselected time or total count is reached. This mode provides instant images for storage and display (Cromey D. W., 2013).

In the list mode, digitized X- and Y-signals are received in sequence in time and are stored as individual events (Saha G. B., 2013). Since data acquisition is completed, these data can be sorted to form images in many ways to suit a specific medical need (Sivarajah U. et al., 2017). First, data can be manipulated by changing the size of the matrix and also the time of acquisition per frame (Adluru G. et al., 2009). Moreover, physiologic markers, such as the start of a cardiac cycle, can be incorporated in the list mode acquisition. In general, the list mode acquisition provides flexibility (Gullberg G. T. et al., 2010). The need of larger memory space, the requirement for longer processing time and the unavailability of images during or immediately after the completion of the study are between the major disadvantages of this technique (Figure 3.2) (Madsen M. T., 2007).

3.2 PLANAR IMAGING

3.2.1 Static Acquisition

The basic nuclear medicine acquisition is the static planar view. A static acquisition is the collection of data in one view of a region of interest for a preset time or preset

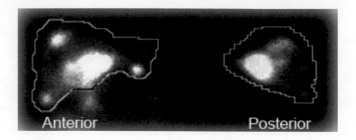

FIGURE 3.3 Planar liver images with Regions of Interest (ROIs) drawn for the activity calculation.

total counts (Dorbala S. et al., 2018). Data are commonly acquired in the frame mode, and normally the size of the matrix is specified before the beginning of the examination (Jolliffe I. T. and Cadima J., 2016). The choice of a matrix size depends on the FOV of the imaging system and the pixel size in order to give the desired image resolution (Chan C. et al., 2016). Usually, a pixel size of 2–3 mm is considered appropriate enough for a good image resolution. Because of the high-count densities in static views, data acquisition in byte mode perhaps overflows in individual pixels and, for this reason, the word mode is usually employed (Cradduck T. D. and Busemann-Sokole E., 1985).

Digital images represent the count density in regions of interest (ROIs) in a target (Rosenkrantz A. B. et al., 2015). The number of counts that should be acquired in an image depends on the size of the region is to be identified and its apparent contrast with the surrounding background (Wickham F. et al., 2017). Generally, the large and high-contrast objects are easily detectable at low count densities, whereas those small and low-contrast objects are difficult to delineate due to the statistical noise (Alsleem H. et al., 2014). Count density should be optimum in order to achieve desirable contrast. Practically, in static acquisitions the same acquisition parameters, such as pixel size, matrix size, and acquisition time, are selected to be constant for all the views (Faber T. L. and Folks R. D., 1994). A liver scan is a very common example of a static study (Figure 3.3).

3.2.2 DYNAMIC ACQUISITION

In dynamic studies, a continuous series of images is acquired over time, in order to watch the time course of the radiopharmaceutical, because in this way important information about the functionality of an organ can be provided (Gullberg G. T. et al., 2010). The patient's position cannot be changed during the image acquisition, but therefore the matrix size and the frame rate can be changed (Saha G. B., 2013). The frame rate varies, depending on the examination. Moreover, it is possible to apply different frame rate, during the same application (Debattista K. et al., 2017). For example, many frames can be acquired during the first few minutes, to obtain further details about the absorbance ability of an organ and later, the rate can be slower than the initial, because the concentration changes more slowly during washout, redistribution, and clearance (Faber T. L. and Folks R. D., 1994). Furthermore, the selection of frame rate depends on the kinetics of the radionuclide through the organ of interest (Ebrahimnejad Gorji K. et al., 2019). Matrix or pixel size may also be changed during dynamic acquisitions. The acquisition of image data is regulated, so that, while one frame is being collected, the previous frame is stored in the external storage device (Cromey D. W., 2013). A renogram is an example of a commonly acquired dynamic study (Figure 3.4) (Faber T. L. and Folks R. D., 1994).

3.2.3 GATED ACQUISITION IN CARDIAC STUDIES

The gated technique is characterized by continuous acquisition of data in multiple sequential images in each cardiac cycle by gating between successive cycles. For this reason, the modern version of gated study is widely known as multiple gated acquisition or MUGA (Mitra D. and Basu S., 2012). In particular, in the MUGA study, the data are acquired in synchronization with the R-wave[1] of the cardiac cycle (Figure 3.5) (Mitra D. and Basu S., 2012).

The normal heart beat is about 1 beat/second. Thus, the R-R interval which is the time elapsed between two successive R-waves of the QRS complex of an electrocardiogram is about 1 s (Lanfranchi P. A. and Somers V. K., 2011). Prior to the acquisition, the R-R interval is examined to determine the heart rate. The R-R interval is divided into several segments or frames depending on the number of preferable frames (Postema P. G. and Wilde A. A. M., 2014). During acquisition, the computer "directs" counts to the first frame when the R-wave from a QRS complex is detected. After completion of counting in all frames, a new R-wave is detected (Faber T. L. and Folks R. D., 1994). Each collected frame is created by hundreds of cardiac cycles, so the resulting images are representative of an average heartbeat and for this reason, this kind of gated acquisition is called frame mode (Faber T. L. and Folks R. D., 1994). This sequence of counts continues until a sufficient number of counts have been accumulated in each frame (Vaquero J. J. and Kinahan P., 2015). A typical example of an obtained plot of the time-activity curve (TAC) is shown in Figure 3.6. The ejection fraction is calculated by Equation 3.1:

$$EF(\%) = \frac{A_d - A_s}{A_d} \cdot 100 \qquad (3.1)$$

where A_d and A_s are the end-diastolic and end-systolic activities (Chengode S., 2016).

This method is effective, in the case of a regular heartbeat. Otherwise, in cases with cardiac arrhythmia, the R-R interval is altered, and the data become corrupted from R-wave to R-wave (Conn N. J. et al., 2018). Irregular heartbeat causes the average heartbeat image to contain some abnormal contractions, compromising image quality (Hornberger L. K. and Sahn D. J., 2007). Moreover, if the heart rate changes, each gated frame starts to hold larger or smaller portions of the cardiac cycle than initially intended (Zhang J. et al., 2008). The modern systems for gated acquisition

---

[1] In a QRS complex of an electrocardiogram (ECG), the Q phase is the first negative depolarization of the ventricular depolarization complex. The R-wave is the first positive deflection which follows the P-wave, and the S-wave is the first negative wave, after the R-wave. T-wave represents the ventricular repolarization (Noble R. J. et al., 1990).

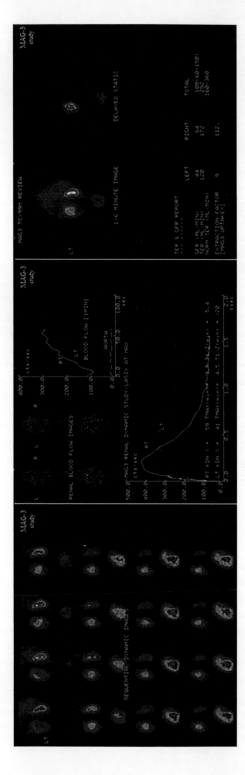

FIGURE 3.4　Sequential dynamic images of kidneys. (99m)Tc-MAG(3) study.

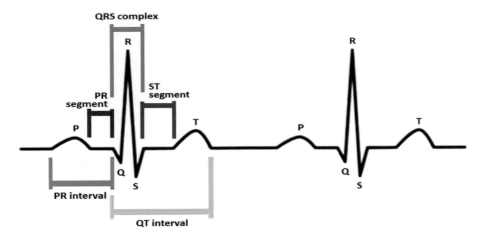

FIGURE 3.5 The QRS complex of the cardiac cycle.

have been designed to reject the "wrong" heartbeat cycle, by continuously monitoring heart rate and adjusting frame acquisition times to account for changes (Gullberg G. T. et al., 2010).

In list mode, time markers are also placed to give some indication of the time of acquisition of each count (Dorbala S. et al., 2018). List mode acquisition allows to sort out and reject the bad heartbeat data, in a post-acquisition reformatting (Saha G. B., 2013). The drawback of list mode acquisitions is that computer storage requirements are very high (Watabe H. et al., 2006). Particularly, in the frame mode, only the running summation of counts is stored, while the list mode stores every count and its location separately. For this reason, while planar gated studies are usually acquired in the list mode, gated tomographic acquisitions are generally acquired in the frame mode (Faber T. L. and Folks R. D., 1994).

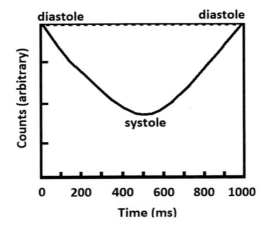

FIGURE 3.6 Time-activity plot from a gated study with segments of a QRS cardiac cycle (Saha G. B., 2013).

3.3 SPECT-PET ANGULAR PROJECTIONS – SINOGRAM

Conventional γ-cameras provide two-dimensional (2D) planar images of three-dimensional (3D) objects (Chen N. et al., 2018). Thus, depth, which is the third dimension, is obscured due to the superimposition of all data along this direction. In planar images, data can be obtained in different projections, such as posterior, anterior, lateral, and oblique, in order to gain some information about the depth of a structure of interest (Van Audenhaege K. et al., 2015). If enough planar views are acquired at various angles, then these projections can be mathematically reconstructed to create a 3D stack of slices, estimating the distribution of radioactivity at every point in the body of the patient (Van Audenhaege K. et al., 2015). This approach is referred to as a tomogram. Thus, precise assessment of the depth of a structure in an object is made by tomographic scanners (Vaquero J. J. and Kinahan P., 2015).

The basic principle of tomographic imaging in nuclear medicine is based on the detection of radiation at different angles around the patient (Shukla A. K. and Kumar U., 2006). Generally, it is called emission computed tomography (ECT) and it is based on algorithms, providing images at distinct depths (slices) of the object (Van Audenhaege K. et al., 2015). Every slice of a tomographic stack can be created from each row of the planar projection image set (Green M. V. et al., 2017). A 3D slice-by-slice tomogram allows every portion of the object to be completely seen and analyzed (Wu B. et al., 2010). In correspondence with the planar views, each tomographic slice has an associated matrix size and pixel size (Lyra M., 2009). Additionally, tomographic slices have a thickness, equal to the pixel size in the vertical direction of the original projection (Lyra M., 2009). Tomographic acquisition usually takes much more time than a common planar acquisition (Ljungberg M. and Sjögreen Gleisner K., 2016).

In nuclear medicine, two types of ECT have been widely used in clinical practice, based on the type of radionuclides used: SPECT and positron emission tomography (PET) (Van Audenhaege K. et al., 2015).

3.3.1 SINGLE PHOTON EMISSION COMPUTED TOMOGRAPHY (SPECT)

SPECT is routinely utilized in nuclear medicine for various organ imaging (Lyra M., 2009). It commonly uses γ-emitting radionuclides such as ^{99m}Tc, ^{123}I, ^{67}Ga, and ^{111}In. The fundamental of SPECT is to obtain a snapshot of the γ-emitter distribution in any slice of the body, as accurately as possible, using projections of this image acquired by a rotating γ-camera from several angles of view (Lyra M. et al., 2011).

A typical SPECT system consists of a γ-camera (with one up to three NaI (Tl) detector heads, mounted on the gantry), an online computer system for data acquisition and processing and a display system (Saha G. B., 2013). The detector head allows rotation around the long axis of the patient for data collection over 180 or 360°. The 180° data collection is commonly used but many investigators prefer 360° data acquisition because they suggest that it minimizes the attenuation effects and the effects due to variation of resolution with depth (Saha G. B., 2013). The 180° acquisition, using a dual-head camera with heads mounted interdimensionally, requires only one detector head for data collection. Thus, the data obtained by the other head essentially can be discarded. Otherwise, there is the possibility of the use of the arithmetic mean $\frac{(A_1 + A_2)}{2}$

or the geometric mean $\left(A_1 \cdot A_2\right)^{1/2}$ of the counts, A_1 and A_2, acquired by the two heads, in order to allow correction for attenuation of photons in tissue (Kheruka S. C. et al., 2012). Sometimes 180° collection by a dual-head camera with heads mounted at 90° angles to each other can shorten significantly the required imaging time. For 360° data acquisition, dual-head cameras with heads mounted at 90° or 180° angles to each other and triple-head cameras with heads oriented at 120° to each other are commonly used, offering short imaging time (Dorbala S. et al., 2018).

The data are collected in the form of pulses for each angular position and stored for reconstruction of the images and of the planes of interest. Pulses are formed by the PM tubes from the light photons produced by the interaction of γ-ray photons from the object (Peterson T. E. and Furenlid L. R., 2011). Then they are amplified, verified by X, Y position and PH analyses, and finally stored. Collected data can generate transverse (short axis), sagittal (vertical long axis) and coronal (horizontal long axis) images. Actually, multi-head γ-cameras can collect data in several projections at the same time, and thus can reduce the time of imaging (Bocher M. et al., 2010).

Data collection can be achieved in continuous motion or in "step-and-shoot" mode. In the first case, the detector rotates continuously at a constant speed around the patient (Winant C. D. et al., 2012). In the step-and-shoot mode, the detector moves around the patient at selected incremental angles. Each projection view is acquired at each angular stop (DiFilippo F. P., 2008). There is a short pause of a few seconds between views to allow for the automatic rotation of the camera head to the next stop, so the camera makes a single rotation around the patient.

Data acquisition in SPECT is schematically illustrated in Figure 3.7, in which it is clearly depicted, the detector rotating around the object of interest, allowing the observation of the pattern of γ-emission in the FOV for many angles (Peterson T. E.

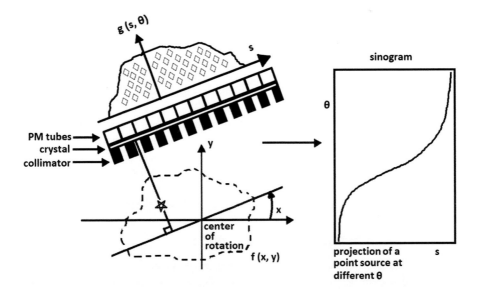

FIGURE 3.7 Tomographic acquisition. At each angle, data is considered as the projection of radioactivity distribution onto the detector.

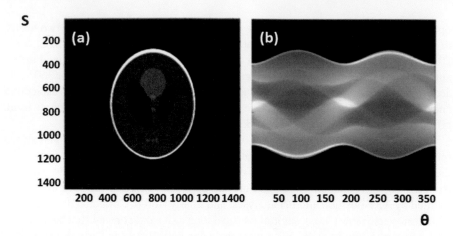

FIGURE 3.8 (a) Shepp–Logan phantom slice. Shepp–Logan is a standard model of a human head which was created by Larry Shepp and Benjamin F. Logan (Shepp L. A. and Logan B. F., 1974). It was developed in order to be used for the design and testing of image reconstruction algorithms. (b) Corresponding sinogram. Each row of sinogram is projection of slice at given angular position of detector.

and Furenlid L. R., 2011). The function $g(s, \vartheta)$ stands for the number of scintillations detected at any location s along the detector, when the detector head is at an angular position ϑ. The quantity $f(x, y)$ is defined as the estimated number of photons emitted at any point (x, y) of the transverse slice in the FOV (Bruyant P. P., 2002). The function g is the projection of f onto the crystal, therefore $g(s, \vartheta)$ is the total number of radioactive counts, recorded in any time interval, at point s since the detector is at angle ϑ. A major disadvantage of this method is that photons, emitted at different depths, along the same direction, could potentially produce scintillations, wrongly, detected at the same location in the crystal (Bruyant P. P., 2002). Fortunately, by acquiring projections for many distinct angular positions of the detector, this problem can be diminished.

At the end of the data acquisition process, each point of the detector, for each angular position, contains the total number of scintillations, or counts. So, data acquisition is the record of projections and actually, gives as a result, a set of angular projections (Bailey D. L., 2003). The set of projections of a single slice is widely known with the term "sinogram" (Figure 3.8). In a sinogram, which is a 2D image, the horizontal axis represents the count location on the detector, and the vertical axis corresponds to the angular position of the detector. Each point of the sinogram corresponds to the number of counts. There is a separate sinogram image for each slice location along the long axis of the patient (Bruyant P. P., 2002).

3.3.2 POSITRON EMISSION TOMOGRAPHY (PET)

PET is utilized in nuclear medicine for various organ imaging, using β^+-emitting radionuclides such as ^{11}C, ^{13}N, ^{15}O, ^{18}F, ^{68}Ga, and ^{82}Rb. The sensitivity of PET is substantially higher than in conventional, single-photon nuclear imaging (Feng H. et al., 2019). Fundamentally, PET is based on the detection in coincidence of the two

511-keV annihilation radiations, originating from the previous mentioned β^+-emitting sources (Shukla A. K. and Kumar U., 2006). In fact, positrons are annihilated in the patient's body and produce two 511-keV annihilation photons which are emitted in opposite directions (Shukla A. K. and Kumar U., 2006). These photons are detected in an electronic time interval, the so-called "coincidence time window", by two detectors connected in coincidence and must be along the straight line connecting the centers of the two detectors called the line of response (LOR) (Vaquero J. J. and Kinahan P., 2015). The 511-keV photons are converted into light photons in the detector. A pulse is formed by the PM tube, and pulse-height analysis follows, in accordance with the principles of a conventional γ-camera (Peterson T. E. and Furenlid L. R., 2011). The detectors are placed in the array of several rings in order to put the organ of interest in the FOV. Data acquisition is held over 360° simultaneously around the patient's body (Gullberg G. T. et al., 2010). As the two opposite photons are detected in a straight line, interdimensionally, there is no need of a collimator to limit the FOV; this technique is called electronic collimation (Peterson T. E. and Furenlid L. R., 2011). The coincident events are called prompts, including true, random, scattered, and multi-coincidence events. These events are parastatically illustrated in Figure 3.9 (Saha G. B., 2013). A true coincident event arises when two 511-keV photons from a single annihilation event are detected by a detector pair along the LOR, while random coincidences occur when two unrelated 511-keV photons derived from two separate annihilation events are detected by a detector, as a pair, occurred within the same time window (Shukla A. K. and Kumar U., 2006). Random events increase the background, worsening image contrast. Compton scattering of annihilation photons inside the patient's body is the main cause of scatter coincidences, degrading the contrast and decreasing the image quality (Cherry S. R. et al., 2012).

In a common full ring system, the data are collected by all detector pairs, simultaneously (Cherry S. R. et al., 2012). In partial ring systems, the detector is rotated around the patient in angular increments for data collection. Data acquisition consists of three important steps. First of all, the location of the detector pair in the detector ring must be determined for each coincident event (Dorbala S. et al., 2018). Secondly, the pulse height of the photon detected is checked in order to exist in the pulse energy window set for 511 keV. Finally, the position of the LOR is determined in terms of polar coordinates for the following storage in the computer system (Saha G. B., 2013).

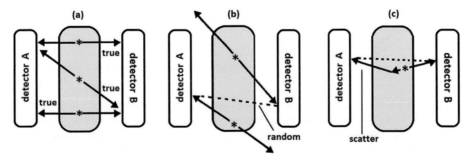

FIGURE 3.9 (a) True coincidence events, (b) Random coincidence events, (c) Scattered coincidence events.

In a PET system, block detectors are cut into small detectors and coupled with four PM tubes, which are arranged in arrays of rings (Bailey D. L., 2003). Each detector is connected to N/2 detectors, where N is the number of small detectors in the ring (Peterson T. E. and Furenlid L. R., 2011). The pulses which are produced in PM tubes determine the locations of the two detectors (Saha G. B., 2013). The position of each detector is estimated by a weighted centroid algorithm, which can estimate a weighted sum of individual PM tube pulses (Peterson T. E. and Furenlid L. R., 2011). These pulses are then normalized with the total pulse obtained from all PM tubes. The X and Y positions of the detector element in the ring are obtained by Equations 3.2 and 3.3 (Peterson T. E. and Furenlid L. R., 2011; Saha G. B., 2013).

$$X = \frac{(B+C)-(A+D)}{A+B+C+D} \tag{3.2}$$

$$Y = \frac{(C+D)-(A+B)}{A+B+C+D} \tag{3.3}$$

where A, B, C, and D are the pulses from the four PM tubes, which are then summed up to give a Z pulse, which is checked by the pulse height analyzer (PHA) to be within the energy window set (Saha G. B., 2013). If this pulse is outside the window of 511 keV, it is rejected. Otherwise it is accepted for further process and storage.

Data storage in the computer system is the last step of data acquisition. The coincidence events in PET systems are stored in the form of a sinogram, similarly to SPECT systems (Figure 3.10) (Fahey F. H., 2002). R and φ coordinates are used,

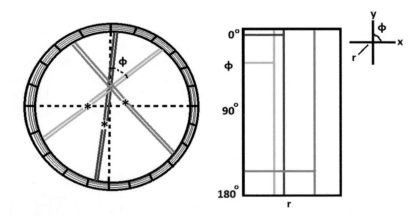

FIGURE 3.10 PET data acquisition. Positrons are annihilated in the patient's body and produce two 511-keV annihilation photons which are emitted in opposite directions. These photons are detected in the "coincidence time window", by two detectors connected in coincidence along the line of response (LOR). The photons are converted into light photons in the detector. The detectors are placed in the array of several rings in order to put the organ of interest in the field of view. Data acquisition is held over 360° around the patient's body. The coincidence events in PET systems are stored in the form of a sinogram. (From Cherry S. R. et al., 2012.)

where *r* is the distance of the annihilation event from the center and *d* is the projection angle from y-axis (Fahey F. H., 2002).

Thus, PET data are acquired directly into a sinogram in a matrix of appropriate size in the computer memory (Kadrmas D. J., 2008). The sinogram is a 2D histogram of the LORs in the (*r*, φ) coordinate system. Hence, each LOR corresponds to a particular pixel in the sinogram, characterized by the coordinates *r* and φ (Kadrmas D. J., 2008). The number of total counts in each pixel represents the number of coincidence events, which are detected during the selected time by the two detectors along the LOR (Fahey F. H., 2002).

Data acquisition is held in frame mode or in list mode and can be applied to both static and dynamic imaging. In static imaging, a frame can be obtained consisting of a set of sinograms acquired over the length of the scan (Gullberg G. T. et al., 2010). In dynamic imaging, the data are collected in multiple frames of sinograms. In dynamic imaging, a gated method can also be employed. The static scans can estimate the gross tracer uptake, while dynamic scans provide information about the tracer distribution as a function of time (Dorbala S. et al., 2018).

3.4   FACTORS AFFECTING ACQUISITION

The process of data projection acquisition is affected by a number of factors. These factors are usually related to count density, pixel size selection, field of view, matrix size, zoom factor, statistical noise, attenuation, scattering, collimator/detector – blurring, artefactual nonuniform resolution, blurring due to patient's motion, artefactual nonuniform intensity. Each of these factors needs to be corrected prior to image reconstruction (Frey E. C. et al., 2012).

3.4.1   COUNT DENSITY

Count or information density is an important factor which estimates the number of counts per pixel. It has a significant impact on image quality, as it can improve the image contrast (Frey E. C. et al., 2012). A minimum number of counts need to be collected for reasonable contrast in the image. The higher the counts in an image, the better are the contrast. Count density depends on the administered dose and the uptake in the organ of interest (Madsen M. T. et al., 1994). Contrast is usually improved by increasing administered activity and also when there is a differential uptake between the area of interest and surrounding tissues (Saha G. B., 2013). According to a practical rule an optimum count density is considered to be about 1000 counts/cm^2.

Digital images essentially represent the count density in ROIs in an object (Alsleem H. et al., 2014). So, large and high-contrast objects can be easily detectable at low count densities, whereas small and low-contrast objects are difficult to be delineated due to statistical noise. A low-density image can decrease spatial resolution as well as lesion detectability (Foster B. et al., 2014). Increase in count density without increasing acquisition times can be achieved by using a smaller acquisition matrix, by injecting a larger dose of the radiopharmaceutical, or by using a high-sensitivity collimator (Frey E. C. et al., 2012).

3.4.2 Field of View – Pixel Size – Matrix Size – Zoom Factor

The size of a matrix is selected by the operator, depending on the type of task to be performed. The matrix size is determined from the number of the columns (m) and the number of rows (n) of the image matrix ($m \times n$). It is approximated to the FOV (Gong K. et al., 2017). Generally, as the matrix dimension increases the resolution gets better. Nuclear medicine image matrices, nowadays, range from 64×64 to 1024×1024 pixels (Madsen M. T. et al., 1994).

Each pixel in a digital image corresponds to a specific location in the detector. The number of the counts which can be stored in a pixel depends on the depth of the pixel, which is represented by a byte. The pixel size depends on the matrix size selection and it is considered as a very important factor that affects the spatial resolution of a digital image (Peterson T. E. and Furenlid L. R., 2011). The pixel size is calculated by dividing the FOV, by the number of pixels across the matrix. For a standard FOV, an increase of the matrix size decreases the pixel size and the ability to see details is improved. Sometimes, a zoom factor is applied during data acquisition, in order to improve spatial resolution, as it reduces the pixel size. Thus, the pixel size d can be calculated as:

$$d = \frac{FOV}{z \cdot N} \tag{3.4}$$

where z is the zoom factor and N is the number of pixels across the matrix.

For example, a zoom factor of 2, reduces the pixel size by half, improving the spatial resolution. In this way, the counts per pixel are reduced thus increasing the noise on the image (Peterson T. E. and Furenlid L. R., 2011). The pixel size and zoom factor selection is limited by the spatial resolution of the imaging device. In ideal conditions, the pixel size must be less than 1/3 of the expected spatial resolution of a SPECT system, measured at the center of rotation (Equation 3.5) (Difilippo F. P., 2008).

$$d \leq \frac{FWHM}{3} \tag{3.5}$$

FWHM is the full width at half-maximum of the line spread function of the imaging system. A pixel size larger than this limit will degrade the image (Saha G. B., 2013).

3.4.3 Image Noise – Statistical Noise – Noise in Projection Data

Noise is an integral part of images in nuclear medicine decreasing the detectability of lesions. It is presented in two forms, the systematic and the random or statistical or quantum noise (Vicente E. M. et al., 2017).

Systematic noise is related to a fixed variation in count rate all across the image, impairing the image contrast (Dorbala S. et al., 2018). This type of noise may be caused by system artifacts or by the radioactivity distribution in different overlapping organs (Frey E. C. et al., 2012). Moreover, nonuniformities in the γ-camera and tomographic reconstruction of images contribute to systematic noise (Dorbala S. et al., 2018).

Statistical noise is due to the random nature of emission and/or detection of photons. Statistical noise is a major component of image contrast caused by statistical variations of the background counts and is given by the standard deviation of the background count B (Frey E. C. et al., 2012). Statistical noise follows a Poisson distribution.

Thus,

$$Noise = \sqrt{B} \qquad (3.6)$$

It can be determined as a fraction of the total background counts contributed to image contrast, which is called the contrast noise, C_N (Vaquero J. J. and Kinahan P., 2015). Therefore,

$$C_N = \frac{\sqrt{B}}{B} = \frac{1}{\sqrt{B}} \qquad (3.7)$$

Hence, noise originates from random variations of the background count density which is composed of scattered radiations inside the patient, through the detector, and septal penetration of photons in the collimator (Frey E. C. et al., 2012). These categories of counts contribute to the final noise in the image, worsening the contrast and the image quality (Vaquero J. J. and Kinahan P., 2015).

Another more reliable parameter to evaluate the image quality is the signal-to-noise ratio (SNR) (Chang T. et al., 2012).

$$SNR = \frac{N}{\sqrt{N}} \qquad (3.8)$$

where N is the number of detected photons per pixel. If the *SNR* is high, the diagnostic information of an image is appreciated regardless of the noise level (Lyra M. et al., 2011). Structured noise is derived from nonuniformities in the scintillation camera and overlying structures in patient body (Polemi A. M. et al., 2016).

Usually, noise is a bigger trouble in SPECT than in planar imaging, because counts per pixel are less than in planar imaging and also the reconstruction process in SPECT amplifies the noise (Dorbala S. et al., 2018). Of course, there are other advantages in SPECT, such as better contrast, since, in planar imaging, overlapping structures cause more contrast degradation. Noise is also affected by the choice of a matrix (Bruyant P. P., 2002).

3.4.4 Attenuation – Scattering

The effect of photon attenuation is essentially the reduction of photon fluence due to photon interaction (Lee T. C. et al., 2016). So, photons are attenuated in body tissue, while passing through a patient. The photon energy, the thickness of tissue, and the linear attenuation coefficient of the photons in this tissue can affect the degree of attenuation (Chen Y. and An H., 2017). If I_0 is the number of photons emitted from an organ and I is the number of photons detected by the γ-camera, then (Frey E. C. et al., 2012):

$$I = I_0 \cdot e^{-\mu x} \tag{3.9}$$

where μ is the linear attenuation coefficient of the photon in tissue and x is the depth of tissue traversed by the photon (Frey E. C. et al., 2012).

Attenuation causes a gradual decrease of count density from the edge to the center of the image and there is the need of correction in order to avoid artifacts (Saha G. B., 2013).

In SPECT imaging, the usual techniques which are employed to estimate I_0 is either obtaining two counts in opposite projections and taking their arithmetic mean or taking their geometric mean, as it was previously mentioned (Peterson T. E. and Furenlid L. R., 2011). PET data acquisition is also affected by photon attenuation, variation in detection efficiency of the detectors, scatter coincidences, partial volume effect, and dead time (Shukla A. K. and Kumar U., 2006). Chang method and transmission method are usually used to correct the attenuation effect due to radiation scattering (Berker Y. and Li Y., 2016). Chang's mathematical formulation is a widely used method for attenuation correction (Saha K. et al., 2016). It can be applied on most γ-cameras that do not require an external transmission source, while the transmission measurements use an external positron emitting source (Seo Y. et al., 2008).

3.4.5 COLLIMATOR/DETECTOR BLURRING – ARTEFACTUAL NONUNIFORM RESOLUTION

Collimator resolution is a limiting factor when absorptive collimators are used for spatial localization (Difilippo F. P., 2008). Collimator hole diameters must be relatively large in order to obtain reasonable collimator efficiency, but in this way, there is blurring of the image (Dorbala S. et al., 2018). Emitted photons travel along the central axis of the collimator hole as well as the off-axis due to the collimator aperture (Van Audenhaege K. et al., 2015). Off-axis rays, from another projection angle can be regarded as central-axis rays (Motomura N. et al., 2006). So, raw data can be thought of as an ensemble of central-axis rays from within an arc equal to that created by the collimator aperture (Motomura N. et al., 2006).

If only central-axis photons could be acquired, these data could compensate for collimator blurring. Thus, collimator blurring can be compensated for, by using fine angular sampling (Dorbala S. et al., 2018).

The point spread function (PSF) is applied to describe the response of an imaging system to a point source or point object, estimating the intense of blurring (Figure 3.11) (Beekman F. J. et al., 1999).

The collimator's resolution depends on source-to-detector distance. The obvious solution could be the projection of a point source (Figure 3.12) (Van Audenhaege K. et al., 2015). In practice, however, the projection is totally different. For parallel hole collimators, resolution is dominated by collimator blurring, and the FWHM is proportional to the distance to the collimator (Van Audenhaege K. et al., 2015).

Four-quadrant round bar phantoms are used to offer precise determination of intrinsic resolution, collimator spatial resolution, field size, and linearity. We offer

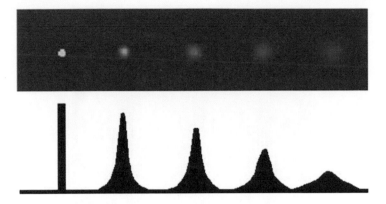

FIGURE 3.11 The point spread function (PSF) is applied to describe the response of an imaging system to a point source or point object, representing the intense of blurring.

a range of sizes manufactured to the highest quality standards (Figure 3.13) (Van Audenhaege K. et al., 2015).

The quality of images produced by the tomographic imaging systems is, generally, measured in various image quality parameters. Image resolution is usually defined in physical units or in pixels with known pixel size and is the most common index used for quality control of an image (Zanzonico P., 2008). Image resolution is also described in its uniformity and symmetry across the span of the reconstructed images (Kim K. et al., 2019). A high resolution is fine enough to present

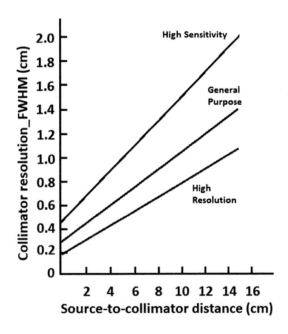

FIGURE 3.12 Collimator resolution versus source-to-collimator distance for three different collimators. (From Hine G. J. et al., 1978.)

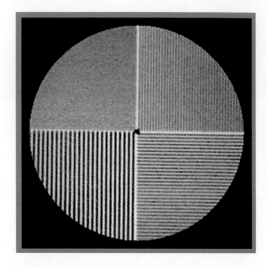

FIGURE 3.13   Four-quadrant round bar phantoms image obtained using a parallel-hole collimator.

even the smallest features in the image with enough sharpness (Cromey D. W., 2013). Uniformity is also necessary, as it is important to observe the same value across the whole image at the same time in all directions (Van Audenhaege K. et al., 2015).

All the factors, that can degrade PET or SPECT system's final response, including collimator/detector effect, make spatially-variant, anisotropic and ultimately degrade quality of the final reconstructed images, produced by these systems (Ljungberg M. and Pretorius P. H., 2018). For instance, statistical image estimators, based on quadratic priors, induce nonuniform smoothing and, consequently, smooth out regions inside the object with varying intensity, thus, producing anisotropic smoothing (Bai B. et al., 2013). This nonuniform smoothing leads to nonuniform reconstructed resolution properties in the final image. An anisotropic response of the system may produce shape deformations and degrade the quantitation system capability, which is very critical for diagnosis (Dutta J. et al., 2013).

3.4.6   Patient Motion Blurring – Artefactual Nonuniform Intensity

Significant patient motion during imaging can cause artifacts or blurring in an image, reducing the image contrasts (Frey E. C. et al., 2012). This fact primarily results from the overlapping of adjacent areas by the movement of different organs. Restraining the patients or having them in a comfortable position may mitigate this blurring (Khalil M. M. et al., 2011). This attempt can include straps and comfortable pillows. But sometimes, this motion is unintentional, like cardiac activity. To detect patient motion, the cine display and/or the sinogram should be reviewed prior to releasing the patient (Lyra V. et al., 2016). The resulted artifacts due to patient motion can partly be corrected by automated correction algorithms that shift the projection views to align the organ of interest, but repeat acquisition is absolutely recommended if these algorithms are not successful (Berker Y. and Li Y., 2016).

During dosimetry calculations performing, even if the repositioning is very careful, misalignment due to patient repositioning on the couch for the different imaging sessions will occur (Lodge M. A. et al., 2011). Image registration or alternatively drawing separate ROIs for each of the time point measurements can help to compensate this problem (Vaquero J. J. and Kinahan P., 2015). If only one transmission study is performed, it is very important for the patient to remain in the same position for the transmission scan since the correction is usually made on a pixel by pixel basis (Frey E. C. et al., 2012). Otherwise, the transmission scan needs to be registered to the whole-body measurement by appropriate software (Seo Y. et al., 2008).

In clinical routine, breath holding, or respiratory gating may improve the thoracic images. Artefacts which are related to heart motion can be reduced by using a cardiac gating technique (Parker J. G. et al., 2009). In this mode, data are acquired in synchrony with the heart beat or with the breathing cycle, so that all images are acquired at the same time during the motion cycle (Chun S. Y., 2016). This technique helps reduce blurring and other possible image artifacts induced by body motion (Frey E. C. et al., 2012).

3.5 MATLAB ASSISTANCE IN NUCLEAR MEDICINE ACQUISITION

Nuclear medicine imaging is an incontestable vital tool for diagnosis, as it provides in noninvasive manner the internal structure of the body to detect eventually diseases or abnormalities of tissues (Khalil M. M. et al., 2011). The presence of speckle noise in imaging affects edges and fine details which limit the contrast resolution (Chan C. et al., 2016). Enhancement of nuclear medicine images using filtering technique may be a promising approach, utilizing MATLAB in data analysis, in order to evaluate contrast enhancement pattern in different nuclear medicine images such as gray color and to evaluate the usage of new nonlinear approach for contrast enhancement of soft tissues (Abdallah Y. M. Y. and Wagiallah E., 2012). For this reason, the data are analyzed by using MATLAB software, to enhance the contrast within the soft tissues, the gray levels in both enhanced and unenhanced images and noise variance (Seo Y. et al., 2008). Top-hat filtering and deblurring images using the blind deconvolution algorithm are among the main techniques of enhancement that are usually used (Li H. et al., 2009).

MATLAB, as a high-performance interactive software package for scientific and engineering computation, allows matrix computation, implementation of algorithms, simulation, plotting of functions and data, signal and image processing (Lyra M. et al., 2011). It enables quantitative analysis and visualization of nuclear medical images of several modalities, such as SPECT, PET, or hybrid systems (Abdallah Y. M. Y. and Wagiallah E., 2012).

3.5.1 CONTRAST ENHANCEMENT

The command which is used to implement contrast processing is the imadjust (Chiu C. C. and Ting C. C., 2016). Then, the contrast in an image can be enhanced or degraded if needed. Particularly, the appropriate command is the following (MathWorks https://la.mathworks.com):

$$\mathbf{J} = \text{imadjust}\left(\mathbf{I}, \left[\text{low\_in high\_in}\right], \left[\text{low\_out high\_out}\right], \text{gamma}\right); \quad (3.10)$$

a)   b)

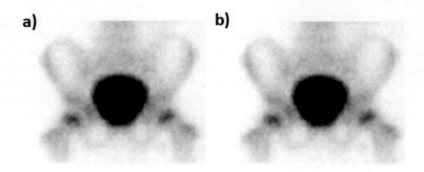

FIGURE 3.14   Bone scan (a) before contrast adjustment and (b) after contrast adjustment. (From Van den Wyngaert T. et al., 2016.)

where *J* is the new image, *I* is the initial image and gamma factor depicts the shape of the curve, describing the relationship between the values of *I* and *J*. If the gamma factor is omitted, it is considered to be 1 (Figure 3.14) (Lyra M. et al., 2011).

Furthermore, another very useful command leads to the inversion of colors, especially in grayscale images, where an object of interest can be efficiently outlined. Generally, the function for color inversion is the following (Zhu Y.-M. and Nortmann C. A., 2011):

$$J = \text{imadjust}\left(I, \begin{bmatrix} 0 & 1 \end{bmatrix}, \begin{bmatrix} 1 & 0 \end{bmatrix}, \text{gamma}\right);$$ (3.11)

$$\text{or } J = \text{imcomplement}(I);$$ (3.12)

3.5.2   Filtering in MATLAB

Image Processing Toolbox can be utilized for image data filtering. For linear filtering, MATLAB provides the fspecial command to generate some predefined common 2D filters (Lyra M. et al., 2011). It can generate Gaussian, Laplacian, Laplacians of Gaussian. The application of the mean (average) filter in a SPECT slice for different convolution kernel sizes (3 × 3, 9 × 9, 25 × 25 average filter) is quite usual (Figure 3.15) (Lyra M. et al., 2011). The series of commands may include the following:

$$
\begin{aligned}
&\text{h = fspecial ('average', }[3\ 3]);\\
&\text{b = imfilter (a, h);}\\
&\text{figure, imshow (b);}\\
&\text{i = fspecial ('average', }[9\ 9]);\\
&\text{c = imfilter (b, h);}\\
&\text{figure, imshow (c);}\\
&\text{j = fspecial ('average', }[25\ 25]);\\
&\text{d = imfilter (c, h);}\\
&\text{figure, imshow (d);}
\end{aligned}
$$
   (3.13)

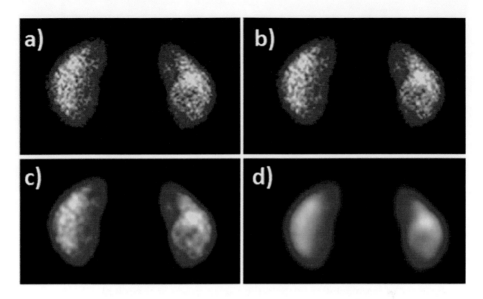

FIGURE 3.15   Mean filter applied on kidney images: (a) original image, (b) average filter 3×3, (c) average filter 9×9, (d) average filter 25×25. (From Lyra M. et al., 2011.)

3.5.3   IMAGE SEGMENTATION

The process through which an image is divided into constituent parts, in order to isolate and study separately areas of special interest is called image segmentation. This process may assist in detecting critical parts of a nuclear medicine image that are not easily displayed in the original image (Abdallah Y. M. Y. and Wagiallah E., 2012). Particularly, in nuclear medicine, segmentation techniques are employed to detect the extent of a tissue, an organ, a tumor, ambiguous boundaries of specific structures, and areas with high-radiopharmaceutical concentration (Fass L., 2008).

The common segmentation techniques are based on either the discontinuities or the similarities of structures inside an image (Kaur D. and Kaur Y., 2014). In nuclear medicine the discontinuity segmentation is the most frequently encountered technique (Li L. et al., 2017). It is based on the use of a threshold, since it is helpful in the removal of the useless information from the image, such as the background activity as well as the appearance of details which are not easily detected (Foster B. et al., 2014). The command edge is used for this reason. Furthermore, a threshold is applied to detect some edges above defined grayscale intensity, since the image is become binary. The general function which is used is the following (MathWorks. https://la.mathworks.com):

$$[BW] = \text{edge} \left(\text{image, 'method', threshold} \right) \tag{3.14}$$

where [BW] is the produced binary image, "image" is the initial one, "method" is the method of edge detection and "threshold" is the applied threshold. *Sorbel, prewitt,*

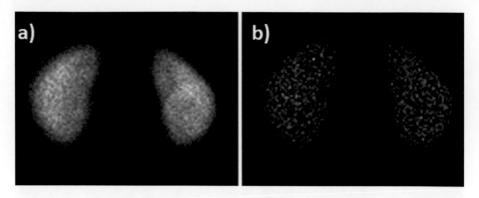

FIGURE 3.16   Images obtained through the gradient magnitude process: (a) original kidney images, (b) image post-implementation of filter and gradient magnitude.

and *canny* are widely used methods in nuclear medicine (Lyra M. et al., 2011). Other applications of segmentation include the use of gradient magnitude (Figure 3.16). The general algorithm for this function is the following (Lyra M. et al., 2011):

$$
\begin{aligned}
&\text{I} = \text{imread('kidneys.jpg');}\\
&\text{Figure, imshow(I)}\\
&\text{hy} = \text{fspecial('sobel');}\\
&\text{hx} = \text{hy';}\\
&\text{Iy} = \text{imfilter(double(I), hy, 'replicate');}\\
&\text{Ix} = \text{imfilter(double(I), hx, 'replicate');}\\
&\text{gradmag} = \text{sqrt(Ix.}^2 + \text{Iy.}^2);\\
&\text{figure, imshow(gradmag,[])}\\
&\text{se} = \text{strel('disk', 20);}\\
&\text{K} = \text{imopen(I, se);}\\
&\text{figure, imshow(K)}
\end{aligned}
\tag{3.15}
$$

REFERENCES

Abdallah Y. M. Y., Wagiallah E. (2012) Enhancement of nuclear medicine images using filtering technique, IJSR, 3 (8): 2319–7064.

Adluru G., McGann C., Speier P., Kholmovski E. G., Shaaban A., Dibella E. V. (2009) Acquisition and reconstruction of undersampled radial data for myocardial perfusion magnetic resonance imaging, Journal of magnetic resonance imaging: JMRI, 29 (2): 466–473.

Akkurt I., Gunoglu K., Arda S. S. (2014) Detection efficiency of NaI(Tl) detector in 511–1332 keV energy range, Sci Technol Nucl Ins, 2014: 186798.

Alsleem H. Paul U., Mong K. S., Davidson R. (2014) Effects of radiographic techniques on the low-contrast detail detectability performance of digital radiography systems, Radiol Technol, 85 (6): 614–622.

Bai B., Li Q., Leahy R. M. (2013) Magnetic resonance-guided positron emission tomography image reconstruction, Semi Nucl Med, 43 (1): 30–44.

Bailey D. L. (2003) Chapter 3, Data acquisition and performance characterization in PET, 41–62, reproduced from Valk P. E., Bailey D. L., Townsend D. W., Maisey M. N. Positron Emission Tomography: Basic Science and Clinical Practice. Springer-Verlag London Ltd, London, 69–90.

Beekman F. J., Slijpen E. T., de Jong H. W., Viergever M. A. (1999) Estimation of the depth-dependent component of the point spread function of SPECT, Med Phys, 26 (11): 2311–2322.

Berker Y., Li Y. (2016) Attenuation correction in emission tomography using the emission data – a review, Med Phys, 43 (2): 807–832.

Bocher M., Blevis I. M., Tsukerman L., Shrem Y., Kovalski G., Volokh L. (2010) A fast cardiac gamma camera with dynamic SPECT capabilities: design, system validation and future potential, Eur J Nucl Med Mol Imaging, 37 (10): 1887–1902.

Bruyant P. P. (2002) Analytic and iterative reconstruction algorithms in SPECT, J Nucl Med, 43: 1343–1358.

Chan C., Dey J., Grobshtein Y., Wu J, Liu Y.-H., Lampert R., Sinusas A. J., Liu C. (2016) The impact of system matrix dimension on small FOV SPECT reconstruction with truncated projections, Med Phys, 43 (1): 213–224.

Chang T., Chang G., Clark J. W., Diab R. H., Rohren E., Mawlawi O. R. (2012) Reliability of predicting image signal-to-noise ratio using noise equivalent count rate in PET imaging, Med Phys, 39 (10): 5891–5900.

Chen N., Zuo C., Lam E. Y., Lee B. (2018) 3D Imaging based on depth measurement technologies, Sensors (Basel, Switzerland), 18 (11): 3711.

Chen Y., An H. (2017) Attenuation Correction of PET/MR imaging, Magn Reson Imaging C, 25 (2): 245–255.

Chengode S. (2016) Left ventricular global systolic function assessment by echocardiography, Ann Card Anaesth, 19 (Supplement): S26–S34.

Cherry S. R., Sorenson J. A., Phelps M. E. (2012) Physics in Nuclear Medicine, 4th edition. Saunders, Elsevier, Philadelphia, PA.

Chiu C. C., Ting C. C., (2016) Contrast enhancement algorithm based on gap adjustment for histogram equalization, Sensors (Basel, Switzerland), 16 (6): 936.

Chun S. Y. (2016). The use of anatomical information for molecular image reconstruction algorithms: attenuation/scatter correction, motion compensation, and noise reduction, Nucl Med Mol Imaging, 50 (1): 13–23.

Conn N. J., Schwarz K. Q., Borkholder D. A. (2018) Nontraditional electrocardiogram and algorithms for inconspicuous in-home monitoring: comparative study, JMIR Mhealth Uhealth, 6 (5): e120.

Cradduck T. D., Busemann-Sokole E. (1985) Computers in nuclear medicine, Radiographics, 5 (1): 51–82.

Cromey D. W. (2013) Digital images are data: and should be treated as such, Methods Mol Biol, 931: 1–27.

Debattista K., Bugeja K., Spina S., Bashford-Rogers T., Hulusic V. (2017) Frame rate vs resolution: a subjective evaluation of spatiotemporal perceived quality under varying computational budgets, Comput Graph Forum, 37: 363–374.

Difilippo F. P. (2008) Design and performance of a multi-pinhole collimation device for small animal imaging with clinical SPECT and SPECT-CT scanners, Phys Med Biol, 53 (15): 4185–4201.

Dorbala S., Ananthasubramaniam K., Armstrong I.S., Chareonthaitawee P., DePuey E. G., Einstein A. J., Gropler R. J., Holly T. A., Mahmarian J. J., Park M.-A., Polk D. M., Russell III R., Slomka P. J., Thompson R. C., Wells R. G. (2018) Single photon emission computed tomography (SPECT) myocardial perfusion imaging guidelines: instrumentation, acquisition, processing, and interpretation, J Nucl Cardiol, 25: 1784–1846.

Dutta J., Ahn S., Li Q. (2013) Quantitative statistical methods for image quality assessment, Theranostics, 3 (10): 741–756.

Ebrahimnejad Gorji K., Abedi Firouzjah R., Khanzadeh F., Abdi-Goushbolagh N., Banaei A., Ataei Gh. (2019) Estimating the absorbed dose of organs in pediatric imaging of ^{99m}Tc-DTPATc-DTPA radiopharmaceutical using MIRDOSE software, J Biomed Phys Eng, 9 (3): 285–294.

Faber T. L., Folks R. D. (1994) Computer processing methods for nuclear medicine images, J Nucl Med Techno, 22: 145–162.

Fahey F. H. (2002) Data acquisition in PET imaging, J Nucl Med Technol, 30: 39–49.

Fass L. (2008) Imaging and cancer: a review, Mol Oncol, 2 (2): 115–152.

Feng H., Wang X., Chen J., Cui J., Gao T., Gao Y., Zeng W. (2019) Nuclear imaging of glucose metabolism: beyond 18F-FDG, Contrast Media Mol Imaging, 2019: 7954854.

Fernandes A. R., Oliveira A., Pereira J., Coelho P. S. (2017) Nuclear medicine and drug delivery, Advanced Technology for Delivering Therapeutics, IntechOpen, London, UK.

Foster B., Bagci U., Mansoor A., Xu Z., Mollura D. J. (2014) A review on segmentation of positron emission tomography images, Comput Biol Med, 50: 76–96.

Frey E. C., Humm J. L., Ljungberg M. (2012) Accuracy and precision of radioactivity quantification in nuclear medicine images, Semin Nucl Med, 42 (3): 208–218.

Gong K., Zhou J., Tohme M., Judenhofer M., Yang Y., Qi J. (2017) Sinogram blurring matrix estimation from point sources measurements with rank-one approximation for fully 3-D PET, IEEE T Med Imaging, 36 (10): 2179–2188.

Green M. V., Seidel J., Williams M. R., Wong K. J., Ton A., Basuli F., Choyke P. L., Jagoda, E. M. (2017) Comparison of planar, PET and well-counter measurements of total tumor radioactivity in a mouse xenograft model, Nucl Med Biol, 53: 29–36.

Gullberg G. T., Reutter B. W., Sitek A., Maltz J. S, Budinger T. F (2010) Dynamic single photon emission computed tomography—basic principles and cardiac applications, Phys Med Biol, 55 (20): R111–R191.

Hine G. J., Paras D., Warr C. P. (1978) Recent advances in gamma-camera imaging, Proc SPIE, 152: 123.

Hornberger L. K., Sahn D. J. (2007) Rhythm abnormalities of the fetus, Heart, 93 (10): 1294–1300.

Jolliffe I. T., Cadima J. (2016) Principal component analysis: a review and recent developments. Philosophical transactions. Series A, Proc Math, Phys, Eng Sci, 374 (2065): 20150202.

Jungmann J. H., Smith D. F., MacAleese L., Klinkert I., Visser J., Heeren R. M. A. (2012) Biological tissue imaging with a position and time sensitive pixelated detector, J Am Soc Mass Spectr, 23 (10): 1679–1688.

Kadrmas D. J. (2008) Rotate-and-slant projector for fast LOR-based fully-3-D iterative PET reconstruction, IEEE T Med Imaging, 27 (8): 1071–1083.

Kaur D., Kaur Y. (2014) Various image segmentation techniques: a review, IJCSMC, 3 (5): 809–814.

Khalil M. M., Tremoleda J. L., Bayomy T. B., Gsell W. (2011) Molecular SPECT imaging: an overview, Int J Mol Imaging, 2011: 796025.

Kheruka S. C., Hutton B. F., Naithani U. C., Aggarwal L. M., Painuly N. K., Maurya A. K., Gambhir S. (2012) A new method to correct the attenuation map in simultaneous transmission/emission tomography using Gd/Ga radioisotopes, J Med Phys, 37 (1): 46–53.

Kim K., Kim D., Yang J., El Fakhri G., Seo Y., Fessler J. A., Li Q. (2019) Time of flight PET reconstruction using nonuniform update for regional recovery uniformity, Med Phys, 46 (2): 649–664.

Lanfranchi P. A., Somers V. K. (2011) Cardiovascular Physiology in Principles and Practice of Sleep Medicine, 5th edition, Elsevier Inc.

Lee T. C., Alessio A. M., Miyaoka R. M., Kinahan P. E. (2016) Morphology supporting function: attenuation correction for SPECT/CT, PET/CT, and PET/MR imaging, Q J Nucl Med Mol Im, 60 (1): 25–39.

Li H., Mawlawi O. R., Zhu R. X., Zheng Y. (2009) Blind deblurring reconstruction technique with applications in PET imaging, Int J Biomed Imaging, 2009: 718157.

Li L., Wang J., Lu W., Tan S. (2017) Simultaneous tumor segmentation, image restoration, and blur kernel estimation in PET using multiple regularizations, computer vision and image understanding: CVIU, 155: 173–194.

Ljungberg M., Pretorius P. H. (2018) SPECT/CT: an update on technological developments and clinical applications, Br J Radiol, 91 (1081): 20160402.

Ljungberg M., Sjögreen Gleisner K. (2016) Personalized dosimetry for radionuclide therapy using molecular imaging tools, Biomedicines, 4 (4): 25.

Lodge M. A., Mhlanga J. C., Cho S. Y., Wahl R. L. (2011) Effect of patient arm motion in whole-body PET/CT, J Nucl Med: official publication, Society of Nuclear Medicine, 52 (12): 1891–1897.

Lyra M. (2009) Single Photon Emission Tomography (SPECT) and 3D Images Evaluation in Nuclear Medicine, Image Process. DOI: 10.5772/7056.

Lyra M., Ploussi A., Georgantzoglou A. (2011) Chapter 23, MATLAB as a tool in nuclear medicine image processing, In: MATLAB – A Ubiquitous Tool for the Practical Engineer, INTECH Open Access Publisher, 477–500.

Lyra V., Kallergi M., Rizos, E., Lamprakopoulos G., Chatziioannou S. N. (2016) The effect of patient anxiety and depression on motion during myocardial perfusion SPECT imaging, BMC Med Imaging, 16 (1): 49.

Madsen M. T. (1994) Computer acquisition of nuclear medicine images, J Nucl Med Technol, 22: 3–11.

Madsen M. T. (2007) Recent advances in SPECT imaging, J Nucl Med, 48 (4): 661–673.
MathWorks. https://la.mathworks.com

Mitra D., Basu S. (2012) Equilibrium radionuclide angiocardiography: its usefulness in current practice and potential future applications, World J Radiol, 4 (10): 421–430.

Motomura N., Nambu K., Kojima A., Tomiguchi S., Ogawa K. (2006) Development of a collimator blurring compensation method using fine angular sampling projection data in SPECT, Ann Nucl Med, 20 (4): 337–340.

Noble R. J., Hillis J. S., Rothbaum D. A. (1990) Chapter 33, Electrocardiography, In: Walker H. K., Hall W. D., Hurst J. W. (Eds), Clinical Methods: The History, Physical, and Laboratory Examinations, 3rd edition. Boston, MA: Butterworths.

Parker J. G., Mair B. A., Gilland D. R. (2009) Respiratory motion correction in gated cardiac SPECT using quaternion-based, rigid-body registration, Med Phys, 36 (10): 4742–4754.

Peterson T.E., Furenlid L. R. (2011) SPECT detectors: the anger camera and beyond, Phys Med Biol, 56 (17): R145–R182.

Polemi A. M., Niestroy J., Stolin A., Jaliparthi G., Wojcik R., Majewski S., Williams M. B. (2016) Design and characterization of a low profile NaI(Tl) gamma camera for dedicated molecular breast tomosynthesis, Proc SPIE Int Soc Opt Eng, 9969: 99690O.

Postema P. G., Wilde A. A. (2014) The measurement of the QT interval, Curr Cardiol Rev, 10 (3): 287–294.

Rosenkrantz A. B., Mendiratta-Lala M., Bartholmai B. J. (2015) Clinical utility of quantitative imaging, Acad Radiol, 22: 33–49.

Saha G. B. (2013) Physics and Radiobiology of Nuclear Medicine, 4th edition. New York Dordrecht Heidelberg London: Springer.

Saha K., Hoyt S. C., Murray B. M. (2016) Application of Chang's attenuation correction technique for single-photon emission computed tomography partial angle acquisition of Jaszczak phantom, J Med Phys, 41 (1): 29–33.

Sciammarella M., Shrestha U. M., Seo Y., Gullberg G. T., Botvinick E. H. (2019). A combined static-dynamic single-dose imaging protocol to compare quantitative dynamic SPECT with static conventional SPECT, J Nucl Cardiol: official publication of the American Society of Nuclear Cardiology, 26 (3): 763–771.

Seo Y., Mari C., Hasegawa B. H. (2008) Technological development and advances in single-photon emission computed tomography/computed tomography, Semin Nucl Med, 38 (3): 177–198.

Shepp L. A., Logan B. F. (1974) The Fourier reconstruction of a head section, IEEE T Nucl Sci, NS-21 (3): 21–43.

Shukla A. K., Kumar U. (2006) Positron emission tomography: an overview, J Med Phys, 31 (1): 13–21.

Sivarajah U., Kamal M. M., Zahir I., Weerakkody V. (2017) Critical analysis of big data challenges and analytical methods, J Bus Res, 70: 263–286.

Slomka P., Hung G.-U., Germano G., Berman D. S. (2016) Novel SPECT technologies and approaches in cardiac imaging, Cardiovasc Innov Appl, 2 (1): 31–46.

Van Audenhaege K., Van Holen R., Vandenberghe S., Vanhove C., Metzler S. D., Moore S. C. (2015) Review of SPECT collimator selection, optimization, and fabrication for clinical and preclinical imaging, Med Phys, 42: 4796–4813.

Van den Wyngaert T., Strobel K., Kampen W. U., Kuwert T., van der Bruggen W., Mohan H. K., Gnanasegaran G., Delgado-Bolton R., Weber W. A., Beheshti M., Langsteger W., Giammarile F., Mottaghy F. M., Paycha F. [On behalf of the EANM Bone & Joint Committee and the Oncology Committee] (2016) The EANM practice guidelines for bone scintigraphy, Eur J Nucl Med Mol Imaging, 43: 1723–1738.

Vaquero J. J., Kinahan P. (2015) Positron emission tomography: current challenges and opportunities for technological advances in clinical and preclinical imaging systems, Annu Rev Biomed Eng, 17: 385–414.

Vicente E. M., Lodge M. A., Rowe S. P., Wahl R. L., Frey E. C. (2017) Simplifying volumes-of-interest (VOIs) definition in quantitative SPECT: beyond manual definition of 3D whole-organ VOIs, Med Phys, 44 (5): 1707–1717.

Watabe H., Matsumoto K., Senda M., Iida H. (2006) Performance of list mode data acquisition with ECAT EXACT HR and ECAT EXACT HR+ positron emission scanners, Ann Nucl Med, 20 (3): 189–194.

Wickham F., McMeekin H., Burniston M., McCool D., Pencharz D., Skillen A., Wagner T. (2017) Patient-specific optimisation of administered activity and acquisition times for ^{18}F-FDG PET imaging, EJNMMI Res, 7: 3.

Winant C. D., Aparici C. M., Zelnik Y. R., Reutter B. W., Sitek A., Bacharach S. L., Gullberg G. T. (2012) Investigation of dynamic SPECT measurements of the arterial input function in human subjects using simulation, phantom and human studies, Phys Med Biol, 57 (2): 375–393.

Wu B., Klatzky R. L., Stetten G. (2010) Visualizing 3D objects from 2D cross sectional images displayed in-situ versus ex-situ, J Exp Psychol Appl, 16 (1): 45–59.

Zanzonico P. (2008) Routine quality control of clinical nuclear medicine instrumentation: a brief review, J Nucl Med, 49 (7): 1114–1131.

Zhang J., Fletcher J. G., Scott Harmsen W., Araoz P. A., Williamson E. E., Primak A. N., McCollough C. H. (2008) Analysis of heart rate and heart rate variation during cardiac CT examinations, Acad Radiol, 15 (1): 40–48.

Zhu Y.-M., Nortmann C. A. (2011) Pixel-feature hybrid fusion for PET/CT images, J Digit Imaging, 24 (1): 50–57.

4 Methods of Imaging Reconstruction in Nuclear Medicine

Maria Argyrou

CONTENTS

4.1 INTRODUCTION

Tomographic imaging in Nuclear Medicine was invented in order to overcome the problems of the standard two-dimensional images acquisition. Structures deep in the human body can be obscured by image superposition of overlying and underlying structures. By obtaining two-dimensional projection images of a three-dimensional object, from different angles and after the proper interpretation, we can estimate the distribution of the injected radioactivity. After the acquisition, mathematical algorithms are used to reconstruct images of selected planes within the object from projection data.

The mathematics underlying reconstruction tomography was published by J. Radon in 1917 (Radon 1986), but practical applications in medical imaging were developed later, in 1970s in the case of x-ray computed tomography. Although the instrumentation in the two emission tomography facilities (SPECT and PET) is different, the same mathematics can be used to reconstruct medical images in either of the two modalities.

Before presenting the basic reconstruction methods, it is useful to be familiar with some mathematical concepts, notation, and terminology (Cherry, Sorenson and

Phelps 2012). The data are assumed to be collected with a standard γ-camera fitted with a conventional parallel-hole collimator. The detector is assumed to accept radiation through a narrow cylinder defined by the geometric extension of the hole in front of the collimator. This cylinder defines the line of response. The counts recorded for each collimator hole are proportional to the total radioactivity contained within its line of response and are referred to as the line integral for the line of response. The entire set of line integrals across the detector is called a projection profile.

A typical SPECT camera is mounted on a gantry so that the detector can acquire projection profiles from many angles around the body. PET systems use stationary arrays of detector elements arranged in a ring or hexagonal pattern around the body. In both cases, a set of projections are recorded at equally spaced angular intervals. Then, mathematical algorithms are invoked to relate the projection data to the two-dimensional distributed activity within the projected slice.

Let N projections are recorded at equally spaced angles between 0 and 180 degrees. It is convenient to introduce a polar coordinate system (r, s) that is stationary with respect to the γ-camera detector (Figure 4.1). If the camera is rotated by an angle φ with respect to the (x, y) coordinated system of the scanned object, the transformation between the two coordinated systems is read as follows:

$$r = x \ cos\varphi + y \ sin\varphi \tag{4.1}$$

$$s = y \ cos\varphi - x \sin \varphi \tag{4.2}$$

Equations (4.1) and (4.2) aim to determine the contribution of radioactivity at (x, y) in the object at the measured profile integral at rotation angle φ in a distance r from the origin.

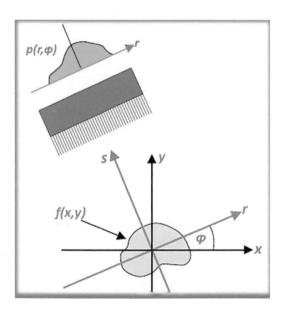

FIGURE 4.1   The rotated coordinate system (r, s) with respect to the (x, y) coordinate system. (Modified from Cherry, Sorenson and Phelps 2012.)

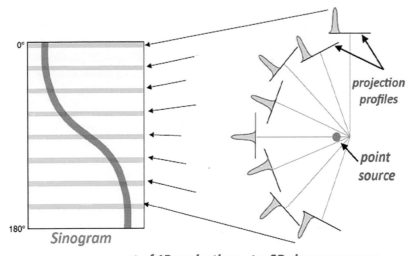

0°

180°

Sinogram

projection profiles

point source

set of 1D projections   to   2D sinogram space

FIGURE 4.2   A sinogram of a point source. Each row in the display corresponds to an individual projection profile displayed from top to bottom. (Modified from Cherry, Sorenson and Phelps 2012.)

It is preferred to display a full set of projection data in the form of a two-dimensional matrix $p(r, \varphi)$ which is known as sinogram (Figure 4.2). Each row in the display corresponds to an individual projection profile. Successive rows from top to bottom represent successive projection angles. The sinogram is a useful tool to examine several cases of artifacts in SPECT and PET images.

The question that arises is how to infer for the distribution of radioactivity inside the scanned region for a given sinogram. The most important reconstruction methods described in the following sections aim to give a realistic answer, applicable for the daily clinical use.

4.2   ANALYTIC RECONSTRUCTION

These methods are based on a continuous modeling and the reconstruction process consists of the inversion of measurement equations. The main concept is that the medical image is the Radon Transformation of the radioactivity distribution in the region of interest. The most frequently used algorithm is the Filtered Back Projection (FBP). Analytic reconstruction algorithms are efficient and elegant but they are unable to handle complicated factors such as blurring and noise.

4.2.1   FILTERED BACK PROJECTION

Simple back projection is the basic approach for reconstructing an image from the profiles. Assuming a point source object, projection profiles are acquired from different angles. Estimation for the source distribution in the plane of interest is obtained by projecting the data from each element in a profile, back across the image grid. The measured counts in a particular profile element are divided uniformly amongst the pixels. This operation, called back projection, can be expressed mathematically as follows, taking into account the contribution of N profiles. Thus, the activity distribution within the slice scanned is the sum of back projection of all the profiles acquired.

$$f'(x, y) = \frac{1}{N} \sum_{i=1}^{N} p(x\ cos\varphi_i + y\ sin\varphi_i,\ \varphi_i) \tag{4.3}$$

where φ_i denotes the projection angle and $f'(x, y)$ represents the approximation to the true radioactivity distribution $f(x, y)$.

This process results in an image resembling the true source distribution (Figure 4.3). However, blurring occurs because counts are projected out of the true source location. Although the increment of projection angles as well as the number of samples along the profile could improve image quality, the blurring effect cannot be eliminated.

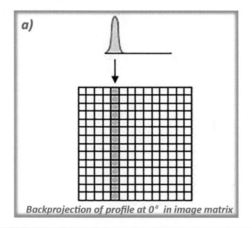

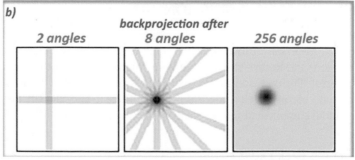

FIGURE 4.3 Back projection of one intensity profile across the image at an angle corresponding to the profile (a). This is repeated for all projection profiles to build up the backprojected image (b). (Modified from Cherry, Sorenson and Phelps 2012.)

In particular, the closer to the point source center, the more intense the blurring effect is. There is a mathematical expression which reveals the relationship between the true and the reconstructed image:

$$f'(x,y) = f(x,y) * \frac{1}{r} \qquad (4.4)$$

where the symbol "*" represents the process of convolution. Simple back projection can be used in simple source distributions. For more complicated forms, sophisticated algorithms are created e.g. Direct Fourier Transformation or Filtered Back Projection should be invoked. In the next paragraph an outline of FBP is presented.

The FBP algorithm is based on the Projection Slice Theorem. According to it, the Fourier Transform (FT) of a profile is equal to the value of the FT of the object measured through the origin and along the same angle, φ, in k-space.

The algorithm can be described by the following steps:

a. Projection profiles at N projection angles are acquired.
b. For each profile, the one-dimensional Fourier Transformation is computed, thus the Fourier Transformation values for a line across k-space are obtained (according to the projection slice theorem mentioned earlier).
c. Then a mathematical function, the ramp filter, is applied to each k-space profile to eliminate the blurring effect. This function multiplies each Fourier Transformed projection by the absolute value of the radial k-space coordinate at each point in the Fourier Transformation. As a consequence, the value of the FT is amplified linearly and proportionally to its distance from the origin of k-space.

$$P'(k_r,\ \varphi) = |k_r|\ P(k_r,\ \varphi) \qquad (4.5)$$

where $P(k_r, \varphi)$ is the unfiltered FT.
d. In the next step, the inverse FT of each filtered FT profile is computed in order to obtain a filtered projection profile:

$$p'(r,\ \varphi) = F^{-1}\left[P'(k_r,\ \varphi)\right] \qquad (4.6)$$

e. And finally, a simple back projection is performed using the filtered profiles. That process induces several artifacts on the reconstructed images, such as the star artifact, and blurring. For that reason a variety of reconstruction filters have been developed in order to improve the final image quality. The role of reconstruction filters is discussed in Chapter 5.

The difference between the two back-projection algorithms is that in the filtered, the profiles are modified by a reconstruction filter before they are back projected across the image. It has to be noticed that the back-projection process is taken place in spatial domain, while data filtration is done in the frequency domain.

4.3   ALGEBRAIC RECONSTRUCTION

4.3.1   ITERATIVE RECONSTRUCTION

The principle of iterative reconstruction algorithm is to find a solution regarding to the final image by successive estimates (Bruyant PP 2002). The main steps of these algorithms are the following (see Figure 4.4):

a. An initial estimate $f^*(x,y)$ of the true image $f(x,y)$ is made. This estimate is usually simple, for instance a uniform image.
b. Projections that would have been recorded from the initial estimate are calculated by forward projection. During this process, the intensities along the potential ray paths are summed up, for all the projections through the estimated image.
c. The calculated projections from step b) are compared to the actual profiles recorded from the source distribution.
d. The difference of this comparison is used to modify the estimated image in order to achieve a closer agreement.
e. This process is repeated until the difference reaches an acceptable small level. A proper design of the image updating and comparing procedure would offer a progressively convergence toward the true image.

The iterative reconstruction algorithm is based first of all on the method for comparing estimated and actual profiles and secondly on the image updating method. Comparing is achieved through the cost function, which measures the difference

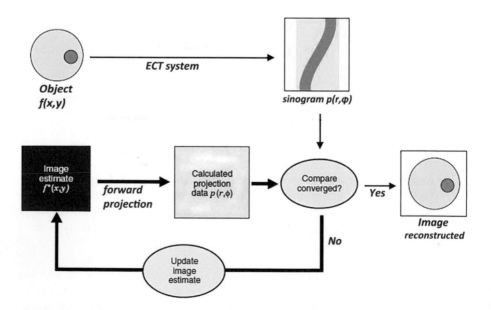

FIGURE 4.4   Iterative reconstruction steps illustration. (From Cherry, Sorenson and Phelps 2012.)

between the two kinds of profiles. Then, the update function uses the output of the cost function to update the estimated image.

In general, the goal of iterative process is to produce rapid and accurate convergence of the estimated toward the true image. For that reason, one should deal with the issue of statistical noise. For example, some algorithms give more weight to profiles (or sinograms) which contain higher number of counts and thus lower levels of statistical noise. There are also algorithms which incorporate some sort of "prior information" such as the expected shape or smoothness of the image.

Iterative algorithms are more intensive computationally, as compared to the analytical algorithms mainly the FBP. Most of them require several iterations in order to approach effectively the true image. The iteration consists of a separate back-projection procedure, which is the most time consuming part of an FBP algorithm, but it is performed just once. Furthermore, iterative algorithm can encompass specific characteristics of the imaging system. Calculation of forward projection profiles for the estimated image takes into account contribution from all image pixels and not only from these along a ray path. This is against the rapidity of calculations.

The expectation-maximization (EM) algorithm is based on statistical considerations to compute the most likely source distribution that would have created the observed projections (Vandenberghe and etal. 2001). It assigns greater weight to high-count elements of a profile. This is the main difference as compared to back-projection algorithms in which a uniform statistical weight is assigned to all elements.

The true source distribution $f(x, y)$ is presented as a vector $V(i)$ $(i=1,\dots m \times m)$ in a discretized image of $m \times m$ pixels. Assuming that there are N angles of projection and M samples of projection, the acquisition data can be represented by a one-dimensional vector $u(j)$, $j = 1\dots.(N \times M)$

Then

$$u(j) = \sum_{i=1}^{m \times m} P(i,j) V(i) \qquad (4.7)$$

where $P(i, j)$ is the probability of detecting a photon originating in voxel I, in detection bin j. These probabilities form a matrix of dimensions $(m \times m)(N \times M)$. This is a set of linear equations which can be solved for $V(i)$. It is difficult though to directly inverse this set of equations because of matrix dimensions. Moreover, this process could be time and memory consuming and could lead to instabilities. There is a class of iterative reconstructions methods which are based on maximization of a likelihood function. These are the Expectation Maximization algorithms. They use a statistical model of measurement process, taking into account the noise. The best-known representative is the ML-EM algorithm which is described in the next section.

4.3.2 Maximum Likelihood-Expectation Maximization (ML-EM)

In this method, projection data are considered as Poisson variables with a mean equal to the profile perpendicular to the projection bin, through the activity distribution. For a large number of photons, the measured data are similar to the value of the line integral. For low-count statistics, however, measured data can have large mean variations. This is why analytic algorithms perform poorly in low-count acquisition. The measured data can be expressed as follows:

$$V(j) = \sum_{i=1}^{mxm} P(i,j) u(i) \tag{4.8}$$

The goal of reconstruction algorithm is to find the distribution $u(i)$ which can have generated the measured projection data $V(j)$ with the largest probability. The probability function is called likelihood function and is derived from Poisson statistics.

$$L(P) = \prod_{j=1}^{m \times m} e^{-V(j)} \frac{V(j)^{u(j)}}{u(j)!} \tag{4.9}$$

In order to maximize this likelihood, the Expectation Maximization algorithm is implemented which yields the following update concerning the image estimation:

$$V_{k+1}(i) = V *_k (i) \sum_{j=1}^{M \times N} P(i,j) \frac{V(j)}{\sum_{l'=1}^{m \times m} P(l',j) V_k(l')} \tag{4.10}$$

This approach has two main disadvantages. First of all, the reconstructions become noisier for high iteration numbers, because of the Poisson statistics. In fact, projection data are similar to the noisy measurements because the projector is a smoothing operator. Several solutions have been proposed to deal with this but the most prevalent in clinical use is the post-filtering process after reconstruction. The smoothness can be chosen by appropriate cutoff frequency of the filter, without repeating the reconstruction.

Secondly, the reconstruction process although very effective, they are slow for daily routine. The time needed for an FBP reconstruction is approximately half the time needed for one iteration by the ML-EM algorithm. The most prominent method to speed up these algorithms is to split up the measurements data into different subsets and to use only one subset for each iteration. The method is called Ordered Subsets Expectation Maximization (OS-EM) and it is described in the next section.

4.3.3 Ordered Subsets-Expectation Maximization (OS-EM)

The OS-EM is the most frequently used iterative reconstruction algorithm in both SPECT and PET (Hudson and Larkin 1994). In ML-EM iteration requires a forward projection of the previous estimation into all projections. In OS-EM the projection data are divided into ordered subsets, each of them containing an equal number of projections. The OS level is defined as the number of these subsets. The standard EM algorithm is then applied to each of the subsets in turn. The resulting reconstruction becomes the starting value for use with the next subset. Iteration in OS-EM is considered as a single pass through all the specified subsets. As the image is refined, a larger number of projection angles are included. The number of projection profiles to be computed diminishes, thus the time per iteration falls rapidly. This approach can be used to speed up simple forward projection as well as more complicated iterative algorithms. For instance, if a SPECT acquisition contains projections from 60 angles, one step of the ML-EM method requires forward projections under 60 angles, whereas in OS-EM data can be divided into 6 subsets of 10 angles, so one iteration step only requires forward projection under 10 different angles. Thus, by using the OS-EM, results in a six-times quicker algorithm. It has been shown that the image quality for the same number of iterations in ML-EM is comparable if the number of subsets is relatively small.

In SPECT, subsets may correspond to groups of projections and it is better to select subsets in a way that pixel activity contributes equally to each subset. In PET, OS-EM can best be applied after full collection of all tube counts. Then, tubes can be binned in parallel families defining projection and OS-EM applies, resulting in reducing iterations as in SPECT. The order in which subsets are processed is arbitrary though careful choice of subsets could improve image quality. For example, new information could be introduced immediately by choosing first the projection which corresponds to the direction of greatest variability in the image, a second projection perpendicular to the first, the next projections in the middle of these and so on.

In concluding, OS-EM offers a high quality of reconstruction in reduced time and thus helping to process SPECT and PET data in real time. This method is also suited to an arbitrary set of projections as they can be obtained during acquisition when a patient movement is detected. It can also be applied with algorithms different than EM and with other collection schemes imposed by camera technologies.

4.4 FOURIER RECONSTRUCTION

A method to avoid 1/r blurring is FT reconstruction. It is an alternative method for representing spatially varying data in spatial frequency space or "k-space". In the case of two-dimensional image profile $f(x, y)$, the FT can be expressed as follows:

$$F\left(k_x, k_y\right) = \mathcal{F}\left[f\left(x, y\right)\right] \tag{4.11}$$

where k_x, k_y are the spatial frequencies in k-space and $F\left(k_x, k_y\right)$ are the corresponding amplitudes for different spatial frequencies. FTs can be calculated quickly and conveniently and a lot of image software packages include FT routines. According to Fourier slice theorem, the FT of a projection of a profile $p(r, \varphi)$ is equal to the value of the FT of the object along the same angle in k-space.

$$\mathcal{F}\left[p(r, \varphi)\right] = F\left(k_r, \varphi\right) \tag{4.12}$$

where $F\left(k_r, \varphi\right)$ represents the value of FT at a radial distance k_r, along a line at angle φ in k-space.

In other words this theorem provides a means for obtain two-dimensional data in k-space from one-dimensional measurements in object space. The reconstruction process is as follows:

a. Projection profiles are acquired in object space at N projection angles φ.
b. The one-dimensional FT of each profile is computed.
c. The values of these transformations are assigned to appropriate locations in k-space, in polar coordinates. These values are closely spaced near the origin and widely spaced farther away. The increased representation of data near the origin explains why the "1/r blurring" occurs in simple back-projection methods.
d. Using the values inserted in polar coordinates, the values of kx, ky are interpolated in the k-space.
e. Perform an inverse FT to compute the object image by using the interpolated values in k-space.

Data free of noise, perfect interpolation, and line integrals representing precisely the sum of activity along a line could lead to an exact representation of the object.

However, interpolation from polar to rectangular coordinates in k-space is computationally intensive and can lead to artifacts in the image. More elegant approaches such as the FBP which is described in the next session could be invoked.

4.5 SELECTION OF RECONSTRUCTION METHOD

Reconstruction methods presented earlier apply to single-slice images. However, SPECT and PET detectors acquire data simultaneously for multiple sections through the body. Image slices are gathered together to form a three-dimensional dataset which can be resliced to obtain images of planes other than those that are directly imaged. Thus three-dimensional volumetric images are generated.

4.5.1 SPECT SLICES RECONSTRUCTION

Reconstruction methods in SPECT were analyzed when a parallel-hole collimator is used. Tomographic reconstruction can be performed when a fan beam or pinhole collimator is used. In this case, non-parallel ray projections are provided and therefore data cannot be inserted directly into reconstruction algorithms used in the case of parallel-hole collimator. One approach is to re-resort the fan beam data into parallel-beam data so that the ordinary reconstruction described in previous section can be used. As an alternate, a reformulated and refined FBP algorithm to include non-parallel data could be invoked. A fan beam collimator can completely cover a three-dimensional tissue volume in a single rotation.

In the case of converging collimator, a 180+θ rotation is required whereas for a diverging collimator the rotation angle should be (180−θ) degrees (θ is half the fan angle for the collimator).

When a cone beam or pinhole collimator is used, all profiles intersect at the center and the process of resorting is not possible by performing a single rotation. A simple approach is to translate the collimator along the rotation axis. The dataset acquired can be resorted into a complete set if parallel projections. Alternatively, approximations and interpolations can be used to convert cone-beam data into fan-beam data. However, these methods work best for small cone angle. Iterative algorithms can be used for direct reconstruction of three-dimensional data but this process is highly time consuming.

4.5.2 PET Slices Reconstruction

PET scanners consist of multiple detector rings. Two-dimensional imaging consists of data collected along Lines of Response within a specified imaging plane (given detector ring) and can be reconstructed either with FBP or iteratively into a transverse image. Projection data can also be acquired at oblique angles between detectors; therefore a three-dimensional reconstruction algorithm is required (Figure 4.5). A simple approach is to "re-bin" the three-dimensional dataset so that each oblique projection ray is placed within the projection data for a particular non-oblique two-dimensional transverse slice (Alessio and Kinahan 2005). This can be accomplished easily by the assignment of each ray to its average axial location. Thus, an oblique projection ray would be positioned as if it were a projection from directly opposed pair located halfway between them. The result of this process is a series of sinograms of parallel-ray projections, each corresponding to different axial locations through the object. Consequently, each sinogram can be reconstructed by two-dimensional FBP or iterative algorithms described earlier.

The main drawback of this method, known as single-slice re-binning, is that significant mispositioning errors occur for events originating close to the scanner edge as

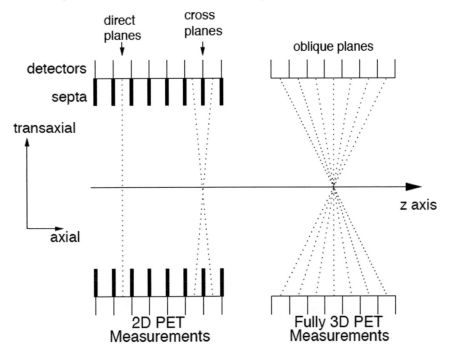

FIGURE 4.5 Three-dimensional versus two-dimensional PET data acquisition. In two-dimensional PET imaging, detectors collect direct and cross planes whereas in fully three-dimensional mode, scanners also collect oblique planes. (From Alessio and Kinahan 2005.)

well as for projections at large angle with respect to the transverse plane. To overcome these problems, more sophisticated re-binning algorithms have been developed.

As an alternate, the three-dimensional formulation of iterative algorithms is proposed, thus, the exact orientation of each line of response is taken into account. Because an extra dimension is added to projection data, the dimensions of matrices involved in the equations described in Section 4.2 increase and the projection steps must be performed though a volume rather than across a two-dimensional slice. Great progress has been made toward matrix size reduction using symmetry arguments and sparse storage techniques. Continuous multiprocessor hardware improvement aims to produce faster and efficient fully three-dimensional algorithms.

4.6   RECONSTRUCTION OF NUCLEAR MEDICINE IMAGES BY MATLAB

In MATLAB there are some examples demonstrating the basic principles of medical image reconstruction using Radon and Inverse Radon Transformation. The reader is referred to MathWorks site (https://www.mathworks.com/) for the definition for *Radon* and Inverse Radon (*iradon*) functions. For their use, the Image Processing Toolbox should be installed. An example is shown in Figure 4.6. The test image is

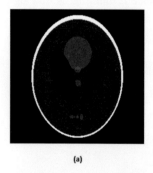

(a)

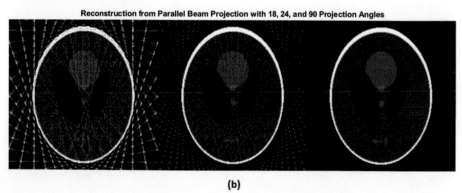

Reconstruction from Parallel Beam Projection with 18, 24, and 90 Projection Angles

(b)

FIGURE 4.6   In (a) The Shepp–Logan phantom test image is shown. In (b) the reconstruction process results using Radon function are presented for Parallel Beam geometry and a variable amount of projection angles. (https://www.mathworks.com/.)

the Shepp–Logan head phantom which can be generated using the *phantom* function. The *radon* function is invoked to create projections for a different number of projection angles and *iradon* function is called to return the output image from the above projections. The number of projecting angles plays an important role to the presence of image artefact as can be seen in Figure 4.6.

NiftyRec (Pedemonte et al. 2010) is software for tomographic reconstruction providing reconstruction tools for emission and transmission computed tomography. It uses MATLAB and Python interfaces which include FB projection, MLEM, and OSEM reconstruction algorithms optimized for achieving high performance in a few seconds (Figure 4.7).

The Michigan Image Reconstruction Toolbox (MIRT) is a collection of open source algorithms for image reconstruction adapted to MATLAB. This software was developed at the University of Michigan by Jeff Fessler and his group and includes iterative and non-iterative algorithms for tomographic imaging (PET, SPECT, X-ray CT) (see Fessler's webpage).

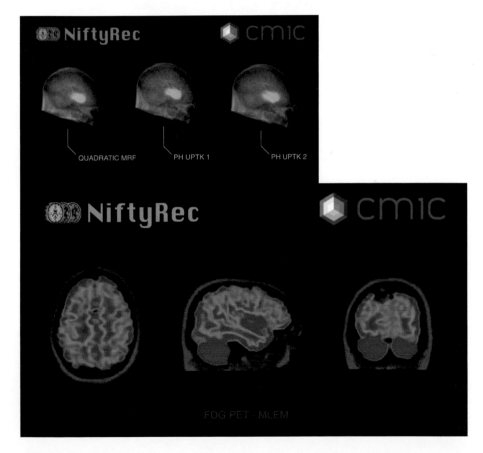

FIGURE 4.7 Top: An example of SPECT reconstruction is shown with Bayesian model expressing hidden states of function and anatomy. Bottom: An MLEM reconstruction of FDG PET with a Bayesian model based on finite states of function and anatomy.

REFERENCES

Alessio, Adam M., and Paul E. Kinahan. 2005. "PET Image Reconstruction." To appear in Henkin et al.: *Nuclear Medicine* 2nd Ed.

Bruyant P.P. 2002. "Analytic and iterative reconstruction algorithms in SPECT." *J Nucl Med* 43 (10): 1343–1358.

Cherry, Simon R., James A. Sorenson, and Michael E. Phelps. 2012. *Physics in Nuclear Medicine, 4th ed.* Elsevier Saunders.

Fessler, Jeffrey A. https://web.eecs.umich.edu/~fessler/code/.

Hudson H.M., and R.S. Larkin. 1994, "Accelerated image reconstruction using ordered subsets of projection data." *IEEE Trans Med Imaging* 13 (4): 601–609.

Pedemonte, S. et al. 2010, "GPU accelerated rotation-based emission tomography reconstruction." IEEE Nuclear Science Symposium & Medical Imaging Conference. DOI: 10.1109/NSSMIC.2010.5874272.

Radon, J. 1986. "On the determination of functions from their integral values along certain manifolds." *IEEE Trans Med Imaging* 5 (4): 170–176.

Vandenberghe S. et al. 2001, "Iterative reconstruction algorithms in nuclear medicine." *Comput Med Imaging Graph* 25 (2): 105–111.

5 Image Processing and Analysis in Nuclear Medicine

Antonios Georgantzoglou

CONTENTS

5.1   INTRODUCTION

Primary nuclear medicine images often suffer from certain artefacts and noise, reducing as such their diagnostic value. The reasons mainly include the interaction of radiation with the tissues and the imperfections of the imaging modalities. Therefore, post-acquisition digital image processing has been introduced to suppress these imperfections up to a certain degree. Image processing is a process that aims at reducing or even eliminating undesired features and simultaneously attempts to expose and enhance useful information. The introduction of modern processing techniques has further assisted in increasing the diagnostic value of nuclear medicine images. Image processing may include many different techniques such as filtering, edge detection, segmentation, texture analysis, and pattern recognition (Beck, 2004). In most cases, these techniques are not applied solely but rather as a serial or parallel pipeline of processes.

In this chapter, we attempt to describe various concepts of image analysis and how these principles are applied to nuclear medicine. Therefore, it gives both an overview of general image analysis tools that have been specifically used or developed for nuclear medicine applications. The chapter is arranged in the following order: first, a short overview is given on the factors that affect the image quality so the reader can appreciate the value of image analysis. The operations of filtering and thresholding follow as tools for image enhancement and basic segmentation. More sophisticated segmentation methods are then described, with application in positron emission tomography (PET) and single photon emission computed tomography (SPECT). The next area coves image registration application between PET, SPECT and computed tomography (CT), magnetic resonance imaging (MRI). The last part of this chapter is dedicated on how MATLAB (The MathWorks, Inc., Natick, MA) can be used in nuclear medicine.

5.2   IMAGE DEGRADING FACTORS

Image quality depends on the spatial resolution and contrast. Spatial resolution is the shortest distance between two neighbouring objects that a system can image as separate entities while contrast defines the difference in intensity between two neighbouring objects.

Several factors degrade the quality of the formed image in nuclear medicine. These factors originate from: (a) physical properties (i.e. natural background radioactivity, photon attenuation, photon scattering), (b) imaging devices (i.e. noise, choice of collimator, intrinsic resolution of the system, count rate), (c) the patient (i.e. patient motion). A detailed description of these factors can be found in International Atomic Energy Agency (2014), therefore only a short description of each factor follows:

Background Radioactivity: this originates from natural radioactivity, nearby patients also injected with a radiopharmaceutical or even residual radioactivity.

Background radioactivity can degrade studies that involve low number of counts, misrepresenting the quantitative result.

Radioactive Decay: radioisotopes undergo decay depending on their half-time or decay constant (Equation 5.1):

$$N = N_0 \times e^{-\lambda t} \tag{5.1}$$

where N is the number of photons at time t, N_0 is the initial number of photons at $t=0$ and λ is the decay constant. This factor should be considered in time-dependent radioactivity measurements and the results should be corrected for this decay.

Noise: noise originates from the random character for the decay process and the instrumentation. It has less contribution in studies with high radioactive activity or long acquisition time.

Photon Attenuation: when photons transverse matter, they interact, and they lose energy. These photons may be absorbed by the human body, reducing the amount of radiation that reaches the detector. Equation 5.2 gives the number of photons N that are counted the detector:

$$N = N_0 \times e^{-\Sigma_i \mu_i d_i} \tag{5.2}$$

where N_0 is the true number of photons, d is the thickness of tissue, and μ is the attenuation coefficient (Patton & Turkington, 2008).

Photon Scattering: photons that scatter in matter can be still detected by the imaging system. However, scattered photons give wrong information on the position of the initial photon emission inside the body, leading to poor contrast. Figure 5.1 describes the photon scattering effect on an internal structure of the patient.

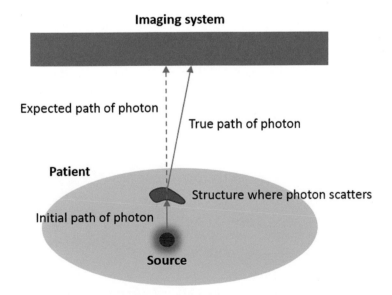

FIGURE 5.1   Photon scattering on an internal patient structure.

Collimator: the choice of collimator in γ-camera depends on the energy of the radionuclide used in a study. High-resolution collimators may suffer from low count-rate issues and high-sensitivity collimators may suffer from poor resolution.

Star Effect: if radiation strikes a collimator with thin septa in small angles and penetrates them, it may contribute to the detected signal; penetration may appear as a star object in the image, preventing the imaging of a high-activity area (Gunter, 2004). Figure 5.2 illustrates this effect.

Resolution and Partial Volume Effect (PVE): the spatial resolution of a system defines the minimum distance that two radioactive sources are imaged as separate entities. The resolution is estimated by the full-width at half-maximum (FWHM) of imaged activity (Soret et al., 2007). PVE occurs when one or more objects interfere between the examining tissues and the imaging system (Erlandsson et al., 2012). Figure 5.3 shows a graphical representation of the loss of resolution and signal of the acquired image.

Count Rate: imaging systems require a certain amount of time to register an event, during which a new event cannot be recorded, limiting the system's sensitivity. Count rate loss (CRL) is given by Equation 5.3:

$$CRL = \frac{TC - MC}{TC} \times 100\% \qquad (5.3)$$

where *TC* is the true counts and *MC* is the measured counts.

Patient Motion: the diagnostic outcome of a nuclear medicine test often depends on imaging for a prolonged period of time. During this time, image quality is affected by the inevitable patient motion and the motion of the internal organs such as the heart, the lungs, and the bowel.

5.3   IMAGE IMPROVEMENT

The inevitable presence of image degrading factors has led to increasing demand for image improvement. The most well-known technique for enhancing the quality of the acquired image is filtering. Filtering includes several different mathematical tools which, if applied to the acquired image, remove the undesired frequencies and, concurrently, maintain the frequencies that contribute to the diagnostic outcome. The way that filtering works is by smoothing or sharpening images (Faber & Folks, 1994).

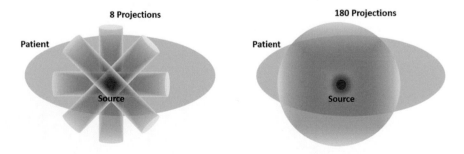

FIGURE 5.2   Star artefact. When few projections are acquired, a star-like artefact is generated in the image which can be significantly reduced when 180 projections are acquired.

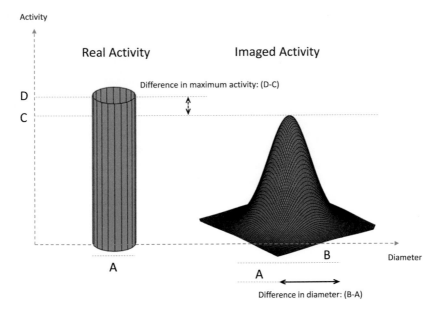

FIGURE 5.3 Resolution and signal loss in an acquired image of radioactivity. Left: Representation of an actual point source. Right: Representation of the measured activity, with increased diameter and decreased height.

Several filters have been introduced, each of them having a different goal. There are filters that can be used for noise suppression, signal enhancement, or star artefact elimination (Van Laere et al., 2001). Although, the goal of filtering is to improve the diagnostic value of an image, this process can reduce the amount of fine details; therefore, special attention should be given (Lyra & Ploussi, 2011). An analytical description of filters has been already published (Gilland et al., 1988; Lyra & Ploussi, 2011; Lyra et al., 2014), therefore only a short overview will follow.

5.3.1 Ramp Filter

One of the most widely used filters in nuclear medicine is the **Ramp Filter**. It is routinely used in Filtered Backprojection (FBP) to eliminate the star artefact (Germano, 2001). It is a high-pass filter that sharpens the image but amplifies the noise (Equation 5.4) (Lyra & Ploussi, 2011):

$$H_R\left(k_x, k_y\right) = \left(k_x^2 + k_y^2\right)^{1/2} \tag{5.4}$$

where k_x, k_y are the spatial frequencies across the x and y axes, respectively.

5.3.2 Filters for Noise Reduction

The application of the Ramp filter results in amplification of noise. Noise reduction is achieved by suppressing the high-frequency signal. Relevant filters are called low-pass filters, simply because they maintain low frequencies.

Butterworth Filter: it is one of the most used filters in nuclear medicine, given by Equation 5.5:

$$B(f) = \frac{1}{\left(1 + \left(f/f_c\right)^{2n}\right)} \tag{5.5}$$

where f is the frequency signal, f_c is a critical frequency, and n is the order of magnitude of filtering. The critical frequency f_c is the frequency where the filter stops its application while n defines the steepness of the filter. An advantage of this filter is that it provides a satisfactory trade-off between noise suppression and fine detail preservation (Lyra & Ploussi, 2011).

Hanning Filter: this filter contains only a cut-off frequency parameter, given by Equation 5.6:

$$H(f) = \begin{cases} 0.5 + 0.5 \ cos\left(\dfrac{\pi f}{f_m}\right), & 0 \le f \le f_m \\ 0, & otherwise \end{cases} \tag{5.6}$$

where f is the frequency signal and f_m is a critical frequency beyond which all frequencies are eliminated. It is efficient in suppressing image noise but in the expense of fine detail preservation (Lyra & Ploussi, 2011). Figure 5.4 shows the Ramp and Hanning filters that are used in FBP.

Hamming Filter: this filter is very similar to Hanning filter as it also contains only a cut-off frequency parameter, given by Equation 5.7:

$$H(f) = \begin{cases} 0.54 + 0.46 \ cos\left(\dfrac{\pi f}{f_m}\right), & 0 \le f \le f_m \\ 0, & otherwise \end{cases} \tag{5.7}$$

where f is the frequency signal and f_m is a critical frequency beyond which all frequencies are eliminated. The difference between the two filters lies in the amplitude of the cut-off frequency (Lyra & Ploussi, 2011). Figure 5.5 shows the minor difference between the Hanning and Hamming filters.

Parzen Filter: this filter provides maximum image smoothing with the expense of resolution degradation (Lyra & Ploussi, 2011), given by Equation 5.8:

$$H(f) = \begin{cases} 2|f|\left(1 - \dfrac{|f|}{f_m}\right)^3, & \dfrac{f_m}{2} < |f| < f_m \\ 0, & |f| \ge f_m \end{cases} \tag{5.8}$$

where f_m is a critical frequency beyond which all frequencies are eliminated.

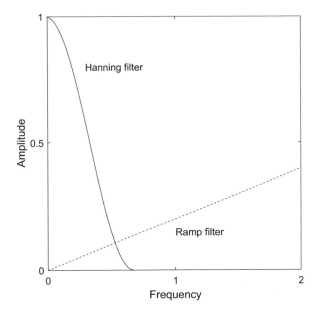

FIGURE 5.4 Filters for FBP. Hanning filter for cut-off frequency 0.7 and Ramp filter with a bandwidth of 0.4. The Hanning filter was designed using the MATLAB function *hann* of Signal Processing Toolbox™ (The MathWorks, Inc., 2019b) and the Ramp filter was designed using the MATLAB function *abs*.

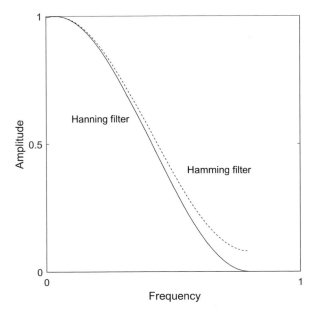

FIGURE 5.5 Comparison between Hanning and Hamming filter. The filters were designed using the MATLAB functions *hann* and *hamming*, respectively.

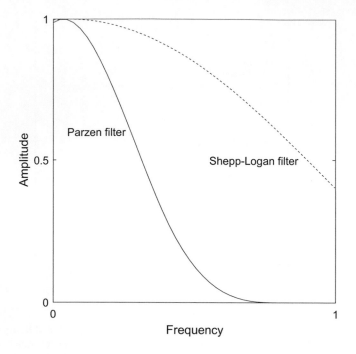

FIGURE 5.6 Difference between Parzen and Shepp–Logan filter. The Parzen filter was designed using the MATLAB function *parzenwin* whereas the Shepp–Logan was designed with a MATLAB custom-written script.

Shepp–Logan Filter: this filter maintains the fine details in the image because it applies a minimum smoothing (Lyra & Ploussi, 2011), given by Equation 5.9:

$$S(f) = \frac{2f_m}{(\pi \left(\frac{sin|f|\pi}{2f_m} \right)}$$ (5.9)

where f is the frequency signal and f_m is a critical frequency. Figure 5.6 shows the difference between the Parzen and Shepp–Logan filters.

5.3.2.1   Other Filtering Approaches

Other filters have also been used in nuclear medicine. Takalo et al. (2015) developed a low-pass filter to reduce noise in SPECT images. The filter is based on autoregression modelling which generates a dual-module image, the predictable and the prediction error image (Equation 5.10):

$$X_{pred}(n_1, n_2) = -\sum_{k_1}\sum_{k_2} a(k_1, k_2) X_{orig}(n_1 - k_1, n_2 - k_2) + w(n, n_2)$$

$$X_{err} = X_{orig} - X_{pred}$$ (5.10)

where $a(k_1,k_2)$ are the prediction weighting factors, (n_1,n_2) is the matrix, $w(n_1,n_2)$ is the error.

Noise reduction can also be achieved using other methods. Stefan et al. (2012) presented a wavelet decomposition method to denoise and restore PET images while Lin et al. (2001) used wavelets to denoise PET images and improve the physiological measurements derived from these images.

5.3.3 FILTERS FOR SIGNAL RESTORATION

These filters have a dual role in image analysis: they both suppress the noise and enhance the signal. Their application depends on the modulation transfer function (MTF) of the imaging device which shows its ability to transfer the modulation of the imaged object to image (Beck & Brill, 2004).

Metz Filter: this filter has two components, an inverse filter which assists in resolution recovery and a low-pass filter for noise suppression, given by Equation 5.11:

$$M(f) = MTF(f)^{-1}\left[1-\left(1-MTF(f)^2\right)^x\right] \tag{5.11}$$

where f is the spatial and x defines the cut-off point of the inverse filter application before the low-pass filter starts its application for noise suppression (King et al., 1988).

Wiener Filter: this filter has the same two components as the Metz filter but it depends on the signal-to-noise (SNR) of the acquired image. Equation 5.12 describes the filter:

$$W(f) = MTF^{-1}(f) \times \frac{MTF^2(f)}{\left(MTF^2(f) + \frac{\overline{N}}{P_0(f)}\right)} \tag{5.12}$$

where N is the total image count and P_0 quantifies the spectrum of object (King et al., 1984).

Other Filters: Benameur et al. (2009) used 2D deconvolution to restore SPECT brain images with limited contrast and found superior results compared to use of Metz filter; Mignotte and Meunier (2000) used 3D blind deconvolution to assist in SPECT brain segmentation; Geets et al. (2007) used a bilateral filter, similar to Gaussian filter, for denoising FDG-PET images.

5.3.4 PARTIAL VOLUME CORRECTION

PVE degrades significantly the image resolution. PVE is calculated as the ratio of measured radioactivity divided by the true radioactivity (Morris et al., 2004). The correction of this effect assists in revealing the true radioactivity concentration, a critical step in nuclear medicine diagnosis. PVE is not routinely used in clinical practice; however, several methods have been suggested (Cal-Gonzalez et al., 2018). A review on correction methods in emission tomography can be found in

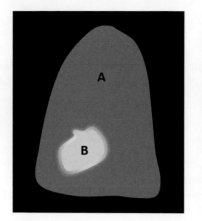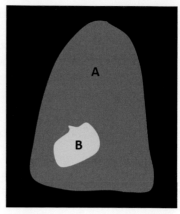

FIGURE 5.7 Graphical demonstration of partial volume effect (PVE) correction. Left: Unclear boundaries between radioactive area and A and B. Right: After PVE correction, radioactive areas A and B are well-separated.

Erlandsson et al. (2012) while an extensive list of methods is also presented in Li et al. (2017). Here, only a short overview of correction strategies will be given. Figure 5.7 presents a graphical representation of PVE correction.

PVE depends heavily on the PSF of the imaging system. Therefore, many correction methods focus on reversing the PSF effect through deconvolution (e.g. Tohka & Reilhac, 2008). The latter utilises the system PSF either during image reconstruction or as an image processing step post-reconstruction. However, undesirable effects include the noise amplification (in the case of correction during reconstruction) or ringing artefacts (sharp changes in signal) due to loss of high-frequency signal at sharp boundaries (for both cases).

To control noise and artefacts, image restoration can be assisted by using anatomical information from other imaging modalities either during or post-reconstruction. Different methods have been suggested for incorporating anatomical information on the image domain, by using the mean intensity value or the voxel intensity value of a single organ or that of multiple regions. The main advantage of voxel-based correction with multiple regions is that it provides partial volume correction over the entire image. However, it should be noted that many methods require the knowledge of true radioactivity and, therefore, combination of methods or iterative utilisation of the same method on multiple regions have been suggested as solutions (Erlandsson et al., 2012).

5.4 THRESHOLDING – EDGE DETECTION

5.4.1 THRESHOLDING

Thresholding is one of the primary processes that are applied on image analysis. It turns a coloured or a grey-scale image into a binary image, or black and white, by assigning that value 0 or 1 to each pixel (Equation 5.13), depending on the threshold value T that has been selected, either by the user or automatically (Solomon & Breckon, 2011):

$$J(x,y) = \begin{cases} 0, \ I(x,y) < T \\ 1, \ I(x,y) \geq T \end{cases} \tag{5.13}$$

where $I(x,y)$ denotes the original image and $J(x,y)$ denotes the binary one. Thresholding is frequently used as a step in an image analysis flow, either immediately after pre-processing or afterwards, in order to eliminate undesired pixel intensities. The original image can then be regenerated without those intensities (i.e. set to 0). Moreover, multi-thresholding can also be applied by introducing multiple thresholds. The image is then transformed not into black and white but into a pseudo-coloured image with each colour defining pixels that had original values within a range.

Thresholding can be applied with two ways: (a) the user defines a threshold value (or many thresholds), usually, (b) an automated method is selected to generate this threshold value(s). Both applications have advantages and disadvantages. The definition of a manual thresholding is a simple process that usually requires few trials but can have very limited application. The use of an automated method eliminates the need for individual image inspection and trial, defining a global application; however, it requires an in-depth understanding of the image dataset to blindly apply to many images. Many automatic threshold methods have been introduced and these can be adaptive to individual datasets, based on different properties of the image such as the histogram or the entropy. Information on several thresholding techniques can be found in Sezgin and Sankur (2004). Here, we provide only an overview of some thresholding methods.

Otsu method: it finds a threshold value that maximises the inter-class variance between the two classes of pixels (black and white) (Otsu, 1979).

Mean method: it selects the mean value of the pixel intensities.

Median method: it selects the median value of the pixel intensities.

Mid-grey method: it uses the middle grey-level value.

Maximum entropy method: it searches for a threshold value that maximises the entropy within the group of pixels below and above this value (Kapur et al., 1985).

Li method: it searches for the threshold value that minimises the difference in entropy between the original and the processed image (Li & Lee, 1993).

5.4.2   EDGE DETECTION

An edge is a discontinuity in an object or image and frequently defines a sharp change in intensity. Edge detection underlines special characteristics of objects. Therefore, this process can find wide application in nuclear medicine image processing assisting in organ delineation. Most often it is used simultaneously with other techniques such as thresholding. As in thresholding, edge detection results in a binary image with 'interesting' objects having white pixels (i.e. pixel value of 1) (Georgantzoglou et al., 2014). Figure 5.8 illustrates the definition of edges.

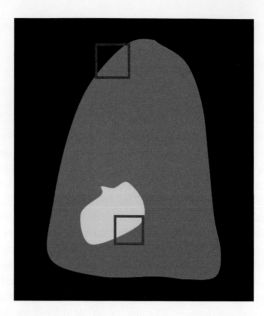

FIGURE 5.8 Edges are detected at sharp changes of intensity between two adjacent areas (black with dark grey, dark grey with light grey) as illustrated by the red boxes.

Edge detection is achieved by mathematical operators that detect discontinuities of the image gradient. They are distinguished into two categories: (a) First-order detectors, (b) second-order detectors. The first- or second-order edge detectors are called as such because they utilise the first- or second-order image derivative (Solomon & Breckon, 2011). In the case of the second-order derivative, the Laplacian operator is applied (Equation 5.14):

$$\nabla^2 I(x,y) = \frac{\partial^2 I}{\partial y} + \frac{\partial^2 I}{\partial x} \tag{5.14}$$

where $I(x,y)$ denotes the image.

5.4.2.1 First-Order Detectors

Roberts: the Roberts edge detector highlights areas of high-frequency, where there is increased likelihood to find an edge. It uses two 2×2 convolution kernels (Equation 5.15) which tend to maximise their response to edge that orient themselves at 45° in relation to image orientation.

$$A_{der} = \begin{bmatrix} 0 & -1 \\ 1 & 0 \end{bmatrix} \quad B_{der} = \begin{bmatrix} -1 & 0 \\ 1 & 1 \end{bmatrix} \tag{5.15}$$

Sobel: the Sobel edge detector also highlights areas of high-frequency. The difference with the Roberts operator is the use of two 3×3 convolution kernels (Equation 5.16), one of which is simply the rotated version of the other by 90°.

Sobel operator respond maximally to edge that orient themselves horizontally or in parallel to image orientation.

$$X_{der} = \begin{bmatrix} 1 & 0 & -1 \\ 2 & 0 & -2 \\ 1 & 0 & -1 \end{bmatrix} Y_{der} = \begin{bmatrix} 1 & 0 & 1 \\ 0 & 0 & 0 \\ -1 & -2 & -1 \end{bmatrix} \tag{5.16}$$

Prewitt: the Prewitt or Compass edge detector (Equation 5.17) is an improved version of Roberts and Sobel operator. It incorporates a higher number of convolution kernels (in equation only two kernels are shown), making the operator sensitive to edges with different orientation.

$$X_{der} = \begin{bmatrix} 1 & 0 & -1 \\ 1 & 0 & -1 \\ 1 & 0 & -1 \end{bmatrix} Y_{der} = \begin{bmatrix} 1 & 1 & 1 \\ 0 & 0 & 0 \\ -1 & -1 & -1 \end{bmatrix} \tag{5.17}$$

5.4.2.2 Second-Order Detectors

Canny: the Canny edge detector first uses a Gaussian kernel to smooth the image. Then, the intensity gradient is taken, and the edge direction is calculated by Equation 5.18, followed by non-maximum suppression and a hysteresis threshold,

$$\theta = tan^{-1} \frac{G_y(x,y)}{G_x(x,y)} \tag{5.18}$$

where G denotes the gradient of the image in (Solomon & Breckon, 2011).

Zero-Cross of Laplacian: the zero-cross edge detector includes two parts: first the Laplacian operator (Equation 5.19, 2D) transforms the image and, then, the detector searches for pixels that present zero crossings (i.e. where the gradient changes).

$$\nabla^2 = \frac{\partial^2}{\partial x^2} + \frac{\partial^2}{\partial y^2} \tag{5.19}$$

Zero-Cross of Laplacian of Gaussian (LoG) or Marr-Hildreth: the difference between this method and the previous one is that fact that the LoG operator (Equation 5.20, 2D) is used to transform the image instead of the Laplacian operator:

$$\nabla^2 G(x,y) = \frac{x^2 + y^2 - 2\sigma^2}{\sigma^4} e^{-(x^2+y^2)/2\sigma^2} \tag{5.20}$$

where σ denotes the width of the Gaussian kernel (Marr & Hildreth, 1980).

5.5 IMAGE SEGMENTATION

Segmentation is the process that divides an image into individual parts, which have different properties, but each part has elements with similar properties. The outcome of segmentation is either the recognition of existence of objects or the actual outlining of these objects. In nuclear medicine, the goal of segmentation is the delineation of healthy organs, tumours, and treatment volumes as well as the quantification of various parameters including anatomical information (i.e. size, shape, extensions) and physiological information (i.e. radiotracer uptake, distribution). The segmentation outcome can be valuable for diagnostic purposes pretreatment, in radiotherapy treatment planning, and for monitoring of tumour response post-therapy.

Manual segmentation has been widely used in nuclear medicine, allowing the radiation oncologists and the nuclear medicine experts to delineate the object(s) of interest. Organ delineation is crucial in radiotherapy treatment planning for various tumours such as head and neck cancer (Scarfone et al., 2004; Delouya et al., 2011), lung cancer (Bradley et al., 2004), gynaecological and rectal (Ciernik et al., 2003) or it can be even used in phantom studies (Grova et al., 2003). It is based on knowledge of anatomical landmarks and the use of atlases.

Although manual segmentation is being currently used in clinical practice, it has certain disadvantages: it is a rather time-consuming process and quite often physician-dependent, leading to lack of reproducibility. Therefore, automated segmentation methods have been introduced. A segmentation method may include one or more processes, integrated in a single processing flow. Nevertheless, initial input depends on the *a priori* knowledge of human anatomy and physiology. Figure 5.9 shows a segmentation flow, with various individual components.

It should be underlined that most of the proposed segmentation techniques are composed of several individual processes which are applied either sequentially or in parallel, while the outcome of these processes may be used as a feedback for optimising parameters of the segmentation algorithm for one or more iterations. Moreover, a segmentation method may include many preprocessing steps such as contrast improvement, edge enhancement, and background subtraction.

In this section, we present an overview of different segmentation tools. Although we have attempted to distinguish different proposed methods based on the core part of the processing algorithm, these methods may include other tools such as filtering or thresholding which have been already discussed. As we consider segmentation a crucial part of image analysis, we present an overview of methods used specifically in SPECT and/or PET segmentation.

5.5.1 ANALYSIS TECHNIQUES FOR SEGMENTATION

Several books and review articles have been published on medical image segmentation. Here, we attempt to outline methods that have been applied in nuclear medicine image analysis. The type and the number of steps that are followed in a segmentation process depend on the type of image data (e.g. standalone tissues or in-between other organs), the quality of the images (e.g. resolution), and the desirable outcome (i.e. segmentation or detection).

Simple **thresholding** can achieve the desirable outcome, frequently assisted by edge detection. Figure 5.10 shows an example of an area segmentation using thresholding. Two homogeneous areas with global pixel intensities of 0.55 and 0.75 (in the pixel intensity range [0,1]), respectively, exist in the left image; a threshold value of 0.7 returns 0 for the pixel intensities of 0.55 and 1 for the pixel intensities of 0.75.

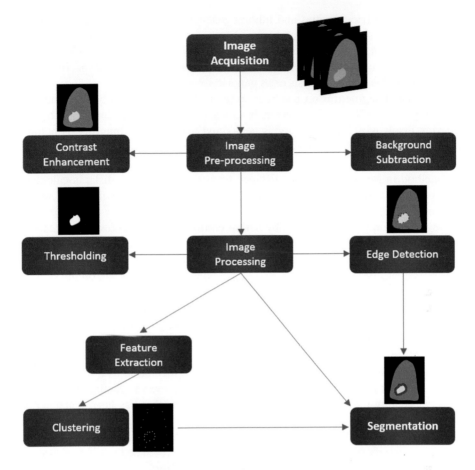

FIGURE 5.9   Flowchart of typical segmentation process.

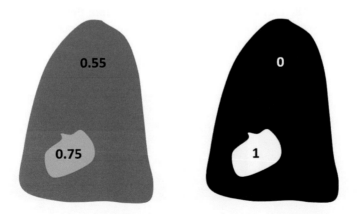

FIGURE 5.10   Segmentation with global thresholding. Left: Simulated image with two homogeneous areas, an 'organ' area (dark grey, global intensity of 0.55) and a 'tumour' area (light grey, global intensity of 0.75). Right: Segmentation using any threshold value.

However, the irregularities and inhomogeneities of human tissues and tumours create complex segmentation problems. Furthermore, the increasing demand for accurate segmentation and parameter quantification has shifted the attention to using more sophisticated techniques. The latter have been either developed specifically for nuclear medicine image analysis or as generic solutions.

Watershed segmentation has been extensively used, especially in segmenting touching objects. The method detects basins and boundary lines in images by considering pixels with high intensity as peaks and pixels of low intensity as being at basal levels. Figure 5.11 shows an illustration of the use of watershed segmentation on a common problem of object detection that of two overlapping objects. Using the method in Meyer (1994), the distance transform of the two disks is calculated and, based on that, the watershed transform of that distance gives the segmentation outcome.

Segmentation can be achieved by **active contours** (Chan & Vese, 2001). There, an initialised curve is evolved by an external force until it satisfies certain criteria. Geodesic active contours evolve based on intrinsic geometric features of the image (Caselles et al., 1997) while parametric active contours evolve based on the satisfaction of certain parameters such as minimisation of the energy functional (Kass et al., 1988; Xu and Prince, 1998). Also, the **region growing** method uses a seed pixel (Pham et al., 2000), which is, then, compared to the neighbouring pixels, or several such seeds (Adams & Bischof, 1994). Figure 5.12 shows an illustration of how active contours technique works.

Learning techniques have also been used in segmentation as a means of classification of structures as healthy tissue or tumour or even to distinguish between adjacent healthy tissues. It considers a data-point map that has been generated by some image analysis technique (e.g. edge detection); concurrently it considers some criteria of similarity or dissimilarity, to cluster those points into groups and, ultimately, to classify these groups or even segment structures. The learning process can be based on either **supervised** or **unsupervised learning**.

Supervised learning uses *a priori* information to classify data-points. In this type of learning, the user considers a training set to train the classifier and extract specific labels to attribute to points or groups. Then, new datasets can be classified based on this training set, attributing the same labels to unknown groups. Examples of supervised learning methods are the following:

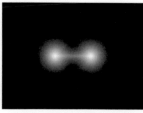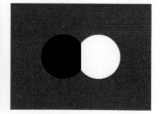

FIGURE 5.11 Watershed segmentation. Left: Simulated image of two circular overlapping objects. Middle: the distance transform of the two objects. Right: The watershed transform of the distance transform leads to object separation and segmentation.

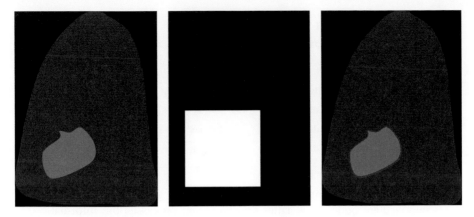

FIGURE 5.12 Example of active contour segmentation. Left: Simulated image with two homogeneous areas, an 'organ' area (dark grey) and a 'tumour' area (light grey). Middle: User-defined binary mask that is used to drive the shrinking of the contour. Right: Segmentation outcome with the red curve outlining the 'tumour'.

Nearest neighbour: each point is assigned to a specific cluster, under the condition that its closest labelled point belongs to that cluster.

Support vector machines: this is a binary classifier; a hyperplane is defined so as to maximise the distance between the two closest points that belong to different clusters (Winters-Hilt & Merat, 2007).

Unsupervised learning processes data without previous human intervention. Clustering is often used as a means of unsupervised learning. It is achieved using a certain criterion of proximity measure to assign data-points to different clusters. Nevertheless, the nature of the data plays an important role on the choice of cluster (Manning et al., 2008). Examples of unsupervised learning techniques are the following (Jain et al., 1999):

Hierarchical clustering: it can have two forms: (a) agglomerative clustering where each point is initially considered one cluster with subsequent further clustering into groups or, (b) divisive clustering where initially all points belong to a single cluster with subsequent partitioning.

k-means clustering: each point is assigned to a cluster provided that its distance to cluster's centroid is minimised (Pham et al., 2000). A common metric that is used to measure the proximity of the point to the cluster's centroid is the Euclidean distance d; however, other distance metrics have also been used. Table 5.1 gives some examples of established metrics.

Fuzzy clustering: contrary to traditional clustering techniques where a point is unambiguously clustered to a specific cluster, fuzzy clustering attributes a score that each point belongs to each cluster, therefore a point does not belong solely to a single cluster. For example, a point may belong 60% to one cluster and 40% to another cluster.

Artificial neural networks (ANNs): an ANN is a parallel framework of interconnected nodes, which resembles the neurons of the human brain. Initial

TABLE 5.1

Metrics Used for Clustering Data

Metric	Equation
Euclidean distance	$d = \sqrt{(x_2 - x_1)^2 + (y_2 - y_1)^2}$
Manhattan distance	$d = \left\|(x_2 - x_1) + (y_2 - y_1)\right\|$
Chebychev distance	$d = \max\left(\left\|x_2 - x_1\right\|, \left\|y_2 - y_1\right\|\right)$
Minkowski distance	$d = \sqrt[q]{(x_2 - x_1)^q + (y_2 - y_1)^q}$

nodes process information and each such node gives output to multiple intermediate nodes. The latter considers and processes the information, giving the same kind of output to the final nodes which present the clustering result. The connections between these nodes give different weights to the nodes leading to learning (Pham et al., 2000).

Supervised and unsupervised learning techniques have certain advantages and disadvantages. Supervised learning offers direct labelling of the new classes that are classified but this requires training data. On the contrary, unsupervised learning does not need training datasets and it is computationally less expensive but its results can be affected by noise and uncertainty of the clustering data (Foster et al., 2014a). Figure 5.13 shows a graphical representation of boundary points detection (middle image, red points) and the same points clustered into organ boundary (right image, green) and tumour boundary (yellow).

5.5.2 PET SEGMENTATION

The main goal of segmentation in PET is the delineation of the areas with radiotracer uptake throughout the body image. A review of segmentation methods in PET can

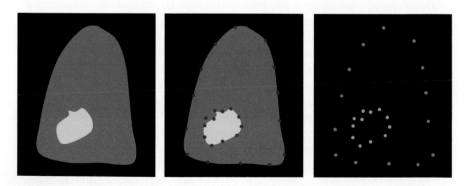

FIGURE 5.13 Graphical representation of points clustering for boundary delineation. Left: Simulated image with two homogeneous areas, an 'organ' area (dark grey) and a 'tumour' area (light grey). Middle: Boundary point detection (red). Right: Clustering of detected points, green for organ delineation and yellow for tumour delineation.

be found in Foster et al. (2014a) and an overview of methods is accounted in Bagci et al. (2013) and Hatt et al. (2018). Therefore, in this section, the author will attempt a short overview of some selected methods.

Manual segmentation has been incorporated in several studies; however, this method remains time-consuming, non-reproducible, and highly subjective to users' eye. The use of thresholding in PET images is a widely accepted method to provide an adequately good separation of foreground and background objects. The reason is that organs in PET images have relatively homogeneous pixel intensities and, therefore, can be distinguished from neighbouring ones (Foster et al., 2014a). Intensity threshold has been used in studies concerning various organs such as head and neck, lung, rectum, and cervix. Daisne et al. (2003) presented a robust automatic segmentation method of a functional part, using information from radioactivity and background. The volume of radioactive spheres was calculated with different values of threshold and the best value was selected as the one that minimised the square difference between true and calculated volume.

Thresholding, together with region growing and morphological processing are simple and fast techniques; however, they can only be applied in images with high contrast between tissues (Suetens et al., 1993). However, the combination of either these techniques or with other image analysis techniques can lead to the desired segmentation outcome. Ciernik et al. (2005) used an algorithm of region growing combined with multi-level thresholding to delineate the tumour in rectal cancer on FDG-PET images.

Analysis can make use of measurements of Signal-to-Noise (SNR) to assist the thresholding process. Erdi et al. (1997) used the signal-to-background ratio from images taken with a Jaszczak phantom, together with the anatomical information from a CT scan to determine optimum threshold value. A similar study utilised an iterative thresholding method and a phantom filled with radioactive sources of ^{18}F-FDG or Na^{124}I to estimate the volume of the sources (Jentzen et al., 2007). The method used threshold-volume curves at varying signal-to-background ratio.

Another combination of techniques includes thresholding with clustering; Gallivanone et al. (2016) presented an automated method for estimation of metabolic tumour volume. The method was based on thresholding for the definition of the volume while a k-means clustering algorithm assisted in the estimation of the background signal.

Clustering is found to have application in PET image segmentation together with other techniques, apart from thresholding. Hagos et al. (2018) described a method for tumour segmentation. The method creates groups of pixels based on similarity in intensity. A principal component analysis extracts statistical features. Finally, a k-means clustering distinguished the clusters as tumours or healthy tissues based. Foster et al. (2014b) proposed a method for tuberculosis detection in small animal PET images to reflect the diffuse character of the radioactive areas. Kernel density estimation via diffusion assists in the construction of histogram of the lungs' pixel intensities. The histogram is then smoothed and an affinity propagation clustering algorithm clusters the image voxels into groups. The lung pathology is then evaluated based on the calculation of several metrics.

Wong et al. (2002) classified different regions based on their time-activity curves with the aim to cluster them, with each cluster including similar regions. A weighting factor was used to enhance the separability of the different clusters due to different tissue radiotracer uptake. Geets et al. (2007) described a gradient-based segmentation method. A watershed segmentation algorithm segmented the gradient-intensity image and the over-segmented pieces were then clustered using the hierarchical clustering to connect parts with similar radioactivity concentration.

Bi et al. (2017) presented a method for automatic classification of FDG uptake in whole-body PET-CT images. The method is based on segmentation using linear spectral clustering and convolutional neural networks (CNN) to cluster regions of FDG uptake. Then, a class-driven feature selection assisted in classifying the different regions, providing the desirable output. CNN has also been used in classification of different regions (i.e. brain, heart, bladder, kidneys) in 2D coronal images (Afshari et al., 2018).

Classification techniques find wide application in structural segmentation in PET. Yu et al. (2009) developed a method for assigning each PET image voxel a label as normal or abnormal. The method was based on decision tree; each step includes the calculation of a feature while k-nearest method classifies the voxels. Another classification example is found in Berthon et al. (2016). This research presents an automatic method for segmentation using a decision tree supervised machine learning algorithm. The method uses a dataset of PET images with known contour as a training set to predict the contour in new cases. Several measurements are considered, such as the tumour volume, tumour peak to background SUV ratio, and a regional texture metric. It has been used to automatically evaluate the contour in head and neck tumour cases (Berthon et al., 2017).

Bibliography in PET image segmentation includes several other techniques. Reuter et al. (1997) utilised edge detection, looking for zero-crossings. Ballangan et al. (2013) presented a method for segmenting lung tumours by representing an image as a graph, where the SUV values have a central role. A segmentation energy term uses an SUV cost function to adapt for the differences in tumour homogeneity while the SUV downhill feature separates the tumour from the neighbouring organs. Li et al. (2017) proposed a process for simultaneous PVE correction, image restoration, and tumour segmentation. The method included total variation semi-blind deconvolution and Mumford-Shah segmentation (Mumford & Shah, 1989). The method achieved preservation of information in tumour edges as well as preservation of smooth radioactive uptake in internal tumour areas.

5.5.3 SPECT Segmentation

Besides PET, thresholding finds wide application in SPECT image segmentation with researchers applying this method throughout many years, even nowadays. Mortelmans et al. (1986) presented a thresholding method for volume determination in SPECT using a global threshold. This threshold is determined by the grey-level histogram and the minimisation of the difference between the true and the measured volume. An iterative threshold technique was presented by Erdi et al. (1995) to delineate the tumour and organs. The method included a background subtraction and an optimisation flow for minimising the difference between the true and the calculated volume of radioactive spheres. Gustafsson et al. (2017) evaluated different methods for tumour segmentation, all of which included a pre-initiated manual ROI delineation. The methods included fixed threshold at 42% of the maximum voxel value, the Otsu threshold method, and the Fourier surface method.

Cardiac SPECT image analysis and segmentation of heart has attracted a lot of interest and standards on how to perform this process has been introduced (Cerqueira et al., 2002). A review of different heart segmentation methods can be found in Kang et al. (2012). Left ventricle (LV) function assessment is the core goal of several publications due to its important role in diagnosis of cardiovascular disease. Soneson et al. (2009) presented a method for automatic segmentation of LV in myocardial perfusion SPECT. This method aims at calculating the ventricle's mass, a measurement that is important for evaluation of hypertension. The method consists of several processes, which combine the Dijkstra's algorithm (Dijkstra, 1959), thresholding and edge detection. Lee et al. (2006) developed a method for calculation of cardiac volumes and ejection fraction. A cylinder model was used to segment the left ventricular myocardium and thresholding was used to eliminate background activity. The endocardial and epicardial boundaries were determined by count profiles fitted to Gaussian curve.

Clustering finds application in SPECT segmentation as well. Boudraa et al. (1993) presented a method for LV segmentation using fuzzy clustering. Using this method, the time behaviour of each image pixel is recorded while Fourier analysis gives the phase parameter. These two parameters are used to cluster the image pixels.

5.5.4 FEATURE EXTRACTION

In modern clinical practice, there is an increasing demand for standardisation of analysis techniques, so the same analysis can be implemented in many image datasets across many medical centres. Feature extraction is a core process in anatomical and functional evaluation of an image dataset, and it allows quantitative imaging, leading to standardisation. Dedicated software packages have been introduced to serve this goal (Lambin et al., 2012). Features can be distinguished into geometric (structural) and textural. Geometric features use measurements such as diameter, perimeter, and orientation to characterise an organ/tissue size, shape, orientation, position (absolute or relative to other organs). Textural features use statistical and spectral measurements such as contrast and homogeneity to characterise radiotracer uptake (absolute or relative) and radiotracer distribution. Examples of textural features are the contrast, correlation, variance, entropy, and angular second moment (Haralick et al., 1973). For the case of PET, an extensive list of feature that can be calculated has already been manifested (Cook et al., 2014); the list includes first- and second-order features which describe both the anatomy and physiology of a tissue, either directly or indirectly.

5.6 IMAGE REGISTRATION

The fusion of different imaging modalities created the so-called multimodality imaging. The latter capitalises on the certain advantages of both nuclear medicine (SPECT, PET) and radiology techniques (CT, MRI), providing a modern and high-impact advance in diagnosis. This technique, however, requires registration of the images originating from two different imaging modalities. Image registration refers to establishment of feature similarity or dissimilarity between two or more images, so they can overlap following geometrical transformation. It is a finite process that requires precision to ultimately boost the diagnostic outcome. A review of registration methods can be found in Maintz and Viergever (1998), Zitová and Flusser (2003), and Brock et al. (2017). In this chapter, we provide a short overview of the process.

Image registration occurs post-acquisition if image datasets have been obtained from separate systems, using specialised software packages. The latter control the whole process, including acquisition, reconstruction, correction, and, finally, representation (Cherry, 2009). The most important part of this process is the transformation of the image datasets from different imaging modalities into a common reference system (International Atomic Energy Agency, 2008). Usually, the radiological image acts as the fixed image and the nuclear medicine image acts as the moving image, which must be registered to the fixed one. Registration is achieved using information from the image datasets or markers that can be found on both modalities (Catana et al., 2008).

Image registration faces specific issues related to (a) the different time that the patient is scanned in the different medical device, (b) the different position that the patient has every time, and (c) the variability across the medical devices. The process includes four interactive parts: (a) the definition of a suitable similarity metric that corresponds the fixed image onto the moving image, (b) the optimisation that finds the optimum transformation parameters for minimisation or maximisation of the selected metric, (c) the transformation which maps the points of the fixed image onto the moving image, (d) the interpolation that creates a version of the moving image to match the fixed image in terms of size, location, and orientation (Brock et al., 2017). Figure 5.14 shows an illustration of a generalised image registration process.

5.6.1   SIMILARITY METRICS

Several similarity metrics have been proposed (Table 5.2), distinguished into two general categories: (a) the geometry or feature-based metrics, and (b) the intensity-based metrics (Brock et al., 2017). Geometry-based metrics refer to point or surface matching; point matching relies on anatomical landmarks or markers while the corresponding metric is the sum of squared distances between the N anatomical points-landmarks P_A and P_B of the image datasets. Surface matching relies on overlapping of surfaces. Here, the metric is the sum or average of squared distances of minimum distance between the points P_A of the first image dataset and the surface S_B of the second image dataset. Intensity-based features use either the images' pixel intensities I_A and I_B or similarity between images $A(x)$ and $B(x)$ (Brock et al., 2007). The more recent mutual information of image entropy aligns voxels that have the high probability to appear in both moving and fixed datasets.

5.6.2   OPTIMISATION

Optimisation aims at maximising the similarity metric. An overview of optimisation processes is listed in Maintz and Viergever (1998). Optimisers can be local or

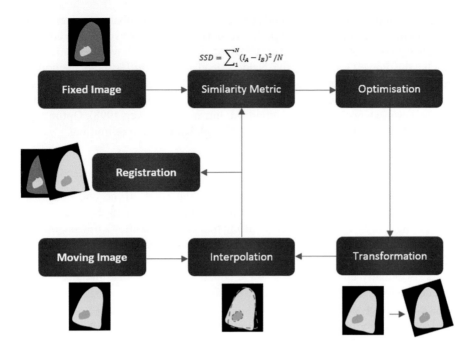

FIGURE 5.14 Flowchart of image registration between a fixed and a moving image.

TABLE 5.2
Similarity Metrics Used for Image Registration

Process	Metric
Geometry-based Metrics	
Point-matching	$P = \sum_{1}^{N} (P_{A'} - P_B)^2 / N$
Surface-matching	$S = \sum_{1}^{N} dist(p_{A'}, S_B)^2 / N$
Intensity-based Metrics	
Sum of squared differences	$SSD = \sum_{1}^{N} (I_{A'} - I_B)^2 / N$
Cross-correlation	$C = \sum_{\vec{x}} B(\vec{x}) \cdot T(A(\vec{x}))$
Correlation coefficient	$CC = \dfrac{\sum_{\vec{x}} (A(\vec{x}) - A)(T(A(\vec{x}) - \bar{B})}{\sqrt{\sum_{\vec{x}} (A(\vec{x}) - \bar{A})^2 \sum_{\vec{x}} (T(B(\vec{x}) - \bar{B})^2}}$
Mutual information	$MI(I_A, I_B) = \sum_{B} \sum_{A} P(I_{A'}, I_B) log_2 [p(I_{A'}, I_B)/p(I_{A'})p(I_B)]$

Source: Brock et al., 2017.

global and single-dimension or multi-dimension solutions. Global optimisers search the entire image for maximisation of the similarity metric; however, this is computationally expensive (Zitová & Flusser, 2003). Many optimisers have been introduced, namely: the Powell's method, the Nelder-Mead or downhill simple method, the Brent's method, the Levenberg-Marquardt method, the stochastic search methods the gradient-descent methods, the simulated annealing method, the geometric hashing and the quasi-exhaustive search method (Maintz and Viergever, 1998). Optimisation methods should be fast and efficient in finding global maxima or minima.

5.6.3 Transformation

Commonly used transformation techniques are the rigid, affine and deformable methods. A rigid transformation allows translation and rotation of the image in x,y,z axes, giving 6 degrees of freedom. An affine transformation allows translation, rotation, scaling, and shear giving 12 degrees of freedom. Deformable models can be either global (e.g. thin-plate splines) or local (e.g. B-Spline). In global models, deformation is globally controlled by a single equation whereas local models can capture local deformations since they employ several control points across a deformable surface (Brock et al., 2017; Patera et al., 2018). Deformable models employ a high number of degrees of freedom since all pixels or voxels transform independently, a process that is computationally expensive (Li & Miller, 2010). Figure 5.15 shows a schematic representation of the 2D affine transformation.

5.6.4 Interpolation

Interpolation is frequently performed by using the mathematical process of convolution. It is a process by which the moving image is resampled by enlargement or shrinking to fit the dimensions of the fixed image. Interpolation algorithms include, among others, the nearest neighbour, the bilinear method, the bicubic method, quadratic splines, B-splines, and Gaussian functions (Zitová & Flusser, 2003). It should be noted that some interpolation methods can yield artefacts while some methods can be computationally expensive.

Affine Transformation

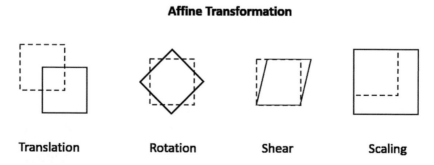

| Translation | Rotation | Shear | Scaling |

FIGURE 5.15 2D affine transformation. It includes translation, rotation, shear, and scaling.

5.7 MATLAB IN NUCLEAR MEDICINE IMAGE ANALYSIS

MATLAB (The MathWorks, Inc., Natick, MA) has been introduced in nuclear medicine image analysis due to its ability to handle images as 2D, 3D, or 4D matrices and its plethora of integrated tools and toolboxes such as the Image Processing Toolbox™ (IPT) (The MathWorks, Inc., 2019a). Moreover, the integrated tool for medical image format DICOM (Bidgood & Horii, 1992) simplifies the process of processing initialisation since patient-specific acquired images can be directly imported to MATLAB's workspace as variables of matrices using the function *dicomread*. Additionally, the function *dicominfo* obtains all properties of the image in MATLAB's workspace for further analysis while the function *dicomewrite* saves the image in DICOM format again. The latter facilitates the return of the processed image to commercial image acquisition software.

The rest of the chapter has been dedicated to describing how image analysis is applied on nuclear medicine images using MATLAB (R2018a). MATLAB code accompanies the described processes to fully assist in direct implementation of the code by the reader. The code is also accompanied by description of each function; it should be noted that it is a good practice to comment on the code in order to make it fully understood by other users. Analysis on nuclear medicine images using MATLAB has been previously described in Lyra et al. (2011).

5.7.1 IMAGE VISUALISATION AND PREPROCESSING BY MATLAB

MATLAB has a dedicated image tool, initiated with the command *imtool*, which can be used to visualise images. Integrated in this tool are several basic features such as zooming, contrast adjustment, pixel information, change of window level and width, distance measurement, cropping, basic information, and change of colourmap. Figure 5.16 shows an overview of the tool's capabilities. The following commands open our test image 'kidneys_image' in the visualisation tool:

```
% Read the dicom image and open it the visualisation toolbox
im = dicomread('kidneys_image.dcm');
imtool(im);
```

Figure 5.17 shows the histogram of the left kidney, generated by the function *imhist* where *n* is the number of bins:

```
% Image histogram with 256 bins of grey-level pixel
intensities
num_bins = 256;
figure; imhist(im, 256);
```

Concerning more advanced tools, IPT uses the dedicated function *imadjust* for contrast adjustment. This function can change the contrast either globally or in an

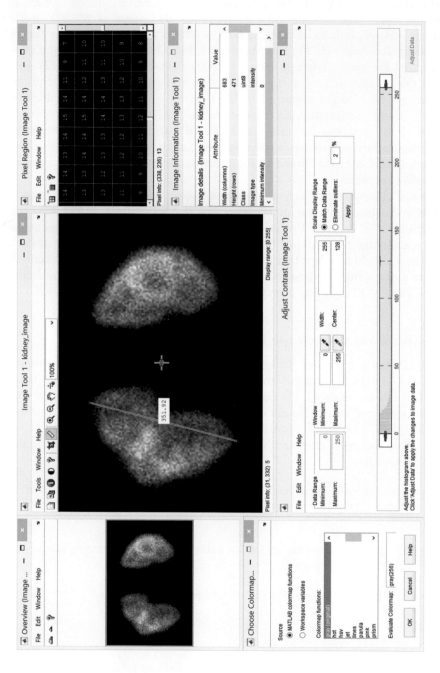

FIGURE 5.16 MATLAB's image tool which allows the user to perform basic preprocessing adjustments and measurements.

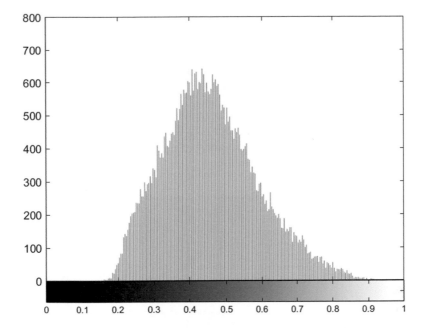

FIGURE 5.17   Histogram of the pixel intensities of the left kidney.

adaptive manner, that can be applied to a series of images. The following command adjusts the contrast manually:

```
% Contrast adjustment with low_in and high_in are the lowest
and highest intensities of the % input image and low_out and
high_out are the lowest and highest intensities of the output
% image
im_contrast = imadjust(im, [low_in high_in], [low_out
high_out]);
```

5.7.2   IMAGE THRESHOLDING BY MATLAB

Thresholding can be applied using the function *graythresh*. As in contrast adjustment, a global or an adaptive threshold can be defined. MATLAB can also generate an automated threshold value using the Otsu method (Otsu, 1979). The following commands apply the Otsu thresholding method and manual thresholding while the *imbinarize* function generates the binary image where pixel intensities above the threshold value obtain value 1 and pixel intensities below the threshold value obtain value 0:

```
% Otsu thresholding method and binarisation
thresh = graythresh(im);
im_bw = imbinarize(im, thresh);
% Manual thresholding with threshold value of 0.2 and
binarisation
```

```
thresh = 0.2;
im_bw = imbinarize(im, thresh);
% Get the original pixel intensities included in the segmented
areas; add them and obtain the
% final image of the segmented areas
im_final = im.*bw;
```

Figure 5.18 shows an example of global image thresholding where some background radioactivity can be eliminated. The top row includes the original kidney image while the bottom row shows the thresholded images with threshold values 0.1, 0.2, and 0.3 in intensity range [0,1], respectively.

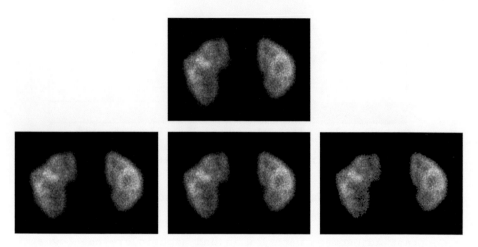

FIGURE 5.18   Top row: Original kidney image. Bottom row: Thresholded images with threshold values 0.1, 0.2, and 0.3, respectively, in the [0,1] range of pixel intensities.

5.7.3   Image Filtering by MATLAB

IPT has a set of functions for application of the Ramp filter. The Ramp filter (and other filters such as the Shepp–Logan, Cosine, Hamming or Hanning filters) can be applied together with FBP reconstruction while the interpolation method can also be chosen (i.e. nearest-neighbour, linear, spline or shape-preserving piecewise cubic method), using the following commands:

```
% Design the Radon transform to be applied on the image using
the angles in degrees used to
% obtain the projections
theta_angles = (0:179);
radon_transf = radon(im, theta);
% Apply filtered backprojection using the Ramp filter and
nearest-neighbour interpolation
% method
im_filt = iradon(radon_transf, theta_angles, 'Ram-Lak',
'nearest');
```

MATLAB offers many filtering options post-reconstruction. The goal of filtering has been already introduced previously in the chapter. Using the function *fspecial*, either a standard, linear, non-linear, or even adaptive filter can be designed by choosing specific parameters. Filters are applied to images using the functions *filter2*, *imfilter (2D-filtering)*, and *nfilter (ND-filtering)*, depending on the individual occasion. Besides, MATLAB offers the chance of creating new functions where other filters can be encoded. Additionally, the user can capitalise on the fact that MATLAB has a wide user community, the members of which share their code, therefore often this fact eliminates the need for creating a new function. The following commands generate and apply a mean filter:

```
% Filter design: mean filter with kernel size 9×9
filt = fspecial('average', [9 9]);
% Filtering application
im_filt = imfilter(im, filt);
% Show the filtered image
figure; imshow(im_filt);
```

5.7.4 Image Edge Detection by MATLAB

As already discussed, edge detection holds a key role in image analysis, with many applications including nuclear medicine. IPT includes a set of functions which perform edge detection. Using the function *edge*, edge detection can be applied. MATLAB offers several options as for the technique to use including Canny, Sobel, Prewitt, and Logarithmic. Edge detection can assist in uncovering boundaries or discontinuities between adjacent organs or between an organ and a tumour but it also searches for internal inhomogeneities. Figure 5.19 shows the application of edge detection on the previously filtered kidney images, using the following commands:

```
% Edge detection with Prewitt detector and threshold 0.1
bw_prewitt = edge(im_filt, 'Prewitt', 0.1);
% Edge detection with Canny detector and threshold 0.5
```

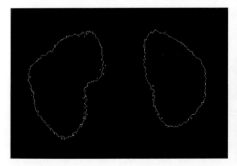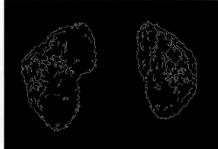

FIGURE 5.19 Edge detection. Left: Detection using the Prewitt edge detector with threshold on edge quality of 0.1. Right: Detection using the Canny edge detector with threshold on edge quality of 0.5.

```
bw_canny = edge(im_filt, 'Canny', 0.5);
% Show the images with the edges
figure; imshow(bw_prewitt);
figure; imshow(bw_canny);
```

5.7.5 Image Contouring by MATLAB

Diagnosis may require qualitative data on the shape and size of organs or the relevant size of an organ compared to a similar one, such us in the case of kidneys. Such information can be obtained using the IPT function *imcontour* that generates the organ's outline and, thus, delineates the outline. Contours connect points that have similar intensities while a multi-threshold technique is applied to generate the so-called isocontours (Faber & Folks, 1994). The following command performs contouring, with three levels of outlining, which are plotted as lines (i.e. '-'):

```
% Generate the contours with 3 levels of outlining; this
function includes plotting of organ
% outlines
im_cont = imcontour(im, 3, '-');
```

Figure 5.20 shows the contouring result of a kidney's image, after being thresholded with a uniform threshold of 0.2 (in intensity range [0,1]) and filtered with an average filter of 15 × 15 pixels.

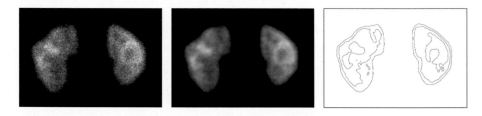

FIGURE 5.20 Left: Original kidney images. Middle: Thresholded with a uniform threshold of 0.2 (in [0,1] intensity range) and filtered image with an average 15 × 15 filter. Right: Contouring outcome.

5.7.6 Image Segmentation and Information Extraction by MATLAB

Image segmentation capitalises on the outcome from other processing techniques such as contrast adjustment, filtering, and edge detection. Nevertheless, segmentation can be applied either on preprocessed images or directly to unprocessed images. IPT includes functions for watershed segmentation (function *watershed*) and active contour segmentation (function *activecontour*). For the latter, there is a choice of using either the Chan-Vese method (Chan & Vese, 2001) or the geodesic active contour technique (Caselles et al., 1997). Active contours can be used to delineate complex organ outlines while watershed transformation can separate two neighbouring organs with ambiguous boundaries. The following commands generate the organ outlining using active contours:

```
% Create the masks for defining the initial seeding areas
mask1 = zeros(size(im));
mask2 = zeros(size(im));
% Define the first organ seeding point
organ1_y = 214; organ1_x = 153;
% Extend the first organ seeding point by 30 pixels in each of
x,y directions
mask1(organ1_y-30:organ1_y+30, organ1_x-30:organ1_x+30) = 1;
% Define the second organ seeding point
organ2_y = 215; organ2_x = 524;
% Extend the second organ seeding point by 30 pixels in each
of x,y directions
mask2(organ2_y-30:organ2_y+30, organ2_x-30:organ2_x+30) = 1;
% Apply active contour segmentation on each organ, with 380
iterations, smoothing factor for % the outline of 0.5 and
contraction bias of -0.5
im_bw1 = activecontour(im, mask1, 380, 'Chan-Vese',
'SmoothFactor', 0.5, …
'ContractionBias', -0.5);
im_bw2 = activecontour(im, mask2, 380, 'Chan-Vese',
'SmoothFactor', 0.5, …
'ContractionBias', -0.5);
% Get the original pixel intensities included in the segmented
areas; add them and obtain the
% final image of the segmented areas
im_temp1 = im.*bw1;
im_temp2 = im.*bw2;
im_final = im_temp1 + im_temp2;
% Get the pixel boundaries to draw, ignoring any holes created
by the segmentation inside the % organ
[bound1, ~] = bwboundaries(bw1, 'noholes');
[bound2, ~] = bwboundaries(bw2, 'noholes');
boundary1 = bound1{1};
boundary2 = bound2{1};
% Show the image and plot the boundaries with red colour
figure; imshow(im_final);
hold on
plot(boundary1(:,2), boundary1(:,1), 'r', 'LineWidth', 1);
plot(boundary2(:,2), boundary2(:,1), 'r', 'LineWidth', 1);
```

Figure 5.21 shows the segmentation outcome (with red line) of an original kidney image (left) after applying an average filter with kernel size 11×11 pixels. Simple thresholding (middle) of 0.2 in intensity range [0 1] results in a rough segmentation with rather curly outlining. Segmentation with active contours using the Chan-Vese method results in smoother outlining since this method gives the opportunity to smooth the outline of the organs (365 iterations, smoothing factor 0.5, and contraction bias −0.5).

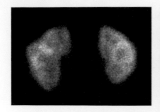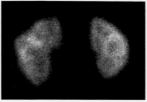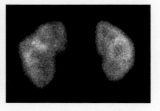

FIGURE 5.21 Left: Original kidney images. Middle: Organ outlining (red line) after segmentation using average filtering (11 × 11 pixels) and thresholding with 0.2 threshold in [0,1] intensity range. Right: Segmentation with active contours using the Chan-Vese method with 365 iterations, smoothing factor 0.5, and contraction bias −0.5.

Segmentation is pivotal in extraction of quantitative information. A list of first-order parameters, common in Radiomics, can be calculated using the function *regionprops* and they describe the anatomy of kidneys. This function uses as input the binary image of each kidney and the parameters to be calculated. The latter are calculated in pixels; nevertheless, if a calibration factor is provided, parameters can be expressed in units of length. The following commands show how to obtain those parameters while Table 5.3 shows the results:

```
% First order parameters for first kidney
stats1 = regionprops(im_bw1, 'Area', 'MajorAxisLength',
'MinorAxisLength', 'Eccentricity', …
'Orientation', 'ConvexArea', 'EulerNumber', 'EquivDiameter',
'Solidity', 'Extent', 'Perimeter');
% First order parameters for second kidney
```

TABLE 5.3

Parameters That Can Be Calculated with MATLAB to Describe the Anatomy of the Kidneys; Parameters with Units Have Units Either as Pixels (p) Or Pixel² (p²)

	Left Kidney	Right Kidney
Area	49438 p²	39292 p²
Major Axis Length	328.5519 p	293.3411 p
Minor Axis Length	202.0875 p	175.8788 p
Eccentricity	0.7885	0.8003
Orientation	70.2268°	−73.3400°
Convex Area	52763 p²	40778 p²
Euler Number	1	1
Equivalent Diameter	250.8912 p	223.6697 p
Solidity	0.9370	0.9636
Extent	0.6493	0.6984
Perimeter	916.78	803.2210

```
stats2 = regionprops(im_bw2, 'Area', 'MajorAxisLength',
'MinorAxisLength', 'Eccentricity', ...
'Orientation', 'ConvexArea', 'EulerNumber', 'EquivDiameter',
'Solidity', 'Extent', 'Perimeter');
```

A list of second-order parameters can be calculated using the function *graycoprops* and describe the texture or physiology of kidneys. This function uses as input the grey-level co-occurrence matrix that is created by the function *graycomatrix*. The latter receives as input the segmented image with the original pixel intensities for each kidney. The following commands show how to obtain those parameters while Table 5.4 shows the results:

```
% Grey-level co-occurrence matrix for first kidney
glcm1 = graycomatrix(im_temp1);
% Texture parameters for first organ
texture1 = graycoprops(glcm1, {'Contrast', 'Homogeneity',
'Energy', 'Correlation'});
% Grey-level co-occurrence matrix for second kidney
glcm2 = graycomatrix(im_temp2);
% Texture parameters for second organ
texture2 = graycoprops(glcm2, {'Contrast', 'Homogeneity',
'Energy', 'Correlation'});
```

MATLAB has been already used for other purposes in nuclear medicine apart from image analysis. An example is the calculation of the resolution of the γ-camera as the half-maximum (FWHM) of the line-spread function (LSF) (Velo & Zakaria, 2017).

5.7.7 INTRODUCING A NEW FEATURE IN MATLAB

Besides the parameters that IPT already includes, custom-made scripts and functions can be written using MATLAB integrated editor to calculate other parameters. Examples can be circularity, rectangularity, and thinness ratio (da Fontoura Costa & Cesar Jr, 2010) as anatomical features while entropy, skewness, and kurtosis (Cook et al., 2014) can be used as functional features.

A new feature can be calculated by writing a new function. A good practice in writing functions and scripts is to add comments to explain every step in designing

TABLE 5.4

Parameters That Can Be Calculated with MATLAB to Describe the Physiology of Kidneys

	Left Kidney	Right Kidney
Contrast	0.1445	0.1548
Correlation	0.9383	0.9384
Energy	0.2005	0.1777
Homogeneity	0.9277	0.9226

```
  calc_circ.m  ✕  +
1    % Function to calculate the circularity of an object. It is calculated based
2    % on the equation:
3    % Circularity = 4 * pi * Area / (Perimeter^2)
4
5    % Created:  31 Mar 2019
6    % Updated:  05 Apr 2019
7
8    %% Beginning of function
9
10   % Input parameters: area and perimeter of object
11   function circularity = calc_circ(area, perimeter)
12
13   % Calulate the circularity
14   circularity = 4 * pi * area / (perimeter^2);
```

FIGURE 5.22 Function to calculate the circularity of an object.

a feature so as it can be understood by other users. Comments are defined by using the symbol % at the beginning of a line. They are not considered when a script runs and they have a characteristic green colour, given by MATLAB due to the % symbol. Also, it is useful to add the dates of creating and last update of the file. Figure 5.22 shows a function that calculates the circularity of an object (Takashimizu & Iiyoshi, 2016) as it has been written in a new script using MATLAB's Editor. The green square at the top-right corner indicates to the user that MATLAB did not detect and coding errors.

5.7.8 IMAGE REGISTRATION BY MATLAB

IPT includes a dedicated set of functions for image registration. The function *imregister* is the one that performs the final registration process; this function takes as input the fixed image, the moving image, the metric, the transformation model, and the optimising method. An automatic default configuration of the metric and the optimiser for intensity-based basic registration can be designed by the function *imregconfig*. The following command gives the automatic configuration:

```
% Automatic configuration of multimodal registration
[optimizer, metric] = imregconfig('multimodal');
```

Transformation of the moving image is a crucial step in registration. The functions *affine2d* and *affine3d* perform the 2D and 3D affine transformations, respectively. The following commands describe the 2D transformation in MATLAB and Figure 5.23 shows the 2D affine transformation after counterclockwise 2D affine transformation by 15°:

```
% Construct the affine2d object for 15° rotation in the
counterclockwise direction around % the central point
theta = 15;
transform = affine2d([cosd(theta) sind(theta) 0; -sind(theta)
cosd(theta) 0; 0 0 1]);
```

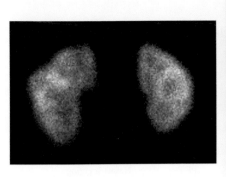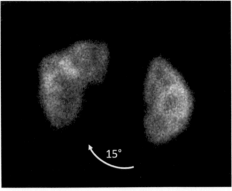

FIGURE 5.23 2D affine transformation by 15° counterclockwise (right) of the kidney image (left) image has orientation L-R.

```
% Transform the original (or processed) image
im_aff = imwarp(im, transform);
% Show image
figure; imshow(im_aff);
```

Regarding the custom characteristics of registration, the function *MeanSquares* creates the mean square error metric while the function *MattesMutualInformation* creates mutual information metric. The function *RegularStepGradientDescent* creates the relevant optimiser. The following commands give the manual registration configuration:

```
% Import the fixed image and the image to register
im_fix = dicomread('image1.dcm');
im_mov = dicomread('image2.dcm');
% Manual configuration of registration parameters: optimiser
and metric
optimizer = registration.optimizer.OnePlusOneEvolutionary
metric = registration.metric.MattesMutualInformation
% Register the images with affine transformation
im_regist = imregister(im_mov, im_fix, 'affine', optimizer,
metric);
% Show the registered image
figure; imshowpair(im_fix, im_regist, ,'Scaling','joint')
```

5.7.9 Software Packages in MATLAB

Several custom-made software packages have been designed in MATLAB to apply different processes in nuclear medicine image analysis. Table 5.5 presents an overview of available packages, the imaging modality for which they have been developed and the process that they perform.

In terms of PET analysis, Stefan et al. (2012) presented a software for denoising PET images using wavelets. Searle et al. (2017) developed the software MIAKAT™

TABLE 5.5

Overview of MATLAB Packages for Nuclear Medicine Image Analysis and Quantification

	Modality	Process
(Stefan et al., 2012)	PET	Denoising
Thyr-Vol (Synefia et al., 2014)	SPECT	Thyroid Volume Calculation
MIAKAT™ (Searle et al., 2017)	PET	General purpose
(Scheinost et al., 2010)	SPECT	Ictal SPECT Scans
QAV-PET (Foster et al., 2014c)	PET	General purpose
VoxelStats (Mathotaarachchi et al., 2016)	PET	Neuro-imaging
CERR (Deasy et al., 2003)	Any	General purpose
PVElab (Neurobiology Research Unit, Righospitalet Copenhagen University Hospital, 2004)	Any	Partial Volume Correction
PRoNTo (Schrouff et al., 2013)	PET	Neuro-imaging

which can perform brain segmentation, motion correction, and modelling. Foster et al. (2014c) developed a software for quantitative analysis of PET images as well as registration of PET with CT images. Mathotaarachchi et al. (2016) developed a software package for voxel-based analysis of brain images. Schrouff et al. (2013) developed a platform for analysis of neurological data including PET brain images, for pattern recognition using multivariate analysis, based on machine learning. As for SPECT analysis, Synefia et al. (2014) developed a software for quantification for the thyroid gland volume. Scheinost et al. (2010) developed a software for ictal SPECT analysis for seizure localisation. Besides, Deasy et al. (2003) have developed the software CERR; although it is intended for radiotherapy research, it can provide a framework for nuclear medicine image analysis to perform various processes such as thresholding, watershed segmentation, contouring, and registration.

REFERENCES

Adams, R. & Bischof, L., 1994. Seeded Region Growing. *IEEE Transactions on Pattern Analysis and Machine Intelligence*, 16(6), pp. 641–647.

Afshari, S., BenTaieb, A. & Hamarneh, G., 2018. Automatic localization of normal active organs in 3D PET scans. *Computerized Medical Imaging and Graphics*, 70, pp. 111–118.

Bagci, U. et al., 2013. Joint segmentation of anatomical and functional images: applications in quantification of lesions from PET, PET-CT, MRI-PET, and MRI-PET-CT images. *Medical Image Analysis*, 17(8), pp. 929–945.

Ballangan, C. et al., 2013. Lung tumor segmentation in PET images using graph cuts. *Computer Methods and Programs in Biomedicine*, 109(3), pp. 260–268.

Beck, R. N., 2004. Imaging Science: Bringing the Invisible to Light. In: M. N. Wernick & J. N. Aarsvold, eds. *Emission Tomography: The Fundamentals of PET and SPECT*. San Diego: Elsevier Academic Press, pp. 1–9.

Benameur, S., Mignotte, M., Meunier, J. & Soucy, J.-P., 2009. Image Restoration Using Functional and Anatomical Information Fusion with Application to SPECT-MRI Images. *International Journal of Biomedical Imaging*, 2009, p. 843160.

Berthon, B. et al., 2017. Head and neck target delineation using a novel PET automatic segmentation algorithm. *Radiotherapy and Oncology*, 122(2), pp. 242–247.

Berthon, B., Marshall, C., Evans, M. & Spezi, E., 2016. ATLAAS: an automatic decision tree-based learning algorithm for advanced image segmentation in positron emission tomography. *Physics in Medicine & Biology*, 61(13), pp. 4855–4869.

Bi L. et al. 2017, Automatic detection and classification of regions of FDG uptake in whole-body PET-CT lymphoma studies. *Computerized Medical Imaging and Graphics*, 60, pp. 3–10.

Bidgood Jr, W. D. & Horii, S. C., 1992. Introduction to the ACR-NEMA DICOM standard. *RadioGraphics*, 12 (2), pp. 345–355.

Boudraa, A. E. O. et al., 1993, Left ventricle automated detection method in gated isotopic ventriculography using fuzzy clustering, *IEEE Trans Med Imaging*, 12 (3): 451–465.

Bradley, J. et al., 2004. Impact of FDG-PET on radiation therapy volume delineation in non-small-cell lung cancer. *International Journal of Radiation Oncology•Biology•Physics*, 59(1), pp. 78–86.

Brill, A. B. & Beck, R. N., 2004. Evolution of Clinical Emission Tomography. In: M. N. Wernick & J. N. Aarsvold, eds. *Emission Tomography: The Fundamentals of PET and SPECT*. San Diego: Elsevier Academic Press, pp. 25–52.

Brock, K. K. et al., 2017. Use of image registration and fusion algorithms and techniques in radiotherapy: Report of the AAPM Radiation Therapy Committee Task Group No. 132. *Medical Physics*, 44(7), pp. e43–e76.

Cal-Gonzalez, J. et al., 2018. Hybrid imaging: instrumentation and data processing. *Frontiers in Physics*, 6, p. 47.

Caselles, V., Kimmel, R. & Sapiro, G., 1997. Geodesic active contours. *International Journal of Computer Vision*, 22(1), pp. 61–79.

Catana, C. et al., 2008. Simultaneous in vivo positron emission tomography and magnetic resonance imaging. *Proceedings of the National Academy of Sciences of the U.S.A.*, 105(10), pp. 3705–3710.

Cerqueira, M. D. et al., 2002. Standardized myocardial segmentation and nomenclature for tomographic imaging of the heart. *Circulation*, 105(4), pp. 539–542.

Chan, T. F. & Vese, L. A., 2001. Active contours without edges. *IEEE Transactions on Image Processing*, 10(2), pp. 266–277.

Cherry, S. R., 2009. Multimodality imaging: beyond PET/CT and SPECT/CT. *Seminars in Nuclear Medicine*, 39(5), pp. 348–353.

Ciernik, I. F. et al., 2003. Radiation treatment planning with an integrated positron emission and computer tomography (PET/CT): a feasibility study. *International Journal of Radiation Oncology•Biology•Physics*, 57(3), pp. 853–863.

Ciernik, I. F. et al., 2005. Automated functional image-guided radiation treatment planning for rectal cancer. *International Journal of Radiation Oncology•Biology•Physics*, 62(3), pp. 893–900.

Cook, G. J. R. et al., 2014. Radiomics in PET: principles and applications. *Clinical and Translational Imaging*, 2(3), pp. 269–276.

da Fontoura Costa, L. & Cesar Jr, R. M., 2010. *Shape Analysis and Classification: Theory and Practice*. Boca Raton: CRC Press.

Daisne, J.-F. et al., 2003. Tri-dimensional automatic segmentation of PET volumes based on measured source-to-background ratios: influence of reconstruction algorithms. *Radiotherapy and Oncology*, 69(3), pp. 247–250.

Deasy, J. O., Blanco, A. I. & Clark, V. H., 2003. CERR: a computational environment for radiotherapy research. *Medical Physics*, 30(5), pp. 979–985.

Delouya, G. et al., 2011. 18F-FDG-PET imaging in radiotherapy tumor volume delineation in treatment of head and neck cancer. *Radiotherapy and Oncology*, 101(3), pp. 362–368.

Dijkstra, E. W., 1959. A note on two problems in connexion with graphs. *Numerische Mathematik*, 1(1), pp. 269–271.

Erdi, Y. E. et al., 1997. Segmentation of lung lesion volume by adaptive positron emission tomography image thresholding. *Cancer*, 80(12 Supplement), pp. 2505–2509.

Erdi, Y. E., Wessels, B. W., Loew, M. H. & Erdi, A. K., 1995. Threshold estimation in single photon emission computed tomography and planar imaging for clinical radioimmunotherapy. *Cancer Research*, 55(23 Supplement), pp. 5823s–5826s.

Erlandsson, K. et al., 2012. A review of partial volume correction techniques for emission tomography and their applications in neurology, cardiology and oncology. *Physics in Medicine and Biology*, 57(21), pp. R119–R159.

Faber, T. L. & Folks, R. D., 1994. Computer processing methods for nuclear medicine images. *Journal of Nuclear Medicine Technology*, 22(3), pp. 145–163.

Foster, B. et al., 2014a. A review on segmentation of positron emission tomography images. *Computers in Biology and Medicine*, 50, pp. 76–96.

Foster B. et al., 2014b. Segmentation of PET images for computer-aided functional quantification of tuberculosis in small animal models. *IEEE Transactions on Biomedical Engineering*, 61(3), pp. 711–724.

Foster, B., Bagci, U., Papadakis, G. Z. & Mollura, D. J., 2014c. QAV-PET: quantitative analysis and visualization of PET images. *Annual International Conference of IEEE Engineering in Medicine and Biology Society*, pp. 1909–1912.

Gallivanone, F., Interlenghi, M., Canervari, C. & Castiglioni, I., 2016. A fully automatic, threshold-based segmentation method for the estimation of the metabolic tumor volume from PET images: validation on 3D printed anthropomorphic oncological lesions. *Journal of Instrumentation*, 11, p. C01022.

Geets X. et al., 2007. A gradient-based method for segmenting FDG-PET images: methodology and validation. *European Journal of Nuclear Medicine and Molecular Imaging*, 34(9), pp. 1427–1438.

Georgantzoglou, A., da Silva, J. & Jena, R., 2014. Image Processing with MATLAB and GPU. In: K. Bennett, ed. *MATLAB: Applications for the Practical Engineer*. Rijeka: InTech, pp. 623–653.

Germano, G., 2001. Technical aspects of myocardial SPECT imaging. *Journal of Nuclear Medicine*, 42(10), pp. 1499–1507.

Gilland, D. R. et al., 1988. Determination of the optimum filter function for SPECT imaging. *Journal of Nuclear Medicine*, 29(5), pp. 643–650.

Grova, C. et al., 2003. A methodology for generating normal and pathological brain perfusion SPECT images for evaluation of MRI/SPECT fusion methods: application in epilepsy. *Physics in Medicine and Biology*, 48(24), pp. 4023–4043.

Gunter, D. L., 2004. Collimator Design for Nuclear Medicine. In: M. N. Wernick & J. N. Aarsvold, eds. *Emission Tomography: The Fundamentals of PET and SPECT*. San Diego: Elsevier Academic Press, pp. 153–168.

Gustafsson, J., Sundlöv, A. & Sjögreen Gleisner, K., 2017. SPECT image segmentation for estimation of tumour volume and activity concentration in 177Lu-DOTATATE radionuclide therapy. *European Journal of Nuclear Medicine and Molecular Imaging Research*, 7(1), p. 18.

Hagos, Y. B. et al., 2018. Fast PET scan tumor segmentation using superpixels, principal component analysis and K-means clustering. *Methods and Protocols*, 1(1), p. 3397.

Haralick, R. H., Shanmugam, K. & Dinstein, I., 1973. Textural features for image classification. *IEEE Transactions on Systems, Man, and Cybernetics*, SMC-3(6), pp. 610–621.

Hatt, M. et al., 2018. The first MICCAI challenge on PET tumor segmentation. *Medical Image Analysis*, 44, pp. 177–195.

International Atomic Energy Agency, 2008. *Clinical Applications of SPECT/CT: New Hybrid Nuclear Medicine Imaging System*, Vienna: IAEA.

International Atomic Energy Agency, 2014. *Quantitative Nuclear Medicine Imaging: Concepts, Requirements and Methods*, Vienna: IAEA.

Jain, A. K., Murty, M. N. & Flynn, P. J., 1999. Data clustering: a review. *ACM Computing Surveys*, 31(3), pp. 264–322.

Jentzen, W. et al., 2007. Segmentation of PET volumes by iterative image thresholding. *Journal of Nuclear Medicine*, 48(1), pp. 108–114.

Kang, D. et al., 2012. Heart chambers and whole heart segmentation techniques: review. *Journal of Electronic Imaging*, 21(1), p. 010901.

Kapur, N., Sahoo, P. K. & Wong, A. K. C., 1985. A new method for gray-level picture thresholding using the entropy of the histogram. *Computer Vision, Graphics, and Image Processing*, 29(3), pp. 273–285.

Kass, M., Witkin, A. & Terzopoulos, D., 1988. Snakes: active contour models. *International Journal of Computer Vision*, 1(4), pp. 321–331.

King, M. A., Penney, B. C. & Glick, S. J., 1988. An image-dependent Metz filter for nuclear medicine images. *Journal of Nuclear Medicine*, 29(12), pp. 1980–1989.

King, M. A., Schwinger, R. B., Doherty, P. W. & Penney, B. C., 1984. Two-dimensional filtering of SPECT images using the Metz and Wiener filters. *Journal of Nuclear Medicine*, 25(11), pp. 1234–1240.

Lambin, P. et al., 2012. Radiomics: extracting more information from medical images using advanced feature analysis. *European Journal of Cancer*, 48(4), pp. 441–446.

Lee, B.-I. et al., 2006. Development of quantification software using model-based segmentation of left ventricular myocardium in gated myocardial SPECT. *Computer Methods and Programs in Biomedicine*, 83(1), pp. 43–49.

Li, C. H. & Lee, C. K., 1993. Minimum cross entropy thresholding. *Pattern Recognition*, 26(4), pp. 617–625.

Li, G. & Miller, R. W., 2010. Volumetric Image Registration of Multi-modality Images of CT, MRI and PET. In: Y. Mao, ed. *Biomedical Imaging*. Rijeka: IntechOpen.

Li, L., Wang, J., Lu, W. & Tan, S., 2017. Simultaneous tumor segmentation, image restoration, and blur kernel estimation in PET using multiple regularizations. *Computer Vision and Image Understanding*, 155, pp. 173–194.

Lin, J.-W., Laine, A. F. & Bergmann, S. R., 2001. Improving PET-based physiological quantification through methods of wavelet denoising. *IEEE Transactions on Biomedical Engineering*, 48(2), pp. 202–212.

Lyra, M. & Ploussi, A., 2011. Filtering in SPECT image reconstruction. *International Journal of Biomedical Imaging*, 2011, p. 693795.

Lyra, M., Ploussi, A. & Georgantzoglou, A., 2011. MATLAB as a Tool in Nuclear Medicine Image Processing. In: C. M. Ionescu, ed. *MATLAB – A Ubiquitous Tool for the Practical Engineer*. Rijeka: IntechOpen, pp. 477–500.

Lyra, M., Ploussi, A., Rouchota, M. & Synefia, S., 2014. Filters in 2D and 3D cardiac SPECT image processing. *Cardiology Research and Practice*, 2014, p. 963264.

Maintz, J. B. A. & Viergever, M. A., 1998. A survey of medical image registration. *Medical Image Analysis*, 2(1), pp. 1–36.

Manning, C. D., Raghavan, P. & Scutze, H., 2008. *Introduction to Information Retrieval*. 1st ed. New York: Cambridge University Press.

Marr, D. & Hildreth, E., 1980. Theory of edge detection, *Proceedings of the Royal Society of London. Series B. Biological Sciences*, 207(1167), pp. 187–217.

Mathotaarachchi, S. et al., 2016. VoxelStats: a MATLAB package for multi-modal voxel-wise brain image analysis. *Frontiers in Neuroinformatics*, 10, p. 20.

The MathWorks, Inc., 2019a. Image Processing Toolbox. [Online] Available at: https://uk.mathworks.com/products/image.html [Accessed 28 December 2019].

The MathWorks, Inc., 2019b. Signal Processing Toolbox. [Online] Available at: https://uk.mathworks.com/products/signal.html [Accessed 28 December 2019].

Meyer, F., 1994. Topographic distance and watershed lines. *Signal Processing*, 38(1), pp. 113–125.

Mignotte, M. & Meunier, J., 2000. Three-dimensional blind deconvolution of SPECT images, *IEEE Transactions on Biomedical Engineering*, 47(2), pp. 274–280.

Morris, E. D. et al., 2004. Kinetic modeling in positron emission tomography. In: M. N. Wernick & J. N. Aarsvold, eds. *Emission Tomography: The Fundamentals of PET and SPECT*. San Diego: Elsevier Academic Press, pp. 499–540.

Mortelmans, L. et al., 1986. A new thresholding method for volume determination by SPECT. *European Journal of Nuclear Medicine*, 12(5–6), pp. 284–290.

Mumford, D. & Shah, J., 1989. Optimal approximations by piecewise smooth functions and associated variational problems. *Communications on Pure and Applied Mathematics*, 42(5), pp. 577–685.

Neurobiology Research Unit, Righospitalet Copenhagen University Hospital, 2004. Neurobiology Research Unit – PVElab. [Online] Available at: https://nru.dk/index.php/misc/category/37-pvelab [Accessed 03 November 2019].

Otsu, N., 1979. A threshold selection method from gray-level histograms. *IEEE Transactions on Systems, Man, and Cybernetics*, 9(1), pp. 62–66.

Patera, A. et al., 2018. A non-rigid registration method for the analysis of local deformations in the wood cell wall. *Advanced Structural and Chemical Imaging*, 4 (1).

Patton, J. A. & Turkington, T. G., 2008. SPECT/CT physical principles and attenuation correction. *Journal of Nuclear Medicine Technology*, 36(1), pp. 1–10.

Pham, D. L., Xu, C. & Prince, J. L., 2000. Current methods in medical image segmentation. *Annual Review of Biomedical Engineering*, 2, pp. 315–337.

Reuter, B. W., Klein, G. J. & Huesman, R. H., 1997. Automated 3-D segmentation of respiratory-gated PET transmission images. *IEEE Transactions on Nuclear Science*, 44(6), pp. 2473–2476.

Scarfone, C. et al., 2004. Prospective feasibility trial of radiotherapy target definition for head and neck cancer using 3-dimensional PET and CT imaging. *Journal of Nuclear Medicine*, 45(4), pp. 543–552.

Scheinost, D. et al., 2010. New open-source ictal SPECT analysis method implemented in BioImage Suite. *Epilepsia*, 51(4), pp. 703–707.

Schrouff, J. et al., 2013. PRoNTo: pattern recognition for neuroimaging toolbox. *Neuroinformatics*, 11(3), pp. 319–337.

Searle, G., Coello, C. & Gunn, R., 2017. Extension of the MIAKAT analysis software package to non-brain and pre-clinical PET analysis. *Journal of Nuclear Medicine*, 58(S1), p. 1305.

Sezgin, M. & Sankur, B., 2004. Survey over image thresholding techniques and quantitative performance evaluation. *Journal of Electronic Imaging*, 13(1), pp. 146–165.

Solomon, C. & Breckon, T., 2011. *Fundamentals of Digital Image Processing: A Practical Approach with Examples in MATLAB*. Chichester: John Wiley & Sons, Inc.

Soneson, H. et al., 2009. An improved method for automatic segmentation of the left ventricle in myocardial perfusion SPECT. *Journal of Nuclear Medicine*, 50(2), pp. 205–213.

Soret, M., Bacharach S. L. & Buvat I., 2007. Partial-volume effect on PET tumor imaging. *Journal of Nuclear Medicine*, 48(6), pp. 932–945.

Stefan, W. et al., 2012. Wavelet-based de-noising of positron emission tomography scans. *Journal of Scientific Computing*, 50(3), pp. 665–677.

Suetens, P. et al., 1993. Image segmentation: methods and applications in diagnostic radiology and nuclear medicine. *European Journal of Radiology*, 17(1), pp. 14–21.

Synefia, S. et al., 2014. 3D SPECT thyroid volume quantification by thyr-vol algorithm in MATLAB. *Physica Medica*, 30(S1), pp. e115–e116.

Takalo, R., Hytti, H., Ihalainen, H. & Sohlberg, A., 2015. Adaptive autoregressive model for reduction of noise in SPECT. *Computational and Mathematical Methods in Medicine*, 2015, p. 494691.

Takashimizu, Y. & Iiyoshi, M., 2016. New parameter of roundness R: circularity corrected by aspect ratio. *Progress in Earth and Planetary Science*, 3, p. 2.

Tohka, J. & Reilhac, A., 2008. Deconvolution-based partial volume correction in Raclopride-PET and Monte Carlo comparison to MR-based method. *Neuroimage*, 39(4), pp. 1570–1584.

Van Laere, K., Koole, M., Lemahieu, I. & Dierckx, R., 2001. Image filtering in single-photon emission computed tomography: principles and applications. *Computerized Medical Imaging and Graphics*, 25(2), pp. 127–133.

Velo, P. & Zakaria, A., 2017. Determining spatial resolution of gamma cameras using MATLAB. *Journal of Medical Imaging and Radiation Sciences*, 48(1), pp. 39–42.

Winters-Hilt, S. & Merat, S., 2007. SVM clustering. *BMC Bioinformatics*, 8(S7), p. S18.

Wong, K.-P., Feng, D., Meikle, S. R. & Fulham, M. J., 2002. Segmentation of dynamic PET images using cluster analysis. *IEEE Transactions on Nuclear Science*, 49(1), pp. 200–207.

Xu, C. & Prince, J. P., 1998. Snakes, shapes, and gradient vector flow. *IEEE Transactions on Image Processing*, 7(3), pp. 359–369.

Yu, H. et al., 2009. Automated radiation targeting in head-and-neck cancer using region-based texture analysis of PET and CT images. *International Journal of Radiation Oncology•Biology•Physics*, 75(2), pp. 618–625.

Zitová, B. & Flusser, J., 2003. Image registration methods: a survey. *Image and Vision Computing*, 21(11), pp. 977 1000.

6 3D Volume Data in Nuclear Medicine

Christos Chatzigiannis

CONTENTS

6.1 INTRODUCTION

One of the greatest advantages when imaging using SPECT is improved contrast, as overlapping structures are separated (Shih et al., 1993) thus reducing structural noise present in standard static Nuclear Medicine images. When an operator evaluates reconstructed SPECT images, transaxial, coronal, and sagittal slices are usually viewed together for an accurate localization of under investigation lesion(s). If another method is used to create a three-dimensional (3D) volume display, the continuity of inner body structures can be reviewed better along with an easier understanding of any spatial relationships between neighboring uptake regions (Wallis and Miller, 1990).

3D rendering is the process of transforming data from a tomography into a 2D image in a way that all depth information is preserved, along with any spatial relationships. The rendering result should help the viewer with the perception of space; ideally reducing the images to be viewed while preserving any anatomical references needed for an accurate interpretation. Although 3D rendering has been used with great success in Computed Tomography (CT) (Vannier, Marsh and Warren, 1984) or Magnetic Resonance Imaging (Vannier et al., 1988), Nuclear Medicine images are especially challenging as they usually have poor quality due to noise (Wallis and Miller, 1990), low spatial resolution (around 10 mm for a state of the art modern gamma camera according to Khalil et al., 2011) and a variety of tracer uptakes within the body organs.

183

Most of the rendering techniques can be divided generally into two categories: Thresholding (Binary) or Percentage Volume (Continuous) (Calhoun et al., 1999). Rendering with any method usually involves three steps: Formation of the volume, categorization, and image projection (Fishman et al., 2006). The first step involves data acquisition and any processing required for the formation of the image. It includes for example interpolation, filtering, attenuation correction, or data smoothing. During the second step, each voxel is assigned a color and if applicable visual properties such as opacity according to the information it contains (counts and depth for example). Finally, the relevant data are displayed in a user-selected orientation.

6.2 SURFACE RENDERING

Surface rendering is a binary rendering technique where initially each voxel is examined and categorized if it is inside or outside the object under investigation. The user sets a threshold (intensity/count) and each voxel is assigned two different shade values depending on its count value. A surface is then defined from the outer boundary of each transaxial slice and is consequently generated. When displayed, it can be reflective with the use of a simulated light source or self-luminous, depending on the preferences of the viewer.

The process of dividing an image into different components is also referred to as segmentation. The thresholding technique has been used for many years for image 3D segmentation (Wallis and Miller, 1990) as it permits the display of not only object surfaces but also assessment of volumes such as the cardiac chamber size during ejection fraction calculations. For ease of application, the threshold level is usually expressed as a percentage of the maximum level (Lyra, 2009). Typical threshold levels for the heart (Figure 6.1), lungs, or the liver are 40%–50%, while for kidneys and the thyroid gland are 30%–50%.

The thresholding technique has a number of limitations and disadvantages such as not accurate categorization of voxels representing mixed uptake from different organs, not proper image segmentation if attenuation correction is not applied properly or inaccurate results when excessive statistical noise is present either from scattered photons (e.g. because of a "large" patient) or reduced scanning time (e.g. because of a "non-compliant" patient). All the above can add up and create confusing artifacts on the displayed surface such as holes on the structure, small fragments "floating" in space, or exaggeration of details (Kuszyk et al., 1996). The image may also contain voxels inside a non-correct classified volume because of blurring or partial-volume effects during reconstruction. Its main advantage is its speed as it requires a small amount of computational power to generate images quickly and ease of implementation.

6.2.1 THRESHOLD-BASED SURFACE ILLUMINATION

During surface rendering and after all enclosed volume voxels are discarded an illumination technique can be implemented to allow users to view the surface more easily and identify any areas of diagnostic interest. An artificial light source is positioned at a desirable direction and distance relative to the data along with an appropriate viewing angle/plane. When all hidden or non-desirable surfaces are

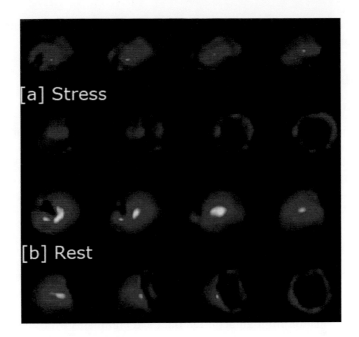

FIGURE 6.1 Example of 3D surface angular images of the heart in (a) stress and (b) rest. (From Lyra, 2009.)

removed the algorithms calculate the amount of light reflected toward the viewing angle on the remaining surface to achieve a 3D-shading effect (Phong, 1975). The method can be modified to decrease the voxel value (or illumination), which is also based on the relative distance of the observer from the voxel. This enhances any depth effect viewed by allowing objects to fade with distance. In addition, the technique of gradient shading (Figure 6.2) can also be used to add a more realistic shade to the final image (Lyra, 2009). When gradient shading, is used the illumination of the surface varies according to the local slope of each surface point. The maximum one occurs when the surface is viewed from upfront while the pixel's

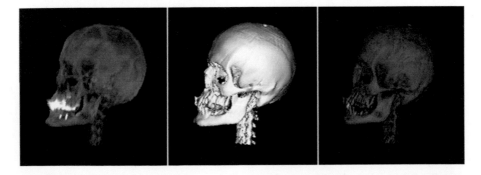

FIGURE 6.2 Render of the skull from CT data using (from left to right) Maximum Intensity Projection, Voxel Gradient Shading, and Volume Compositing. (From Bailey et al., 2014.)

luminosity decreases when the slope of the surface slips away to another angle (Garcia-Panyella and Susin, 2002).

6.3 VOLUME RENDERING

6.3.1 INTRODUCTION

Volume rendering methods do not segment the reconstructed information inside voxels but instead use the whole volume to form a rendered image. Initial research on the implementation of these techniques in Nuclear Medicine was done in the 80s by Gibson (1986), Webb et al. (1987), and Miller, Starren and Grothe (1988) while Wallis et al. (1989) described and summarized the main techniques available in Nuclear Medicine applications.

6.3.2 SUMMED PROJECTION

In this method the matrix of acquired data is re-projected to create a planar image by summing the information (e.g. counts) in the voxels on each data ray (Figure 6.3). This can be performed from any viewing angle. Each view contains the information from all acquired data and eliminates overlying or underling structures from the volume of interest, sub-volumes may also be selected. Re-projection can be done for any chosen viewing angle allowing for a better observation of the volume of interest.

6.3.3 THRESHOLD PROJECTION

This method is similar to Summed Projection with the difference that a single threshold is chosen to define the object of interest. Only the voxels with intensity/counts above the set threshold are subsequently projected.

6.3.4 VOLUMETRIC COMPOSITING

This technique has been employed for use mostly in CT and is a combination of volume and surface rendering (Wallis and Miller, 1990). Several versions of this method have been developed in the past by Levoy (1988) and Drebin (1988). In volumetric compositing each voxel of the reconstructed image is assigned a color (C_i) and an opacity (α) that determines how much of the color from underlying voxels

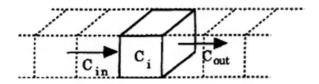

FIGURE 6.3 Light emitted by previous voxels pass through ith voxel with opacity α. This light is combined with the light C_i and emitted as C_{out}. The color emitted C_{out} will be: $C_{out} = C_i + (1 - \alpha)C_{in}$. (From Wallis et al., 1991.)

is transmitted to the viewer. Firstly, a light source and a viewing angle are chosen and then a count range is selected that most probably defines the object of interest, acknowledging the fact that parts of the object may also be outside that range. Voxels with counts within that range are defined to be inside the object and are assigned a high opacity while those away from the predefined range are set as transparent with low opacity. Voxels close to the predefined range are assigned an intermediate opacity depending on the gradient of counts in that location, the absolute difference from the selected range and other user set parameters (edge thickness).

The main advantage of this method is that it smooths out any discontinuities generated during reconstruction and that it produces very good quality renderings (Figure 6.2). Unfortunately, it is much more computationally intensive than previous described techniques and requires users to set input parameters such as maximal opacity, edge thickness, range of counts (intensities), and shading parameters.

6.3.5 Maximum Intensity Projection (MIP)

MIP consists of projecting the maximum intensity value encountered along the trajectory of rays through the data (voxels) on the corresponding screen pixel (Figure 6.4). It is an algorithm well suited for "hot-spot" imaging as the projection value for each ray is taken as the maximum value of the counts in the voxel along that ray (Figure 6.5). To add a more realistic effect and help the user understand the concept of depth, usually a default rotation is implemented in a system displaying MIP images. Moreover, if the MIP images are combined with the summed projection method, *Sum-Weighted Maximum Intensity Projections* can be produced that also depend on the thickness of the object investigated.

6.3.6 Depth Weighting

Each method described before can be modified so that a voxel's luminosity or depth (i.e. the voxel's counts) can decrease with increasing distance/depth. With the use of this method called Depth Weighting, all depth effects along with the image's

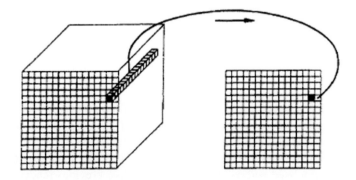

FIGURE 6.4 From a single ray of voxels (left) its maximum is placed on the corresponded rendered image (right) to produce the MIP. If instead of the maximum the sum of the ray is used a summed projection is produced. (From Wallis et al., 1991.)

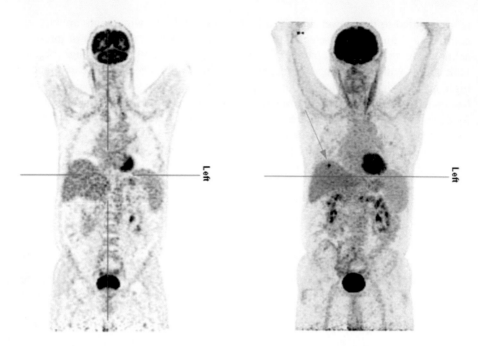

FIGURE 6.5   A PET scan MIP image (right) compared with a standard coronal slice (left). A lesion of increased uptake (arrow) is clearly visible on the MIP image. (From Bailey et al., 2014.)

contrast are enhanced allowing for a better perception of the object as voxels appear to gradually fade with distance. A linear or exponential attenuation can be used with the latter preferred. As an example, if exponential depth weighting is used during the generation of MIP the following equation is used:

$$I_{x,y} = Max\left(e^{-\mu z}D_{x,y,z}\right)$$

where $I_{x,y}$ is the rendered image/surface, D is the 3D matrix with data, Max is the maximum operator and μ is the attenuation coefficient. Typical values for the attenuation coefficient are between 0.024–0.049 cm^{-1}.

6.4   USE OF MATLAB IN RENDERING

Volume rendering or MIPs are the techniques most often used in Nuclear Medicine for the visualization of volumes and investigation of any areas of clinical interest. Rendering can be performed using special workstations, but MATLAB can be an additional and effective tool to view, perform, or visualize a render. In this way the data can be manipulated without any restrictions while previous render and results from other workstations can be validated.

One of the simplest versions of render can be achieved by constructing a simple surface plot from the reconstructed data. After the acquisition of SPECT, iso-contours

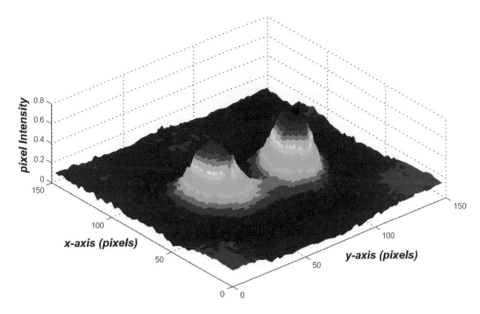

FIGURE 6.6 Surface plot using MATLAB of an I-123 DaTSCAN SPECT. The *x* and *y* axis represent positional information while the *z* axis represents count data (counts). (From Lyra, Ploussi and Georgantzoglou, 2011.)

can be depicted and all of them can be added in order to create a desirable volume image (Lyra et al., 2010a). A surface plot (or mesh plot) can be used to extract information about the consistency of an organ or any loss of functionality. For example to construct a surface plot for a striatum image, a series of images/slices (including the slice with the highest uptake/counts) can be selected and a mesh plot can be visualized by using a few basic MATLAB functions (Lyra et al., 2010b). The plot's angle of view can be further altered allowing for an inspection of any details (Figure 6.6).

REFERENCES

Bailey, D., Humm, J., Todd-Pokropek, A., and van Aswegen, A., 2014, Nuclear Medicine Physics: A Handbook for Teachers and Students (Vienna: IAEA).

Calhoun, P., Kuszyk, B., Heath, D., Carley, J. and Fishman, E., 1999, Three-dimensional volume rendering of spiral CT data: theory and method. RadioGraphics, 19(3), pp. 745–764.

Drebin R., 1988, Volume rendering, Comp Graphics, 22, pp. 65–74.

Fishman, E., Ney, D., Heath, D., Corl, F., Horton, K. and Johnson, P., 2006, Volume rendering versus maximum intensity projection in CT angiography: what works best, when, and why, RadioGraphics, 26(3), pp. 905–922.

Garcia-Panyella, O. and Susin, A., 2002, Left ventricle volume estimation from 3D SPECT reconstruction, Computers in Cardiology, pp. 621–624, DOI: 10.1109/CIC.2002.1166849.

Gibson, C., 1986, Interactive display of three-dimensional radionuclide distributions, Nuclear Medicine Communications, 7(7), pp. 475–488.

Kuszyk, B., Heath, D., Bliss, D. and Fishman, E., 1996, Skeletal 3-D CT: advantages of volume rendering over surface rendering, Skeletal Radiology, 25(3), pp. 207–214.

Levoy M., 1988, Volume rendering: display of surfaces from volume rendered data, IEEE Computer Graphics and Applications, 1988, 29–37.

Lyra M., 2009, Single photon emission tomography (SPECT) and 3D images evaluation in nuclear medicine, In: Chen Y.-S. (Ed). Image Processing, ISBN: 978-953-307-026-1, InTech, Available from: http://www.intechopen.com/books/image-processing/single-photon-emission-tomography-spect-and-3dimagesevaluation-in-nuclear-medicine.

Lyra, M., Ploussi, A., and Georgantzoglou, A., 2011, MATLAB as a Tool in Nuclear Medicine Image Processing, In: Ionescu C. (Ed). MATLAB – A Ubiquitous Tool for the Practical Engineer, InTech, DOI: 10.5772/19999.

Lyra, M., Sotiropoulos, M., Lagopati, N. and Gavrilleli, M., 2010a, Quantification of myocardial perfusion in 3D SPECT images – stress/rest volume differences, Imaging Systems and Techniques (IST), 2010 IEEE International Conference on 1–2 July 2010, pp. 31–35, Thessaloniki, DOI: 10.1109/IST.2010.5548486.

Lyra, M., Striligas, J., Gavrilleli, M. and Lagopati, N., 2010b, Volume quantification of I-123 DATSCAN imaging by MATLAB for the differentiation and grading of Parkinsonism and essential tremor, International Conference on Science and Social Research, Kuala Lumpur, Malaysia, December 5–7, 2010. Available from: http://edas.info/p8295.

Khalil, M., Tremoleda J., Bayomy T. and Gsell W, 2011, Molecular SPECT imaging: an overview, International Journal of Molecular Imaging, 2011, Article ID 796025.

Miller, T., Starren, J. and Grothe Jr, R., 1988, Three-dimensional display of positron emission tomography of the heart, The Journal of Nuclear Medicine, [online] 29 (4), pp. 530–537.

Phong, B., 1975, Illumination for computer generated pictures, Communications of the ACM, 18 (6), pp. 311–317.

Shih, W., Schleenbaker, R., Stipp, V., Magoun, S. and Slevin, J., 1993, Surface and volume three-dimensional displays of Tc-99m HMPAO Brain SPECT images in stroke patients by a three-headed gamma camera, Clinical Nuclear Medicine, 18(11), pp. 945–949.

Vannier, M., Marsh, J. and Warren, J., 1984, Three dimensional CT reconstruction images for craniofacial surgical planning and evaluation, Radiology, 150 (1), pp. 179–184.

Vannier, M., Gutierrez, F., Laschinger, J., Gronemeyer, S., Canter, C. and Knapp, R., 1988, Three-dimensional magnetic resonance imaging of congenital heart disease, RadioGraphics, 8 (5), pp. 857–871.

Wallis, J., Miller, T., Lerner, C. and Kleerup, E., 1989, Three-dimensional display in nuclear medicine, IEEE Transactions on Medical Imaging, 8(4), pp. 297–230.

Wallis, J. and Miller, T., 1991, Three-dimensional display in nuclear medicine and radiology, The Journal of Nuclear Medicine, [online] 32(3), pp. 534–546.

Webb, S., Ott, R., Flower, M., McCready, V. and Meller, S., 1987, Three-dimensional display of data obtained by single photon emission computed tomography, The British Journal of Radiology, 60(714), pp. 557–562.

7 Quantification in Nuclear Medicine Imaging

Nefeli Lagopati

CONTENTS

7.1   QUANTIFICATION IN NUCLEAR MEDICINE

The last decades, there has been significant progress in the development of methods, which allow accurate quantification in nuclear medicine images (Seo Y. et al., 2008). Imaging for nuclear medicine images can allow either detection tasks, which usually include the identification of perfusion defects, or quantitative tasks, such as estimating ejection fraction, organ absorbed dose, or standardized uptake values (SUVs) (Dickson J. et al., 2019). In order to obtain images, suitable for quantitative tasks, there is often requirement for additional processing, resulting in improved resolution and contrast and reduced artefacts (Noller C. M. et al., 2016). The common methods of quantification involve relative quantification techniques, using ratios of image intensity values (IAEA, 2014; Noller C. M. et al., 2016).

Moreover, there are several applications of nuclear medicine, which require images that are quantitatively accurate in an absolute sense (Frey E. C. et al., 2012). Targeted radiotherapy treatment planning and advanced kinetic analysis are among these applications (Thorwarth D., 2015). For an absolute quantification, appropriate equipment, software, and human resources are needed (Zimmerman B. E. et al., 2017).

The term "quantitative measurement" is open to different interpretations. Generally, the main goal of the quantification is to determine relative or absolute numerical values for the uptake or distribution of radionuclides in the patient's body (David S. et al., 2011; Dickson J. et al., 2019).

7.1.1   QUANTITATIVE INDICES

There are two general classes of tasks of quantification: classification and quantitative tasks (Griffis D. et al., 2016). Classification tasks involve placing the patient into one of several discrete classes, based on the estimation of some clinical characteristics (Spratt H. et al., 2013). The most basic classification task is a detection or binary classification task where patients are placed in one of two groups (IAEA, 2014). For instance, in a common practice, in fluorodeoxyglucose (FDG) PET imaging, patients can be identified as having cancerous or benign tumors, based on FDG metabolism in the tumor (Zhu A. et al., 2011). Other more advanced approaches of classification tasks involve more than two diagnostic classes (Erickson B. J. et al., 2017). A characteristic example is rest–stress myocardial perfusion imaging, where patients are identified as having normal perfusion or fixed or reversible perfusion defects (Cremer P. et al., 2014). On the contrary, quantitative tasks involve the extraction of a numerical value or values from data obtained in a nuclear medicine procedure (Sciagrà R., 2012).

Classification tasks, often, involve the qualitative interpretation of information obtained from nuclear medicine procedures, which means that an observer uses his own experience or knowledge to evaluate the information and make the classification decision (IAEA, 2014). Of course, there are many classification tasks can, which can be performed on the basis of quantitative values, obtained from images (Morin O. et al., 2018). For example, it is possible to make a classification in Parkinson's disease, with DaTscan quantitative analysis (Nuvoli S. et al., 2018).

7.1.1.1   Counts, ROIs, VOIs, and SUVs

In planar single photon imaging, the number of photons which are detected by a scintillation camera, falling inside the predefined acceptance energy window, at a position corresponding to a pixel in the image is termed as "counts" (Frey E. C. et al., 2012). In a non-calibrated reconstructed image, the signal intensity of photons assigned to each voxel is often termed as "number of counts" (Frey E. C. et al., 2012; IAEA, 2014).

In a certain area in the image, the number of counts is calculated for a region of interest (ROI) (Figure 7.1) or volume of interest (VOI) (Figure 7.2) defined by the user (Vicente E. M. et al., 2017). The sum of the counts is assumed to be a proportional measure of a clinically relevant factor (Taylor A. T., 2014). Count ratios for several ROIs are usually used to detect differences in activity uptakes (Taylor A. T., 2014). Dynamic studies provide information about the activity uptake over time and can be important in addressing the biokinetics of a particular organ or tissue (Li T. et al., 2017). A normal

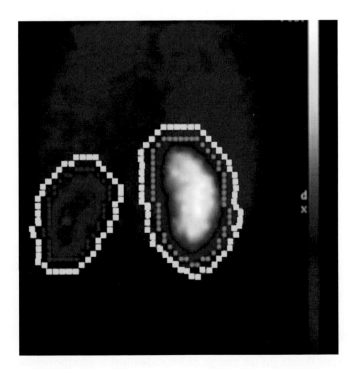

FIGURE 7.1   The definition of ROI in an image of kidneys. By calculating count rate in a defined ROI, it is possible to obtain a renogram which provides quantitative data.

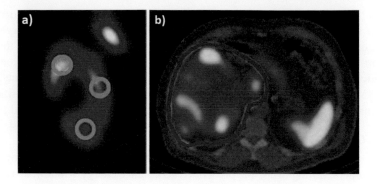

FIGURE 7.2   The definition of VOI is used for activity quantification on SPECT images for: (a) kidneys and (b) liver. (Revised image Marin G. et al., 2018.)

tissue is considered that follows a certain time–activity curve (TAC) and that a change in the shape of the curve reveals clinically relevant changes in the functional behavior of the organ (Ljungberg M. et al., 2016). Counts in ROIs are compared with corresponding data from ROIs in images of patients that have been classified as normal, based on other independent methods (McGoron A. J. et al., 2008).

SUV is perhaps the most commonly used index for the quantification of tumor uptake in PET and PET/CT scans (Kinahan P. E. et al., 2010). SUV is defined as the ratio of the tumor uptake (kBq/mL) divided by the injected activity (kBq) normalized by the dilution volume (mL) at the time of imaging (Wang Y. Y. et al., 2017). With the assumption that the patient has a volumetric density of 1 g per mL, the SUV is usually calculated by Equation 7.1. SUV is a unit-less physical quantity (IAEA, 2014). If the tracer is distributed uniformly throughout the patient, then the SUV is estimated as 1 everywhere (Byrd D. W. et al., 2016). The patient volume is often approximated as equal to the patient weight (IAEA, 2014).

$$SUV = \frac{tumor\ uptake\ \left(\frac{kBq}{ml}\right)}{\frac{injected\ dose\ (kBq)}{patient\ weight\ (g)}} \cdot \frac{(ml)}{(g)} \tag{7.1}$$

7.2   QUANTITATIVE PLANAR IMAGING

Two-dimension planar scintigraphy creates images by detecting the photons emitted in certain directions, which are determined by the collimator which is equipped with γ-camera. Thus, the third dimension, namely the depth, cannot be directly resolved from a single projection image (Geng J., 2013).

7.2.1   COLLIMATOR SELECTION

The appropriate design for the energy of the radionuclide and the selection of a collimator with the appropriate resolution–sensitivity trade-off are the main issues in the selection process of a collimator (Van Audenhaege K. et al., 2015). This decision can affect the image quality, by creating artefacts, degrading spatial resolution, and distance-dependent sensitivity and thus decreasing the quantitative accuracy (Zarinabad N. et al., 2015). The size of the object of interest is of crucial importance to select a collimator (Martin C. J., 2007).

7.2.2 ATTENUATION CORRECTION

Correction for attenuation, which is based on the conjugate view method, is the most common method in planar imaging (Sorenson J. A., 1971). This correction process is based on two opposite measurements in the same energy window. For a point source, located in a uniform medium, the attenuation of the count rate, measured in the two projections is given by Equations 7.2 and 7.3 (IAEA, 2014):

$$C_A = C_0 \cdot e^{-\mu d} \tag{7.2}$$

$$C_P = C_0 \cdot e^{-\mu(T-d)} \tag{7.3}$$

where d is the depth to the source in the anterior view and T is the total thickness of the patient. The geometric mean of the opposite projections can reduce the attenuation to a function of patient thickness T (Equation 7.4):

$$\sqrt{C_A C_P} = C_o \cdot e^{-\mu T/2} \tag{7.4}$$

The activity A is determined from Equation 7.5:

$$A = \frac{C_o}{K} = \frac{\sqrt{C_A C_P}}{K \cdot e^{-\mu T/2}} \tag{7.5}$$

where K is the system sensitivity (cps per MBq) measured in air. This equation is only valid for a point source. To include the source thickness l, Equation 7.6 is suggested (IAEA, 2014):

$$A = \frac{\sqrt{C_A C_P}}{K \cdot e^{-\mu T/2} \cdot \frac{\sinh\left(\frac{\mu l}{2}\right)}{\mu l/2}} \tag{7.6}$$

7.2.2.1 Patient Thickness

An estimation of the thickness of the patient needs to be obtained for a successful attenuation correction. The thickness can be measured manually at various points of the body (Ljungberg M. et al., 2016). Alternatively, the attenuation correction can be applied on the sum of the data within the ROIs (Papanastasiou E. et al., 2017).

7.2.2.2 Transmission Measurement

It is possible to separately measure the thickness of a patient if there is available equipment. A photon flood source on the opposite side of the γ-camera can be mounted and the transmission of the photons can be measured in two steps, one with and one without the patient (Frey E. C. et al., 2012; Ljungberg M. et al., 2016). Thus, a pixel by pixel ratio of the counts in the presence of the patient is divided by the counts in the absence of the patient, giving the attenuation factor, which is needed to be adjusted to account for differences in energy, for flood sources of different energy (IAEA, 2014). The patient has to remain in the same position for the transmission scan. Otherwise, the transmission scan needs to be registered to a whole-body measurement by appropriate software (Minarik D. et al., 2005).

7.2.3 SCATTER

Scatter is responsible for the spatially varying low-frequency blurring, the loss of contrast and, the spurious and spatially varying increase in image intensity, in attenuation compensated images (IAEA, 2014). Scatter quantitative effects are greater for low energy photons, or larger objects, and/or wider energy windows and poorer energy resolutions.

7.2.3.1 Effective Attenuation Coefficient

An effective attenuation coefficient, which is lower in magnitude than the tabulated linear attenuation coefficient based on a narrow-beam geometry, is a simple method to compensate for the additional counts in the projections from scattered photons, by reducing the magnitude of the attenuation correction (Lee Z. et al., 2016). An effective attenuation coefficient depends on the source distribution and surrounding attenuating tissue.

7.2.3.2 Energy Window

The scatter correction methods based on the energy window are simple and effective. The scatter component in the photopeak is estimated, based on measurements in one or more additional scatter windows. The triple energy window (TEW) method is the most common (Papanastasiou E. et al., 2017). In this method, scatter energy windows are placed above and below the photopeak energy windows. A trapezoidal approximation is used for the estimation of the scatter in the photopeak energy window. Wider energy windows can reduce noise in the scatter estimation. There is the need of low pass filtering of scatter to reduce noise (Frey E. C. et al., 2012).

7.2.3.3 Buildup Factor

One of the suggested scatter compensation methods is based on the conjugate view method combined with a depth-dependent buildup function (Sjögreen K. et al., 2002). This iterative method starts with the assumption of the activity in the middle of the volume, where $d = T/2$. Equations 7.7 and 7.8 give the count rate in air, C_O, and the depth from the anterior view, d.

$$C_A = C_O \cdot B(\mu d) \cdot e^{-\mu d} \qquad (7.7)$$

$$C_P = C_O \cdot B(\mu(T-d)) \cdot e^{-\mu(T-d)} \qquad (7.8)$$

where C_A and C_P are the measured count rates in the anterior and the posterior views, respectively. New values of d can be calculated until the C_O is equal for both equations.

7.2.4 RESOLUTION AND PARTIAL VOLUME

Planar imaging provides images, which are affected by depth-dependent spatial resolution, as result of the design of the collimator (Sorenson J. A. et al., 1987). The resolution can be improved by employing methods, based on hardware solutions or postprocessing restoration filtering, where a description of Point Spread Function (PSF) is used to restore the images or methods which include scatter and collimator penetration effects in the PSF (IAEA, 2014). In the case of small targets such as tumors, recovery coefficients can be used for activity quantification. These factors are usually determined from phantom experiments with spherical sources of known sizes, as well as activity contents and similar acquisition conditions (Pereira J. M. et al., 2010).

7.2.5 RECOVERY OF 3D INFORMATION

Radioactivity from other organs and tissues and the blood activity within the target-organ itself can be characterized as non-target activity. Therefore, the proper design of ROIs is of crucial importance (Sjögreen K. et al., 2002). ROIs can be drawn over the whole organs despite overlap. The background activity is obtained from the activity per pixel in a representative ROI and a fraction of patient thickness, occupied by the particular organ. Generally, this fraction is estimated for some organs using two phantoms (He B. et al., 2006). Recovering of 3D information about organ activity requires compensation for overlap in image of the organ of interest and the background. 3D information about the organ shape and its relationship to the background and other organs are required for accurate compensation (Liu A. et al., 1996). The use of a registered 3D CT image combined with a set of 3D VOIs obtained either by segmenting the CT image or from a registered SPECT image is also required. Scatter models as opposed to a buildup function method and modeling of the full Collimator–Detector Response (CDR), including collimator scatter and penetration are promising techniques (Zaidi H., 2006).

7.3 QUANTITATIVE SPECT OR SPECT/CT

The first complete review about the factors affecting quantitative SPECT was published in 1995, by the Society of Nuclear Medicine. Since then, there have been significant advances relevant to the practice and state of the art in quantitative SPECT (Bailey D. L. et al., 2013).

7.3.1 TOMOGRAPHIC RECONSTRUCTION

The filtered back projection is widely used in SPECT reconstruction, often in parallel with Chang attenuation compensation (Figure 7.3). It has universal commercial availability in many general-purpose image processing packages (Lyra M. et al., 2011). The major disadvantage is that it does not allow an absolute quantification, although it is attempted using phantom-based calibration. Relative quantification is generally possible and absolute quantification can be obtained, based on iterative reconstruction, using compensation for the image degrading effects (Oloomi S. et al., 2013). Among

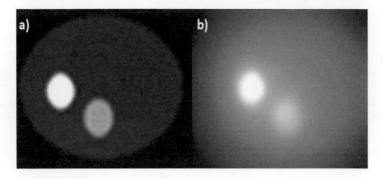

FIGURE 7.3 Filtering process. (a) Model (128 × 128 pixels), (b) Image obtained after backprojection of 128 projections.

the most common iterative algorithms for SPECT reconstruction, OS-EM (Ordered Subsets Expectation Maximization), and ML-EM (Maximum-Likelihood Expectation Maximization) are found (Katua A. M. et al., 2011). ML-EM algorithm is theoretically appealing, and it requires a relatively large number of iterations, in order to produce quantitatively accurate images. OS-EM is also available on recent commercial SPECT systems. It is considered as a state-of-the-art tool for obtaining quantitative SPECT images (IAEA, 2014).

7.3.2 Photon Attenuation

Attenuation significantly affects the quantification in SPECT, by degrading (Figure 7.4). Attenuation compensation is necessary; otherwise, in the absence of this process, absolute estimates of activity will be underestimated, while the relative errors will be smaller, in general. This effect is greater for low energy photons or larger and denser regions of the body (Frey E. C. et al., 2012).

Analytical and statistical attenuation compensation methods have been developed. These methods compensate for the impact of attenuation on accuracy. Analytical

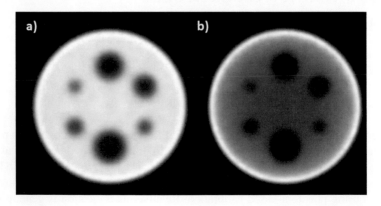

FIGURE 7.4 (a) Image of a phantom. (b) The effect of attenuation in imaging liver. (Revised image from Larsson A., 2005.)

methods can include exact methods and approximate methods (Hobbs R. F. et al., 2010). The exact methods are not widely used, because they provide poor noise properties. They can be developed in short reconstruction times, but they require powerful computers and for this reason the interest around them is limited (Kunyansky L. A., 2001). The Chang attenuation compensation method is available commercially. It is used in cases, where the attenuation distribution in the area of interest is uniform. It performs voxel by voxel attenuation compensation of an image reconstructed without attenuation compensation (Novikov R., 2002). OS-EM and ML-EM are preferred, due to the poor noise properties of exact analytical algorithms and the inferior performance of approximate algorithms (Takabe S. et al., 2016).

Attenuation coefficient in each pixel is required for attenuation correction process. These data can either be estimated from the emission data or measured (Yan Y. et al., 2009). Estimated attenuation maps with the attenuation coefficients from each pixel are often obtained by fitting an ellipse to the boundaries of the object in the emission projection data (Berker Y. et al., 2016). A radionuclide source or a registered CT image is usually utilized for the development of a measured map (Seo Y. et al., 2008). It is usual phenomenon, the produced attenuation maps to be generally noisy with poor resolution and with limited potential to provide anatomical information. But, as the resolution is approximately matched with that of the SPECT image, the poor resolution or noise does not lead to degradation of attenuation compensation (Kheruka S. C. et al., 2011). The source energy and strength are also important, so that sufficient photons can pass through the thickest portion of the patient. Thus, attenuation maps from radioactive sources could be quite effective for attenuation compensation, if the appropriate attention to these factors is given in parallel regular source replacement if it is necessary (Berker Y. et al., 2016).

CT images can also provide attenuation maps. The difference in the energy between the CT and the emission radionuclide has to be considered. A linear scaling is needed in this case (Kheruka S. C. et al., 2011; Berker Y. et al., 2016). Some additional care is required during the reconstruction projection data acquisition by summing photopeaks with significantly different energies (Zhao W. et al., 2018). CT attenuation maps must be well registered with the emission image. For this reason, registration of more than one emission image pixel is required in order to avoid artefacts. Patient position during the scans limits the usage of CT images obtained from a diagnostic CT scan, performed on an independent CT scanner (Frey E. C. et al., 2012).

Recently, the most widespread method used in clinical practice, requires SPECT/CT scanners. Artefacts due to patient motion or misregistration due to imaging table sag, systematic problems in system calibration can decrease the quality of attenuation compensated images, even in these systems (Seo Y. et al., 2008). Properly implemented attenuation compensation can achieve satisfactory compensation for the effects of attenuation (Franceschini E. et al., 2008).

7.3.3 Scatter

In SPECT, scatter fractions can result in spatially varying quantitative errors. The scatter increases approximately linearly with source depth. Thus, the quantitative effects of scatter are more significant for source positions near the center of the object (Frey E. C. et al., 2012). When the narrow beam attenuation coefficient is used for attenuation compensation performing, the scattered photons can affect the image (Lee H.-H. et al., 2013).

For a scatter compensation, a scatter estimation and a compensation method are required. The scatter estimations can be obtained are obtained spatially or based on measurements in additional non-photopeak energy windows (Ghaly et al., 2015). The spatial estimations of the scatter can be based on stationary approximation of the scatter response function, on scatter models or on fast Monte Carlo simulations (Wu G. et al., 2017).

There is a fundamental limitation in compensating for scatter from activity outside the field of view. Therefore, the region reconstructed should extend above and below the ROI. Moreover, an attenuation map must be available in the same region (Dorbala S. et al., 2018).

The model-based methods provide accurate and quick estimation of the spatially varying scatter response function. They are based on the physics of the scattering and use calibrations based on Monte Carlo simulations or experimental measurements of the scatter response function (Saffar M. H. et al., 2013). The scatter in the projection data is usually estimated by a hypothetical activity distribution and an input attenuation map. These methods can model the spatial variance of the scatter response and can give good agreement for regions with uniform attenuation distributions (IAEA, 2014). In general, scatter models require significant computational time. Fortunately, some optimized versions can produce scatter estimates acceptable times for clinical use.

Monte Carlo simulation is the method of choice for cases where downscatter is important. The long computation time and the lack of commercial availability are the main obstacles of this approach (IAEA, 2014).

Scatter compensation can be achieved via pre-reconstruction subtraction or by including the scatter estimation (Dewaraja Y. K. et al., 2006). Pre-reconstruction subtraction is not a desirable choice. It can lead to negative projection pixel values that must be handled carefully and can result in increased reconstructed image noise (Nuyts J. et al., 2013). Well calibrated TEW, model or Monte Carlo-based methods are capable of significantly reducing the effects of scatter on accuracy (Figure 7.5).

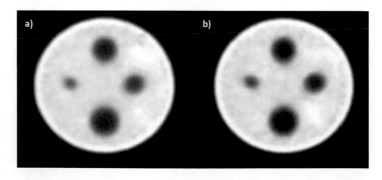

FIGURE 7.5   (a) Image of a phantom, (b) Monte Carlo-based scatter correction.

7.3.4 RESOLUTION AND PARTIAL VOLUME

In common SPECT imaging, spatial resolution is decreased by the spatially variant CDR and significant partial volume effects (PVEs) are created for small objects
(Pretorius P. H. et al., 2009). Since the FWHM of the CDR function increases proportionally to the distance from the collimator, SPECT images present spatially varying
resolution. Thus, PVEs are spatially varying and the accuracy of quantifying small
objects is position dependent (Rahmim A. et al., 2008). Generally speaking, the tangential resolution becomes better away from the center of rotation, and the radial resolution remains almost constant. The contrast near the center of rotation is reduced by
the spatially varying resolution, as a result of increased PVEs (Pramanik M., 2014).

7.3.5 RESOLUTION COMPENSATION

Resolution compensation is accomplished by compensating for the effects of the
CDR in SPECT imaging. Thus, there is an improvement of resolution in the reconstructed images and ameliorated quantitative accuracy for small objects. Iterative
and non-iterative methods are usually chosen (Dewaraja Y. K. et al., 2010).

The use of inverse filters for the implementation of the non-iterative methods
in parallel with other limitations make them not usually available on commercial
systems (Sun Y. et al., 2009). The whole process is relatively easy using general purpose image processing software, but with low efficacy at reducing spatially varying
effects or noise, compared to iterative methods (Shen D. et al., 2017).

Commercially available iterative reconstruction-based methods model the CDR
in the forward and backward projection processes of an iterative reconstruction algorithm, and therefore perform CDR compensation (He B. et al., 2005). More iterations
are required to achieve optimal images. Iterative CDR compensation cannot fully
remove the blurring due to the CDR. Some information on high spatial frequencies
is lost and the resolution remains spatially varying (Rahmim A. et al., 2008).

The modeling of the CDR is very important issue during iterative reconstruction
CDR compensation (Ghaly M. et al., 2016). This model is often limited to modeling the geometric component of the CDR. The geometric CDR can be described by
exact or approximate analytical expressions (Dewaraja Y. K. et al., 2012). For systems with important septal penetration and scatter, such as the radionuclides and collimators, CDR modeling is employed in order to include these effects. The CDRs are
either measured or estimated with Monte Carlo simulations (Chun S. Y. et al., 2013).

Iterative reconstruction-based geometric CDR compensation is available only for
certain imaging protocols. Iterative reconstruction-based CDR compensation has
limited use in current clinical practice (Brambilla M. et al., 2017). It is usually utilized in order to reduce scan time by providing an improved resolution–noise trade-
off (Frey E. et al., 2006).

7.3.6 PARTIAL VOLUME CORRECTION

SPECT images are usually degraded by the effects of the CDR and this leads to
significant PVEs. Spill-in effects are often significant in areas close to objects with

very high activity (Pretorius P. H. et al., 2009; Frey E. C. et al., 2012). The addition of partial volume correction (PVC) is required in order to achieve high quantitative accuracy. This technique is based on the assumption that the activity distribution inside a VOI is uniform (Erlandsson K. et al., 2012). Some PVC methods are based on calibration methods, while others based on postprocessing methods or reconstruction-based methods (Deng X. et al., 2019).

The use of phantom experiments is mandatory for calibration methods which focus on the estimation of recovery coefficients, which are defined as the ratio of the true activity divided by the measured activity (Frey E. C. et al., 2012). These coefficients are a function of the object shape, size, and activity relative to the background. Because of the spatially varying resolution in SPECT images, recovery coefficients also depend on the position in the image (Frey E. C. et al., 2012). In order to reduce the number of variables, the recovery coefficients are usually estimated as a function of object size (Jolliffe I. T. et al., 2016). The shape is assumed to be spherical, the spatial variations are ignored, and the background activity is considered as zero. For this reason, these approaches have limited accuracy. The main advantage is that they do not require any special software (IAEA, 2014).

Post-reconstruction methods measure spill-in and spill-out effects for each VOI or voxel for each patient. Thus, the consideration of the effects of activity in neighboring VOIs or voxels, the shapes of the VOIs, and the spatial variation of resolution is possible (Finocchiaro D. et al., 2019). Some post-reconstruction methods perform PVC for each voxel and others compensate for the PVEs of the activity estimates of each VOI.

7.3.7 RECOVERY OF 3D INFORMATION

3D information is recovered by tomographic reconstruction, in SPECT imaging. SPECT images provide 3D framework, and this can eliminate the issue of overlap of target and non-target activity (Gullberg G. T. et al., 2010).

7.4 QUANTITATIVE PET OR PET/CT

The nature of PET is quite quantitative, as it has the potential to estimate the local concentration of the radiotracer and the relevant physiological parameters but obtaining real quantitative PET images requires careful attention to some factors (Vaquero J. J. et al., 2015).

7.4.1 TOMOGRAPHIC RECONSTRUCTION

The quantitative reliability of PET images is indirectly affected by the tomographic reconstruction, through the spatial resolution and the noise properties of the reconstructed images (Frey E.C. et al., 2012; Vaquero J.J. et al., 2015). The reconstruction algorithm can impact the quantitative accuracy. A reconstruction algorithm can be classified, based on whether it is applied to list mode or sinogram data (Vaquero J.J. et al., 2015). For list mode data, iterative reconstruction algorithms are used, while for sinogram data, iterative, or analytical algorithms are available. Another classification of algorithms is based on whether they come from 2D or 3D acquisition modes. Data which is acquired in 2D acquisition mode is reconstructed using 2D reconstruction algorithms that are applied to sinogram data from each 2D slice. Data acquired in 3D mode is reconstructed by rebinning into 2D sinograms and then using 2D reconstruction or by using 3D reconstruction algorithms (Tong S. et al., 2010).

Each detected event defines a line of response (LOR), which is a line joining the centers of two detectors along which the associated radioactive decay is assumed to have occurred (Vaquero J. J. et al., 2015). The detected event can be recorded in either sinogram or list mode format. Since there is a relatively high number of possible LOR, the list mode format can actually be more compact than the sinogram format (IAEA, 2014).

Both analytical and iterative reconstruction approaches are used in 2D and 3D PET. The use of iterative reconstruction techniques takes advantage of the fine sampling of the list mode data, employing the list mode reconstruction methods (Yu H. et al., 2015). There is no need of a priori sampling to be defined, so this choice can improve the spatial resolution of the reconstructed image. Furthermore, statistical iterative reconstruction algorithms have the advantage of explicitly considering the noise properties of the projection data in the reconstruction process (Nuyts J. et al., 2013).

The iterative reconstruction algorithms which are commonly used in PET are ML-EM and its accelerated versions, OS-EM and RAMLA. Particularly RAMLA includes an explicit regularization parameter to control noise (Bai B. et al., 2013). All the iterative algorithms are offered in 2D versions, appropriate for 2D PET data and in a 3D version, appropriate for 3D PET data. Rebinning is the process which is used when 2D versions are used to reconstruct 3D PET data (Cho S. et al., 2009). Single slice rebinning, multislice rebinning, and Fourier rebinning are proposed as rebinning techniques. All these approaches include approximations and typically result in a loss of axial resolution. For this reason, 3D PET data should be reconstructed with fully 3D reconstruction approaches instead of 2D methods after rebinning (Yu H. et al., 2015). Fully 3D reconstructions provide the opportunity to model physical effects such as scatter, geometric effects, or detector response. Accurate modeling of the 3D effects requires accurate simulation, such as Monte Carlo simulations (Dewaraja Y. K. et al., 2006).

7.4.2 Photon Attenuation

Attenuation is a major bias factor in PET and PET/CT imaging, since, without attenuation compensation, activity values can be significantly underestimated (Chow P. L. et al., 2005). Attenuation introduces distortions in the apparent activity distribution and therefore the relative quantification becomes unreliable (Berker Y. et al., 2016). The transmission measurement allows accurate attenuation correction to be performed in PET, leading to an estimation of an attenuation coefficient map, μ, which characterizes the attenuation properties of the tissues (Berker Y. et al., 2016).

CT is now a standard device in the common commercial scanners, so that attenuation correction is based on CT images. This technique has several benefits over maps derived from a gamma or positron-based transmission device (Seo Y. et al., 2008). The duration of the transmission scan is shorter than the corresponding time of a conventional transmission device (Vaquero J. J. et al., 2015). The attenuation maps are almost noise free, unlike the maps resulting from the conventional transmission device (Lee Z. et al., 2016). Moreover, the CT attenuation maps have better spatial resolution, but this does not necessarily provide an advantage in terms of attenuation correction (Seo Y. et al., 2008). However, an important drawback of the use of CT to derive the attenuation map for attenuation correction is the mismatch which is related to the differences in respiration during CT and PET acquisitions (Mollet P. et al., 2014). CT data usually correspond to a specific phase of the respiratory motion, while PET data includes the averaged respiratory blur (Pépin A. et al., 2014). It leads to artefacts, especially at the interface between the lung and liver, which must be carefully recognized in order to avoid misinterpretation and quantitative errors in this region (Bharwani N. et al., 2013).

Noise from transmission images propagates directly into the PET images. This is a minor issue for CT attenuation correction (Teymurazyan A. et al., 2013). The reduction of noise can be achieved by smoothing the attenuation map reconstructed from the transmission data before reprojection to reduce the propagation of the noise in the attenuation map to the attenuation corrected PET images (Chow P. L. et al., 2005).

One important issue in the case of measured attenuation maps is rescaling if the energy of the transmission source is different from the appropriate. Then the attenuation coefficient values need to be translated to describe the attenuation coefficients at the selected energy (typically 511 keV). There are two mechanisms for attenuation correction, given the rescaled attenuation map (Zaidi H. et al., 2003). The first one is the multiplication of each LOR signal by the appropriate attenuation coefficient factors, which are derived from the transmission measurement (Mollet P. et al., 2014). The second one is the modeling of the attenuation in the transfer matrix, which is used for image reconstruction, based on the estimated attenuation map. The latter provides improved noise properties compared to the former, although both these approaches are theoretically accurate (Berker Y. et al., 2016).

Attenuation correction is available on almost all PET scanners, although the transmission map measurements and the actual correction method differ from one scanner (Bowen S. L. et al., 2016).

7.4.3 Scatter

The energy resolution is limited in PET or PET/CT scanners, and for this reason, coincidences including one or two scattered photons are frequently detected (Shukla A. K. et al., 2006). The detection of scattered events can lead to incorrectly located events in the reconstructed PET image (Vaquero J. J. et al., 2015). The point response due to scattered events is broad in PET and can extend beyond the boundary of the object. Thus, activity is detected in activity-free regions and overall, contrast and quantitative accuracy are reduced due to scatter. For accurate quantification, scatter must be compensated (Guérin B. et al., 2011).

A usual strategy which is commonly used for scatter compensation includes energy window-based methods, requiring the acquisition of the coincidences detected in several energy windows (Guérin B. et al., 2011). In this case, the scatter distribution is estimated by combining the data from different energy windows.

These techniques are appropriate for scanners with high energy resolution (Popescu L. M. et al., 2006). The estimated scatter distribution can be subtracted from the projection data or, alternatively can be modeled during the reconstruction.

Another process which is utilized for scatter compensation is fitting of activity measured outside the patient boundaries in order to estimate the scatter distribution inside the patient boundaries (Aravind V. K. et al., 2012). This approach is based on the fact that coincidences involving scattered photons are detected outside the patient boundaries and can therefore be easily identified as resulting from scatter (Chinn G. et al., 2011).

Finally, scatter compensation can be achieved by estimating the distribution of scattered coincidences using scatter models. The scatter distribution is estimated based on the attenuation map, which describes the scattering medium, and on a preliminary estimate of the true activity distribution (Guérin B. et al., 2011). Then, the image of scattered photons is estimated, using a single scatter simulation model. The contribution from higher order scatter is often estimated, by rescaling the scatter distribution in order to match the portion of the sinogram outside the patient boundaries (IAEA, 2014). These sinograms are calculated and subtracted from the sinograms that are actually measured. This reconstruction leads to a scatter corrected PET image (Santarelli M. F. et al., 2017). More sophisticated models for the estimation of the contribution of the scattered coincidences to the sinograms can also be used. Scatter correction methods are available on all PET or PET/CT scanners and should be systematically used, especially in 3D PET (Vaquero J. J. et al., 2015).

7.4.4 RESOLUTION AND PARTIAL VOLUME

The effects of resolution and partial volume problems are similar to those encountered in SPECT imaging (Rahmim A. et al., 2008). The imperfect spatial response function of the PET scanner can be corrected by modeling the spatial blurring which is introduced by this response function into the system matrix model used in tomographic reconstruction. This correction significantly improves the quantification in small structures (Matej S. et al., 2016).

The common approach consists of multiplying the measured activity value in an ROI or VOI by a recovery coefficient, which is measured as a function of the object size, the surrounding activity, the spatial resolution in the reconstructed images, and the image sampling (Bai B. et al., 2013). Another more sophisticated approach consists of segmenting the images into N non-overlapping regions, with uniform activity. The calculation the spill-out and spill-in between these regions, assuming that the spatial resolution in the reconstructed images is known, is following. This process leads to an $N \times N$ matrix, in which each entry (i, j) corresponds to the fraction of activity, which is emitted from region i that is actually measured in region j (Frey E. C. et al., 2012). By inverting this matrix, the activity values in each compartment are estimated, given the activity value measured in the reconstructed image. This correction first requires a segmentation of the VOI (Erlandsson K. et al., 2012). Furthermore, there is the need of the segmentation of the other compartments of the images, characterized by different activity values and assumes that the activities in these compartments are uniform (Wang H., 2012).

7.5 FACTORS AFFECTING RELIABILITY OF IMAGE-BASED ACTIVITY ESTIMATES

7.5.1 PHYSICAL FACTORS AFFECTING RADIOACTIVITY QUANTIFICATION

7.5.1.1 Count Rate Problem

7.5.1.1.1 Planar Images

The scintillation camera system is responsible for the magnitude of count rate losses. There are many methods for correction of this problem literally (Chiesa C. et al., 2009). Sometimes these correction processes are quite tricky, such as the whole-body scans acquired in continuous acquisition mode, due to the changes of the count rate for each row of pixels in the image existing, while the gamma camera scans over the patient (Hobbs R. F. et al., 2010).

7.5.1.1.2 SPECT

The same problems are also found in SPECT (Seo Y. et al., 2008). Significant count rate losses are an important issue for very high-count rates. The effects are dependent on the camera manufacturer and model for therapeutic doses. There is very limited work focusing on developing compensation methods (Frey E. C. et al., 2012).

7.5.1.1.3 PET

Calculating a calibration factor for a count rate similar to that used in a clinical setting can solve any issue relevant to dead time, in PET systems. A calibration curve, which depends on the count rate and implicitly corrects for dead time effects, is used for the calibration of PET scanners (Frey E. C. et al., 2012). Random coincidences can be systematically compensated for by the manufacturer. This is achievable either by recording the single events or using a delayed time window to estimate the random fraction in each LOR and subtract it accordingly (Hallen P. et al., 2018). Random coincidences can be corrected by incorporation into an iterative reconstruction algorithm.

7.5.1.2 Noise

7.5.1.2.1 Planar Images

The unprocessed planar images usually follow the Poisson distribution. In planar imaging, the noise is a function of administered activity, duration of the acquisition, and camera specific parameters, such as the choice of collimator, the energy window setting, and the thickness of the crystal (IAEA, 2014).

7.5.1.2.2 SPECT

Poisson noise in the projection data also affects the quality of SPECT images. This noise is correlated by the tomographic reconstruction algorithm and it results in a loss of precision for quantitative estimates (Lyra M. et al., 2011). Regularization of image reconstruction by post-reconstruction filtering can reduce variance, by improving the precision (Figure 7.6). Longer acquisition times or larger injected tracer activities can reduce the effects of noise. Higher sensitivity collimators can also reduce noise in the projection data (IAEA, 2014).

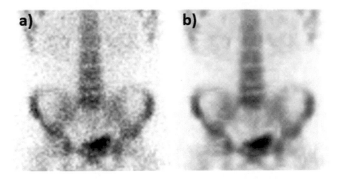

FIGURE 7.6 Projection of noise in a SPECT image of skeletal. (a) Image before noise removed, (b) image after noise removed. Adaptive autoregressive model is used and block size of 5×5 pixels.

7.5.1.2.3 PET

Noise can bias quantitative measurements. It also complicates the fitting of curves when using kinetic analysis. The noise is usually reduced by decreasing the number of updates, using low pass post-reconstruction filtering of the reconstructed images, regulating the reconstruction or increasing the smoothing in the reconstruction filter for filtered back projection methods (Kinahan P. E. et al., 2018). These noise reduction methods are usually associated with a reduction in the spatial resolution of the reconstructed activity distribution. Low noise images do not always lead to low error quantities estimated from the images (Yan J. et al., 2016).

7.5.1.3 Region and Volume Definition

7.5.1.3.1 Planar Images

ROIs are defined for whole organs from anterior or posterior views. There are some methods for correction of the contribution from other tissues and thus, the regions which contain the whole organ can be defined (Momennezhad M. et al., 2016).

7.5.1.3.2 SPECT

The definition of VOIs in the SPECT images is required for the quantification of the activity in an organ or VOI. In SPECT imaging the noise is relatively high, and the resolution is relatively poor (Ljungberg M. et al., 2016). Thus, VOIs which are defined in SPECT often are inaccurate, operator dependent, and imprecise. Generally, automatic and semiautomatic segmentation of SPECT images is considered difficult (Cheimariotis G.-A. et al., 2018). The selection and use of well registered and high-resolution anatomical images for VOI definition could improve manual or automated definition of VOIs. SPECT/CT systems can provide obvious improvement in definition of VOIs, but sometimes there is mismatch between functional and anatomical volumes, resulting in errors (Vicente E. M. et al., 2017). In order to obtain good accuracy, registration should be accurate to more than one voxel. The problems in the definition of organ boundaries, due to variation and the degree of misregistration, introduce uncertainties in organ activities estimation, due to noise (Vicente E. M. et al., 2017).

7.5.1.3.3 PET

The quantitative measurements in PET are affected by the way through which regions and volumes are defined. The estimation of the activity of the object using a VOI inside the object can reduce PVEs, especially in the estimation of tumor SUVs (Foster B. et al., 2014). Using a small VOI away from the edges can reduce the impact of PVEs. Nevertheless, since the activity concentration is averaged over a small number of voxels, the measurement will be less precise (Greve D. N. et al., 2016). Using a larger region, closer to the edges of the object, would tend to make the bias introduced by PVE more severe, but reduce the impact of noise. So, there is no universally preferred method for region definition (Hatt M. et al., 2017). In PET/CT systems, the CT image is often used to define a region before placing it on the PET image. Regions are defined automatically or semiautomatically, by using an intensity threshold, a process which is sometimes quite difficult (Vaquero J. J. et al., 2015). It is important that the SUV which is associated with a tumor depends strongly on the defined region and the calculation method. The definition of the region can impact the estimation of the total activity within a region (Bai B. et al., 2013).

7.5.1.4 Calibration

7.5.1.4.1 Planar Images

It is important to calibrate the system sensitivity in units of cps per MBq, for each type of collimator and for each energy window setting. For this process, the source has to be located at a distance which matches the center of the patient (Zhao W. et al., 2018). The calibration factors need to be measured for corresponding acquisition protocols, depending on the type of acquisition. The acquisition if an image, before voiding in order to compare measured total body activity compared to the administered activity is usual and practically reasonable (Van den Wyngaert T. et al., 2016).

7.5.1.4.2 SPECT

The units of reconstructed SPECT images are typically proportional to the number of decays occurring in a voxel during the acquisition. So, the activity, A_a, in some VOI is equal to the sum of the SPECT voxel values in the VOI, D_a, obtained for SPECT acquisition, a, multiplied by a calibration factor, k (Equation 7.9) (Kappadath S. C., 2011).

$$A_a = k \frac{D_a}{t_a} \tag{7.9}$$

where t_a is the total acquisition duration for the SPECT and the units of k correspond to the activity multiplied by time per SPECT voxel value.

This calibration factor depends on the imaging system, acquisition parameters, and patient, and also on the selected reconstruction and compensation methods. The selection of appropriate attenuation and scatter compensation can remove the effect of patient-dependent factors (Berker Y. et al., 2016). The quantification of SPECT images requires the measurement of a calibration factor which accounts for the effects of the remaining factors (Zhao W. et al., 2018). This calibration factor depends on the imaging system, acquisition parameters and patient, and also on the selected reconstruction and compensation methods. The selection of appropriate attenuation and scatter compensation can remove the effect of patient-dependent factors (Berker Y. et al., 2016). The quantification of SPECT images requires the measurement of a calibration factor which accounts for the effects of the remaining factors (Zhao W. et al., 2018).

The calibration factor, k, is divided into two components, the sensitivity of the collimator and detector for the system used for the imaging and the protocol factors, including residual errors resulting from the image processing, compensation methods, and scaling factors (Anizan N. et al., 2014). The sensitivity, S, is expressed in counts per unit activity, per unit time, and varies over time due to changes in the gain of the detector system. It is easy to be measured and it is calibrated frequently (Saha G. B., 2001). The protocol specific factor or the reconstruction factor, R, is more difficult to be measured but it tends to be less dependent on time (Equation 7.10). R is expressed in counts per SPECT voxel value.

$$k = \frac{R}{s} \tag{7.10}$$

A calibration source is usually used for the measurement of the sensitivity of the collimator detector system. The calibration source has to be small enough, in order to avoid the attenuation contribution, with a small activity, to avoid count rate loss problems, but large enough, to provide a count rate higher than the natural background (Frey E. C. et al., 2012). The sensitivity is measured for the same values of the collimator, radionuclide, and energy window settings, which are used for the study to be quantified, at least daily and ideally for each patient (Cachovan M. et al., 2013). Furthermore, the sensitivity depends on the distance from the collimator, when the septal penetration and sensitivity are non-negligible. For this reason, the distance from the collimator to the detector calibration source is very important to be standardized (Van Audenhaege K. et al., 2015). The source is suspended between the collimators, in order to reduce the effects of attenuation. It is of crucial importance to acquire an image of the natural background. The source is typically suspended at a distance from the collimator approximately equal to the average radius of rotation (IAEA, 2014). The measurement for the natural background should be acquired for the same acquisition time and subtracted from the acquired image. For the selection of ROI over the source, it is important to cover and include all the geometrically collimated photons, excluding as many photons undergoing septal penetration and scatter as possible (IAEA, 2014). The sensitivity is calculated, as the counts inside the ROI in the background, corrected calibration source image, divided by the acquisition time and the activity of the source (Equation 7.11).

$$s = \frac{C - B}{t} \tag{7.11}$$

where t is duration of the acquisition, C goes for the counts in the ROI in the image of the standard source, and B is the number of counts in the ROI for the background image.

7.5.1.4.3   PET

In PET or PET/CT, the calibration process gives the sensitivity of the scanner that is used to convert the count images into an activity image. The efficiency of the variable detector, since there are different responses of the crystals that compose a PET machine is compensated for through a normalization procedure (Lockhart C. M. et al., 2011). The variable photomultiplier tube responses and the geometric effects are compensated, too. The calibration procedure is used to create a normalization map, which is then incorporated directly into the reconstruction process (Bedard N. et al., 2012).

7.5.2   PATIENT FACTORS – PHYSIOLOGY AND FUNCTION-ANATOMY

7.5.2.1   The Effect of the Motion

7.5.2.1.1   Planar Images

Spatial resolution can be affected by the patient motion and for this reason some devices, such as straps and comfortable pillows, have been developed in order to aid the patient in remaining. Even with careful repositioning, incorrect alignment during the different imaging sessions will occur and it affects the dosimetry calculations. Image registration or separate definition of ROIs for each of the time point measurements can mitigate the problem (Brock K. K. et al., 2017). It is important that the patient remains in the same position for the transmission scan, if one and only transmission study is performed, because the correction is usually undergone on a pixel by pixel basis (Pereira J. M. et al., 2010). Otherwise, the transmission scan should be registered to the whole-body measurement by appropriate software. It is possible to hold a transmission study for each time point, but this method may increase the absorbed dose.

7.5.2.1.2   SPECT

Motion is actually one of the most important factors degrading SPECT images. The effects tend to be important for small objects, like tumors, but less important for large objects, such as organs (Rahmim A. et al., 2008).

The beating of the heart generally causes low blurring or a similar size to CDR blurring. So, this is not considered as a major effect (Dorbala S. et al., 2018). These effects can be reduced by cardiac gating, but this type of images generally has higher noise. Respiratory motion also causes artefacts and blurring, but respiratory gating is generally not available on SPECT systems (Dorbala S. et al., 2018). Voluntary patient motion is a significant source of artefacts in SPECT images, mainly due to the long acquisition times. The utilization of patient restraints is recommended to moderate these effects (van Dalen J. A. et al., 2007). Post-acquisition compensation for patient motion is commercially available. These methods have not been validated for quantitative applications and remain unreliable (Dutta J. et al., 2013).

7.5.2.1.3   PET

Physiological motion, such as the motion due to heart rhythm and respiratory function and voluntary motion, like the movements of the patient or coughing, during the scan can affect the quantitative accuracy of PET or PET/CT images (Seo Y. et al., 2008).

The impact of respiratory motion on the image of a tumor in PET is predictable, as it can result in an apparent increase in the volume of tumor and decrease in its uptake. Furthermore, respiratory motion can interfere with attenuation correction when an attenuation map obtained with an X-ray CT is used (Dorbala S. et al., 2018).

On the one hand, due to long acquisition duration in PET, compared to the respiratory period, PET images represent the averaged activity distribution and effects of attenuation on the whole respiratory cycle (Pépin A. et al., 2014). On the other hand, the CT volume is acquired at a specific time during the respiratory cycle, causing some mismatch between the boundaries of moving organs in the CT and PET images (Catana C., 2015). This misregistration can introduce errors in attenuation correction. Thus, accurate registration of CT and PET data before attenuation correction is obligatory (Wagenknecht G. et al., 2013).

Monitoring of the motion of the respiratory system during the PET data acquisition is necessary for the compensation of the respiratory motion. External devices can be used for this reason (Pépin A. et al., 2014). There is a variety of devices which are temperature-sensitive, measuring the temperature of the exhaled air. Others can evaluate any changes in inductance of the coil and can be worn around the patient's chest (Massaroni C. et al., 2019). Moreover, there are real-time position management systems that monitor the motion of the chest wall. Dynamic PET images acquired within a short period of sampling can alternatively be used for the extraction of the respiratory signal (Saw C. B. et al., 2007).

Respiratory gating is utilized for image acquisition. In this case, the acquired data are sorted as a function of the detection time during a respiratory cycle, forming a few frames, each of whom covering part of the respiratory cycle and including less motion blur than the typical in an ungated image (Hamill J. J. et al., 2008). Both phase and amplitude gating, separately or in combination, can be performed for respiratory gating. The phase gating requires the division of each respiratory cycle into a number of time intervals, which have equal duration (Jani S. S. et al., 2013). The amplitude gating is performed with the division of the respiration amplitude into a number of amplitude intervals with the same magnitude of the signal of the respiratory gating. This type of gating is considered to be preferable (Chang G. et al., 2010).

There are two suggested methods for the compensation of respiratory motion using respiratory gated data, the image-based and the raw data-based methods. The first category requires a registration algorithm, in order to realign the gated frames before summing them (Dawood M. et al., 2007). The raw data-based approaches need the data to be modified in order to account for motion between frames before or during the reconstruction. In both cases, rigid or non-rigid registration models can be performed (Özbay E. et al., 2019).

Four-dimensional (4D) reconstruction methods can also be applied for the reconstruction of the series of gated images, with the use of a single reconstruction of the data corresponding to all gates (Jin M. et al., 2009).

Furthermore, involuntary motion can affect the measured data, resulting in inconsistent projection data, because of the long duration of PET acquisition. Interestingly, there are some optical tracking systems which can track the involuntary motion in brain imaging (Catana C. et al., 2011).

7.6 MATLAB AS A TOOL IN NUCLEAR MEDICINE QUANTITATIVE IMAGING

MATLAB is often considered an excellent environment for fast algorithm development. Proper programming practices – vectorization, pre-allocation, and specialization – applications in MATLAB can run as fast as in C language (Bister M. et al., 2007). Fast implementations of bilinear interpolation, watershed segmentation, and volume rendering are among the high-performance processes of MATLAB.

Medical image processing is characterized by the problems of memory usage and low execution speed, which are compounded with ever-increasing sizes of data sets.

7.6.1 BILINEAR INTERPOLATION

Interpolation is a well-covered topic in digital image processing. Bilinear interpolation is typically used to zoom into a 2D image or for rendering, for display purposes. The bilinear interpolation formula of a point (x, y) is given by Equation 7.12:

$$L(x,y) = (\lceil x \rceil - x)(\lceil y \rceil - y)L(\lfloor x \rfloor, \lfloor y \rfloor) + (\lceil x \rceil - x)(y - \lfloor y \rfloor)L(\lfloor x \rfloor, \lceil y \rceil)$$

$$+ (x - \lfloor x \rfloor)(\lceil y \rceil - y)L(\lceil x \rceil, \lfloor y \rfloor) + (x - \lfloor x \rfloor)(y - \lfloor y \rfloor)L(\lceil x \rceil, \lfloor y \rfloor) \quad (7.12)$$

$$+ (x - \lfloor x \rfloor)(y - \lfloor y \rfloor)L(\lceil x \rceil, \lceil y \rceil)$$

The interpolation points are all the values (x, y)=(i/s, j/s) lying within the area of the image. s = scale of resizing and i and j are integers. The following code is used for 2D bilinear interpolation (Bister M. et al., 2007).

```
Function outim = bilinterp2D(inim, scale)
%specialization
[Nx Ny]=size(inim);
Nu=floor(scale*Nx);
Nv=floor(scale*Ny);
U=1:(Nx-1)/(Nu-1):Nx;
Tmpim=zeros(Nu, Ny, class(inim));         % pre-allocation
for i=1:Nu
ul=floor(u(i));
fract=u(i)-ul;
if fract
tmpim(I, :)=(1-fract)*inim(ul, :)+fract*inim(ul+1, :);
%vectorization
else
tmpim(I, :)=inim(u(i), :);
%vectorization
            end
end
v=1: (Ny-1)/(Nv-1):Ny;
outim=zeros (NU, Nv, class(inim));
 % pre-allocation
```

```
for j=1:Nv
vl=floor(v(j));
fract=v(j)-vl;
if fract
outim(:, j)=(1-fract)*tmpim(:, vl)+fract*tmpim(:, vl+1);
%vectorization
else
outim(:, j)=tmpim(:, v(j));
%vectorization
            end
end.
```

7.6.2 WATERSHED SEGMENTATION

Watershed algorithms are very useful, in all their implementations, such as when they are applied on gradient image for object delineation, on original image for definition of ROI, or on distance-transformed images for separation between convex components. They are also associated with filtering and/or merging or with multi-resolution implementation.

It is possible to define a watershed function that is very efficient, as shown in the following code (Bister M. et al., 2007).

```
function [segim, rootim, linkim]=watershed(inim)
%define augmented input image to deal with border problems
[Nx  Ny Nz] = size(inim);
Augim=zeros(Nx+2,  Ny+2, Nz+2, class(inim)) +
(max(inim(:))+1);
% pre-allocating
x=2:Nx+1;
y=2:Ny+1;
z=2:Nz+1;
augim (x, y, z)=inim;
%vectorization
%initialize minim and linkim
minim=inim;
%vectorization
linkim0=uint32(reshape(find(inim>(min(inim(:))-1)),
size(inim));
%vectorization
linkim=linkim0;                                    %vectorization
%look for steepestpath downward from each pixel
for i=-1:+1
  for j=-1:+1
    for k=-1:+1
    shitftim=augim(x+i, y+j, z+k);                  %vectorization
            ind=find(shiftim<minim);
            if length(ind)
            [u v w]=ind2sub([Nx Ny Nz], ind);
            minim(ind)=shiftim(ind);
%vectorization
```

```
        linkim(ind)=linkim0(sub2ind([Nx Ny Nz], u+i, v+j, w+k));
        %vectorization
        end
    end
  end
end
```

7.6.3 VOLUME RENDERING

Volume rendering is also very important in medical image processing. A Code for volume rendering is presented below (Bister M. et al., 2007).

```
Function outim=volrender (inim, x0, y0, z0, pan, tlt, foc, Np,
dist, ROIim)
[Nx Ny Nz]=size(inim);
outim=zeros(Np, Np, 'uint16');
  %pre-allocating
range=atan(13/foc)*180/pi;
step=range*2/(Np-1);
phi=tlt+(-range:step:range)';
%vectorization
the=pan+ (-range:step:range);
%vectorization
xstep=sind(phi)+cosd(the);
%vectorization
ystep=sind(phi)+sind(the);
%vectorization
zstep=cosd(phi)*ones(sizwe(the));
%vectorization
Nr=sqrt(z0^2+y0^2+z0^2)+sqrt(Nx^2+Ny^2+Nz^2);
Sc=max(1, 1+Nr *dist);
Inim=single(inim);
For r=5:Nr
  xf=x0+r+xstep;  x0=floor(xf);  x1=x0+1;
%vectorization
  yf=y0-r+ystep;  y0=floor(yf);  y1=y0+1;
%vectorization
  zf=z0-r+xstep;  z0=floor(zf);   z1=x0+1;
%vectorization
ind=find((x0>0)&(x1<Nx) & (y0>0)&(y1<=Ny)  & (z0>0)&(z1<=Nz));
if length(ind)
xf=xf(ind); x0=x0(ind);  x1=x1(ind);
%vectorization
yf=yf(ind); y0=y0(ind);  y1=y1(ind);
%vectorization
zf=zf(ind); z0=z0(ind);  z1=z1(ind);
%vectorization
ref=sub2ind([Nx Ny Nz], x0, y0, z0);
ref2=find(ROIim(ref));
xf=xf(ref2);   x0=x0(ref2);  x1=x1(red2);
```

```
%vectorization
yf=yf(ref2);    y0=y0(ref2);    y1=y1(red2);
%vectorization
zf=zf(ref2);    z0=z0(ref2);    z1=z1(red2);
%vectorization
ref=ref(ref2);
%vectorization
ind=ind(ref2);
%vectorization
NxNy=Nx*Ny;
val=(xf-x0).*(yf-y0).*(zf-z0)  .*  inim(ref+1+Nx+NxNy) + …
    (x1-xf).*(yf-y0).*(zf-z0)  .*  inim(ref+1+Nx+NxNy) + …
    (xf-x0).*(y1-yf).*(zf-z0)  .*  inim(ref+1    +NxNx) + …
    (x1-xf).*(y1-yf).*(zf-z0)  .*  inim(ref+1    +NxNy) + …
    (xf-x0).*(yf-y0).*(z1-zf)  .*  inim(ref+1+Nx      ) + …
    (x1-xf).*(yf-y0).*(z1-zf)  .*  inim(ref    +Nx    ) + …
    (xf-x0).*(y1-yf).*(z1-zf)  .*  inim(ref+1        ) + …
    (x1-xf).*(y1-yf).*(z1-zf)  .*  inim(ref            );
%vectorization
    outim(ind) =max(outim(ind), uint16(val*sc/(1+r*dist)));
%vectorization
    end
end
```

7.6.4 QUANTITATION OF VOLUME DIFFERENCES

The quantification of the volume differences, which are detected in medical images and are related to different radiopharmaceutical uptake, can provide supplemental information in clinical diagnosis, since it allows a primary classification of the grades of some diseases (European Society of Radiology (ESR), 2015).

A characteristic example is the quantitation of volume differences in 3D SPECT stress/rest images in Myocardium Perfusion Imaging. The DICOM file, for each patient and each phase can be imported to MATLAB (Lyra M., 2009). Then an Index of Quantification (IQ) can be determined to define the global quantitative defect size as a fraction of the myocardial volume area in 3D images that could give confidence in cardiac perfusion efficiency recognition by SPECT (Dorbala S. et al., 2018). The observed 3D images irregularities give visual and quantitative evaluation of myocardium perfusion and the 3D volume visualization increases the reliability of myocardium perfusion diagnosis by SPECT imaging (Lyra M. et al., 2014). The difference relating to the rest and stress data of the 3D images, in voxels, can be calculated, using MATLAB image processing analysis and the quantification and analysis of differences can provide the desirable categorization (Xiong G. et al., 2015). The IQ, uses the myocardium VOI at rest and at stress, considering that there is an indicative diagnostic factor. The index is ranging from normal to maximal severity (actually allows categorization as normal, reversible, and irreversible ischemia (Hung G.-U., 2013).

Respectively, a similar approach can be achieved in order to classify different grades of tremor, Parkinsonism, and Parkinson's disease.

REFERENCES

Anizan N., Wang H., Zhou X.C., Hobbs R.F., Wahl R.L., Frey E.C. (2014) Factors affecting the stability and repeatability of gamma camera calibration for quantitative imaging applications based on a retrospective review of clinical data, EJNMMI Res, 4: 67.

Aravind V.K., Krishnaram V.D., Thasneem Z. (2012) Boundary crossings and violations in clinical settings, Indian J Psychol Med, 34 (1): 21–24.

Bai B., Bading J., Conti P. S. (2013) Tumor quantification in clinical positron emission tomography, Theranostics, 3 (10): 787–801.

Bailey D.L., Willowson K. P. (2013) An evidence-based review of quantitative SPECT imaging and potential clinical applications, J Nucl Med, 54 (1): 83–89.

Bedard N., Hagen N., Gao L., Tkaczyk T.S. (2012) Image mapping spectrometry: calibration and characterization, Opt Eng, 51 (11): 111711.

Berker Y., Li Y. (2016) Attenuation correction in emission tomography using the emission data—a review, Med Phys, 43 (2): 807–832.

Bharwani N., Koh D.M. (2013) Diffusion-weighted imaging of the liver: an update, Cancer Imaging, 13 (2): 171–185.

Bister M., Yap C., Ng Kh., Tok Ch. (2007) Increasing the speed of medical image processing in MATLAB, Biomed Imaging Interv J, 3 (1): e9.

Bowen S.L., Fuin N., Levine M.A., Catana C. (2016) Transmission imaging for integrated PET-MR systems, Phys Med Biol, 61 (15): 5547–5568.

Brambilla M., Lecchi M., Matheoud R., Leva L., Lucignani G., Marcassa C., Zoccarato O. (2017) Comparative analysis of iterative reconstruction algorithms with resolution recovery and new solid state cameras dedicated to myocardial perfusion imaging, Phys Med, 41: 109–116.

Brock K. K., Mutic S., McNutt T. R., Li H., Kessler M. L. (2017) Use of image registration and fusion algorithms and techniques in radiotherapy: report of the AAPM Radiation Therapy Committee Task Group No. 132, Med Phys, 44: e43–e76.

Byrd D. W., Doot R. K., Allberg K. C., MacDonald L. R., McDougald W. A., Elston B. F., Linden H. M., Kinahan P. E. (2016) Evaluation of cross-calibrated 68Ge/68Ga phantoms for assessing PET/CT measurement bias in oncology imaging for single- and multicenter trials, Tomography, 2 (4): 353–360.

Cachovan M., Vija A. H., Hornegger J., Kuwert T. (2013) Quantification of 99mTc-DPD concentration in the lumbar spine with SPECT/CT, EJNMMI Research, 3 (1): 45.

Catana C. (2015) Motion correction options in PET/MRI, Semin Nucl Med, 45 (3): 212–223.

Catana C., Benner T., van der Kouwe A., Byars L., Hamm M., Chonde D. B., Michel C. J., El Fakhri G., Schmand M., Sorensen A. G. (2011) MRI-assisted PET motion correction for neurologic studies in an integrated MR-PET scanner, J Nucl Med, 52 (1): 154–161.

Chang G., Chang T., Clark J. W. Jr, Mawlawi O. R. (2010) Design and performance of a respiratory amplitude gating device for PET/CT imaging, Med Phys, 37 (4): 1408–1412.

Cheimariotis G.-A., Al-Mashat M., Haris K., Aletras A. H., Jögi J., Bajc M., Maglaveras N., Heiberg E. (2018) Automatic lung segmentation in functional SPECT images using active shape models trained on reference lung shapes from CT, Ann Nucl Med, 32 (2): 94–104.

Chiesa C., Negri A., Albertini C., Azzeroni R., Setti E., Mainardi L., Aliberti G., Seregni E., Bombardieri E. (2009) A practical dead time correction method in planar activity quantification for dosimetry during radionuclide therapy, Q J Nucl Med Mol Imaging, 53: 658–670.

Chinn G., Levin C. S. (2011) A maximum NEC criterion for Compton collimation to accurately identify true coincidences in PET, IEEE Trans Med Imaging, 30 (7): 1341–1352.

Cho S., Ahn S., Li Q., Leahy R. M. (2009) Exact and approximate Fourier rebinning of PET data from time-of-flight to non time-of-flight, Phys Med Biol, 54 (3): 467–484.

Chow P. L., Rannou F. R., Chatziioannou A. F. (2005) Attenuation correction for small animal PET tomographs, Phys Med Biol, 50 (8): 1837–1850.

Chun S. Y., Fessler J. A., Dewaraja Y. K. (2013) Correction for collimator-detector response in SPECT using point spread function template, IEEE Trans Med Imaging, 32 (2): 295–305.

Cremer P., Hachamovitch R., Tamarappoo B. (2014) Clinical decision making with myocardial perfusion imaging in patients with known or suspected coronary artery disease, Semin Nucl Med, 44 (4): 320–329.

David S., Visvikis D., Roux C., Hatt M. (2011) Multi-observation PET image analysis for patient follow-up quantitation and therapy assessment, Phys Med Biol, 56 (18): 5771–5788.

Dawood M., Büther F., Lang N., Schober O., Schäfers K.P. (2007) Respiratory gating in positron emission tomography: a quantitative comparison of different gating schemes, Med Phys, 34 (7): 3067–3076.

Deng X., Gao N., Zhang Z. (2019) A calibration method for system parameters in direct phase measuring deflectometry, Appl Sci, 9: 1444.

Dewaraja Y. K., Frey E. C., Sgouros G., Brill A. B., Roberson P., Zanzonico P. B., Ljungberg M. (2012) MIRD Pamphlet No. 23: quantitative SPECT for patient-specific 3-dimensional dosimetry in internal radionuclide therapy, J Nucl Med, 53 (8): 1310–1325.

Dewaraja Y. K., Koral K. F., Fessler J. A. (2010) Regularized reconstruction in quantitative SPECT using CT side information from hybrid imaging, Phys Med Biol, 55 (9): 2523–2539.

Dewaraja Y. K., Ljungberg M., Fessler J. A. (2006) 3-D Monte Carlo-based scatter compensation in quantitative I-131 SPECT reconstruction, IEEE Trans Nucl Sci, 53 (1): 181.

Dickson J., Ross J., Vöö S. (2019) Quantitative SPECT: the time is now, EJNMMI Phys, 6: 4.

Dorbala S., Ananthasubramaniam K., Armstrong I. S., Chareonthaitawee P., Gordon DePuey E., Einstein A. J., Gropler R. J., Holly T. A., Mahmarian J. J., Park M.-A., Polk D. M., Russell R., Slomka P. J., Thompson R. C., Glenn Wells R. (2018) Single photon emission computed tomography (SPECT) myocardial perfusion imaging guidelines: instrumentation, acquisition, processing, and interpretation, J Nucl Cardiol, 25 (5): 1784.

Dutta J., Ahn S., Li Q. (2013) Quantitative statistical methods for image quality assessment, Theranostics, 3 (10): 741–756.

Erickson B. J., Korfiatis P., Akkus Z., Kline T. L. (2017) Machine learning for medical imaging, Radiographics, 37 (2): 505–515.

Erlandsson K., Buvat I., Pretorius P. H., Thomas B. A., Hutton B. F. (2012) A review of partial volume correction techniques for emission tomography and their applications in neurology, cardiology and oncology, Phys Med Biol, 57: R119–R159.

European Society of Radiology (ESR) (2015) Medical imaging in personalised medicine: a white paper of the research committee of the European Society of Radiology (ESR), Insights Imaging, 6 (2): 141–155.

Finocchiaro D., Berenato S., Grassi E., Bertolini V., Castellani G., Lanconelli N., Versari A., Spezi E., Iori M., Fioroni F. (2019) Partial volume effect of SPECT images in PRRT with 177Lu labelled somatostatin analogues: a practical solution, Physica Medica, 57: 153–159.

Foster B., Bagci U., Mansoor A., Xu Z., Mollura D.J. (2014) A review on segmentation of positron emission tomography images, Comput Biol Med, 50: 76–96.

Franceschini E., Yu F. T.H., Cloutier G. (2008) Simultaneous estimation of attenuation and structure parameters of aggregated red blood cells from backscatter measurements, J Acoust Soc Am, 123 (4): 85–91.

Frey E., Tsui B. (2006) Collimator-detector response compensation in SPECT, Quantitative Analysis of Nuclear Medicine Images, New York: Springer.

Frey E. C., Humm J. L., Ljungberg M. (2012) Accuracy and precision of radioactivity quantification in nuclear medicine images, Semin Nucl Med, 42 (3): 208–218.

Geng J. (2013) Three-dimensional display technologies, Adv Opt Photonics, 5 (4): 456–535.

Ghaly M., Links J. M., Frey E. C. (2015) Optimization of energy window and evaluation of scatter compensation methods in myocardial perfusion SPECT using the ideal observer with and without model mismatch and an anthropomorphic model observer, J Med Imaging (Bellingham), 2 (1): 015502.

Ghaly M., Links J. M., Frey E. C. (2016) Collimator optimization and collimator-detector response compensation in myocardial perfusion SPECT using the ideal observer with and without model mismatch and an anthropomorphic model observer, Phys Med Biol, 61 (5): 2109–2123.

Greve D.N., Salat D. H., Bowen S. L., Izquierdo-Garcia D., Schultz A. P., Catana C., Becker J. A., Svarer C., Knudsen G. M., Sperling R. A., Johnson K. A. (2016) Different partial volume correction methods lead to different conclusions: an (18)F-FDG-PET study of aging, Neuroimage, 132: 334–343.

Griffis D., Shivade C., Fosler-Lussier E., Lai A. M. (2016) A quantitative and qualitative evaluation of sentence boundary detection for the clinical domain, AMIA Jt Summits Transl Sci Proc, 2016: 88–97.

Guérin B., El Fakhri G. (2011) Novel scatter compensation of list-mode PET data using spatial and energy dependent corrections, IEEE Trans Med Imaging, 30 (3): 759–773.

Gullberg G. T., Reutter B. W., Sitek A., Maltz J. S., Budinger T. F. (2010) Dynamic single photon emission computed tomography—basic principles and cardiac applications, Phys Med Biol, 55 (20): R111–R191.

Hallen P., Schug D., Weissler B., Gebhardt P., Salomon A., Kiessling F., Schulz V. (2018) PET performance evaluation of the small-animal Hyperion IID PET/MRI insert based on the NEMA NU-4 standard, Biomed Phys Eng Express, 4 (6): 065027.

Hamill J. J., Bosmans G., Dekker A. (2008) Respiratory-gated CT as a tool for the simulation of breathing artifacts in PET and PET/CT, Med Phys, 35 (2): 576–585.

Hatt M., Lee J. A., Schmidtlein C. R., Naqa I. E., Caldwell C., De Bernardi E., Lu W., Das S., Geets X., Gregoire V., Jeraj R., MacManus M. P., Mawlawi O. R., Nestle U., Pugachev A. B., Schöder H., Shepherd T., Spezi E., Visvikis D., Zaidi H., Kirov A. S. (2017) Classification and evaluation strategies of auto-segmentation approaches for PET: report of AAPM task group No. 211, Med Phys, 44 (6): e1–e42.

He B., Du Y., Song X. Y., Segars W. P., Frey E. C. (2005) A Monte Carlo and physical phantom evaluation of quantitative In-111 SPECT, Phys Med Biol, 50: 4169–4185.

He B., Frey E.C. (2006) Comparison of conventional, model-based quantitative planar, and quantitative SPECT image processing methods for organ activity estimation using In-111 agents, Phys Med Biol, 51: 3967–3981.

Hobbs R. F., Baechler S., Senthamizhchelvan S., Prideaux A. R., Esaias C. E., Reinhardt M., Frey E. C., Loeb D. M., Sgouros G. (2010) A gamma camera count rate saturation correction method for whole-body planar imaging, Phys Med Biol, 55: 817–831.

Hung G.-U. (2013) Diagnosing CAD: additional markers from myocardial perfusion SPECT, J Biomed Res, 27 (6): 467–477.

IAEA (2014) Human Health Reports No. 9, Quantitative Nuclear Medicine Imaging: Concepts, Requirements and Methods, IAEA Library Cataloguing in Publication Data, VIENNA.

Jani S. S., Robinson C. G., Dahlbom M., White B. M., Thomas D. H., Gaudio S., Low D. A., Lamb J. M. (2013) A comparison of amplitude-based and phase-based positron emission tomography gating algorithms for segmentation of internal target volumes of tumors subject to respiratory motion, Int J Radiat Oncol Biol Phys, 87 (3): 562–569.

Jin M., Yang Y., Niu X., Marin T., Brankov J. G., Feng B., Pretorius P. H., King M. A., Wernick M. N. (2009) A quantitative evaluation study of four-dimensional gated cardiac SPECT reconstruction, Phys Med Biol, 54 (18): 5643–5659.

Jolliffe I. T., Cadima J. (2016) Principal component analysis: a review and recent developments, Philos Trans A Math Phys Eng Sci, 374 (2065): 20150202.

Kappadath S. C. (2011) Effects of voxel size and iterative reconstruction parameters on the spatial resolution of 99mTc SPECT/CT, J Appl Clin Med Phys, 12 (4): 3459.

Katua A. M., Ankrah A. O., Vorster M., van Gelder A., Sathekge M. M. (2011) Optimization of ordered subset expectation maximization reconstruction for reducing urinary bladder artifacts in single-photon emission computed tomography imaging, World J Nucl Med, 10 (1): 3–8.

Kheruka S. C., Naithani U. C., Maurya A. K., Painuly N. K., Aggarwal L. M., Gambhir S. (2011) A study to improve the image quality in low-dose computed tomography (SPECT) using filtration, Indian J Nucl Med, 26 (1): 14–21.

Kinahan P. E., Byrd D. W., Helba B., Wangerin K. A., Liu X., Levy J. R., Allberg K. C., Krishnan K., Avila R. S. 2018, Simultaneous estimation of bias and resolution in PET images with a long-lived "pocket" phantom system, Tomography, 4 (1): 33–41.

Kinahan P. E., Fletcher W. (2010) PET/CT standardized uptake values (SUVs) in clinical practice and assessing response to therapy, Semin Ultrasound CT MR, 31 (6): 496–505.

Kunyansky L.A. (2001) A new SPECT reconstruction algorithm based on the Novikov explicit inversion formula, Inverse Probl, 17: 293.

Larsson A. (2005) Corrections for improved quantitative accuracy in SPECT and planar scintigraphic imaging, Department of Radiation Sciences, Radiation Physics Umeå University, Sweden. ISBN: 91-7305-938-2.

Lee H.-H., Chen J.-C. (2013) Investigation of attenuation correction for small-animal single photon emission computed tomography, Comput Math Methods Med, 2013: 430276.

Lee Z., Ljungberg M., Muzic Jr. R. F., Berridge M. S. (2016) Usefulness and pitfalls of planar g-scintigraphy for measuring aerosol deposition in the lungs: a Monte Carlo investigation, J Nucl Med, 42: 1077–1083.

Li T., Ao E. C. I., Lambert B., Brans B., Vandenberghe S., Mok G. S. P. (2017) Quantitative imaging for targeted radionuclide therapy dosimetry – technical review, Theranostics, 7 (18): 4551–4565.

Liu A., Williams L. E., Raubitschek A. A. (1996) A CT assisted method for absolute quantitation of internal radioactivity, Med Phys, 23: 1919–1928.

Ljungberg M., Sjögreen G. K. (2016) Personalized dosimetry for radionuclide therapy using molecular imaging tools, Biomedicines, 4 (4): 25.

Lockhart C. M., MacDonald L. R., Alessio A. M., McDougald W. A., Doot R. K., Kinahan P. E. (2011) Quantifying and reducing the effect of calibration error on variability of PET/CT standardized uptake value measurements, J Nucl Med, 52 (2): 218–224.

Lyra M. (2009) Single Photon Emission Tomography (SPECT) and 3D Images Evaluation in Nuclear Medicine, In: Chen Y-S (Ed). Croatia: Image Processing, INTECH.

Lyra M., Ploussi A. (2011) Filtering in SPECT image reconstruction, Int J Biomed Imaging, 2011: 693795.

Lyra M., Ploussi A, Rouchota M, Synefia S. (2014) Filters in 2D and 3D cardiac SPECT image processing, Cardiol Res Pract, 2014: 963264.

Marin G., Vanderlinden B., Karfis I., Guiot T., Wimana Z., Reynaert N., Vandenberghe S., Flamen P. (2018) A dosimetry procedure for organs-at-risk in ^{177}Lu peptide receptor radionuclide therapy of patients with neuroendocrine tumours, Physica Medica, 56, 41–49.

Martin C. J. (2007) Optimisation in general radiography, Biomed Imaging Interv J, 3 (2): e18.

Massaroni C., Nicolò A., Lo Presti D., Sacchetti M., Silvestri S., Schena E. (2019) Contact-based methods for measuring respiratory rate, Sensors (Basel), 19 (4): 908.

Matej S., Li Y., Panetta J., Karp J. S., Surti S. (2016) Image-based modeling of PSF deformation with application to limited angle PET data, IEEE Trans Nucl Sci, 63 (5): 2599–2606.

McGoron A. J., Capille M., Georgiou M. F, Sanchez P., Solano J., Gonzalez-Brito M., Kuluz J. W. (2008) Post traumatic brain perfusion SPECT analysis using reconstructed ROI maps of radioactive microsphere derived cerebral blood flow and statistical parametric mapping, BMC Med Imaging, 8: 4.

Minarik D., Sjögreen K., Ljungberg M. (2005) A new method to obtain transmission images for planar whole-body activity quantification, Cancer Biother Radio, 20: 72–76.

Mollet P., Keereman V., Bini J., Izquierdo-Garcia D., Fayad Z. A., Vandenberghe S. (2014) Improvement of attenuation correction in time-of-flight PET/MR imaging with a positron-emitting source, J Nucl Med, 55 (2): 329–336.

Momennezhad M., Nasseri S., Zakavi S. R., Parach A. A., Ghorbani M., Ghahraman Asl R. (2016) A 3D Monte Carlo method for estimation of patient-specific internal organs absorbed dose for 99mTc-hynic-Tyr3-octreotide imaging, World J Nucl Med, 15 (2): 114–123.

Morin O., Vallières M., Jochems A., Woodruff H. C., Valdes G., Braunstein S. E., Wildberger J. E., Villanueva-Meyer J. E., Kearney V., Yom S. S., Solberg T. D., Lambin P. (2018) A deep look into the future of quantitative imaging in oncology: a statement of working principles and proposal for change, IJROBP, 102 (4): 1074–1082.

Noller C. M., Boulina M., McNamara G., Szeto A., McCabe P. M., Mendez A. J. (2016) A practical approach to quantitative processing and analysis of small biological structures by fluorescent imaging, J Biomol Tech, 27 (3): 90–97.

Novikov R. (2002) An inversion formula for the attenuated X-ray transformation, Ark Mat, 40: 145–167.

Nuvoli S., Palumbo B., Malaspina S., Madeddu G., Spanu A. (2018) 123 I-ioflupane SPET and I-MIBG in the diagnosis of Parkinson's disease and Parkinsonian disorders and in the differential diagnosis between Alzheimer's and Lewy's bodies dementias, Hell J Nucl Med, 21 (1): 60–68.

Nuyts J., De Man B., Fessler J. A., Zbijewski W., Beekman F. J. (2013) Modelling the physics in iterative reconstruction for transmission computed tomography, Phys Med Biol, 58 (12): R63–R96.

Oloomi S., Eskandari H. N., Zakavi S. R., Knoll P., Kalantari F., Saffar M. H. (2013) A new approach for scatter removal and attenuation compensation from SPECT/CT images, Iran J Basic Med Sci, 16 (11): 1181–1189.

Özbay E., Çinar A. (2019) A voxelize structured refinement method for registration of point clouds from Kinect sensors, JESTECH, 22 (2), 555–568.

Papanastasiou E., Moralidis E., Siountas A. (2017) The effect of scatter correction on planar and tomographic 123 semiquantitative I cardiac imaging. A phantom study, Hell J Nucl Med, 20 (2): 154–159.

Pépin A., Daouk J., Bailly P., Hapdey S., Meyer M. E. (2014) Management of respiratory motion in PET/computed tomography: the state of the art. Nucl Med Commun, 35 (2): 113–122.

Pereira J. M., Stabin M. G., Lima F. R. A., Guimarães M. I. C. C., Forrester J. W. (2010) Image quantification for radiation dose calculations – limitations and uncertainties, Health Phys, 99 (5): 688–701.

Popescu L. M., Lewitt R. M., Matej S., Karp J. S. (2006) PET energy-based scatter estimation and image reconstruction with energy-dependent corrections, Phys Med Biol, 51 (2006): 2919–2937.

Pramanik M. (2014) Improving tangential resolution with a modified delay-and-sum reconstruction algorithm in photoacoustic and thermoacoustic tomography, J Opt Soc Am A Opt Image Sci Vis, 31 (3): 621–627.

Pretorius P.H., King M.A. (2009) Diminishing the impact of the partial volume effect in cardiac SPECT perfusion imaging, Med Phys, 36 (1): 105–115.

Rahmim A., Zaidi H. (2008) PET versus SPECT: strengths, limitations and challenges, Nucl Med Commun, 29: 193–207.

Saffar M. H., Oloomi S., Knoll P., Taleshi H. (2013) A new approach to scatter correction in SPECT images based on Klein_Nishina equation, Iran J Nucl Med, 21 (1): 19–25.

Saha G. B. (2001) Performance parameters of gamma cameras, In: Physics and Radiobiology of Nuclear Medicine, New York, NY: Springer.

Santarelli M. F., Positano V., Landini L. (2017) Measured PET data characterization with the negative binomial distribution model, J Med Biol Eng, 37 (3): 299–312.

Saw C. B., Brandner E., Selvaraj R., Chen H., Saiful Huq M., Heron D. E. (2007) A review on the clinical implementation of respiratory-gated radiation therapy, Biomed Imaging Interv J, 3 (1): e40.

Sciagrà R. (2012) Quantitative cardiac positron emission tomography: the time is coming!, Scientifica (Cairo), 2012: 948653.

Seo Y., Aparici C. M., Hasegawa B. H. (2008) Technological development and advances in SPECT/CT, Semin Nucl Med, 38 (3): 177–198.

Shen D., Wu G., Suk H.-I. (2017) Deep learning in medical image analysis, Annu Rev Biomed Eng, 19: 221–248.

Shukla A.K., Kumar U. (2006) Positron emission tomography: an overview, J Med Phys, 31 (1): 13–21.

Sjögreen K., Ljungberg M. (2002) An activity quantification method based on registration of CT and whole-body scintillation camera images, with application to 131I, J Nucl Med, 43: 972–982.

Sorenson J. A. (1971) Methods for Quantitation of Radioactivity in Vivo by External Counting Measurements, University of Wisconsin, Madison, WI.

Sorenson J. A., Phelps M. E. (1987) Chapter 16, WeB, In: Physics in Nuclear Medicine, 2nd edition, Saunders Company, Harcourt Brace Jovanovich, Inc.

Spratt H., Ju H., Brasier A. R. (2013) A structured approach to predictive modeling of a two-class problem using multidimensional data sets, Methods, 61 (1): 73–85.

Sun Y., Davis P., Kosmacek E. A., Ianzini F., Mackey M. A. (2009) An open-source deconvolution software package for 3-D quantitative fluorescence microscopy imaging, J Microsc, 236 (3): 180–193.

Takabe S., Hukushima K. (2016) Typical performance of approximation algorithms for NP-hard problems, J Stat Mech, 2016: 113401.

Taylor A. T. (2014) Radionuclides in nephrourology, Part 2: pitfalls and diagnostic applications, J Nucl Med, 55 (5): 786–798.

Teymurazyan A., Riauka T., Jans H. S., Robinson D. (2013) Properties of noise in positron emission tomography images reconstructed with filtered-backprojection and row-action maximum likelihood algorithm, J Digit Imaging, 26 (3): 447–456.

Thorwarth D. (2015) Functional imaging for radiotherapy treatment planning: current status and future directions—a review, Br J Radiol, 88 (1051): 20150056.

Tong S., Alessio A. M., Kinahan P. E. (2010) Image reconstruction for PET/CT scanners: past achievements and future challenges, Imaging Med, 2 (5): 529–545.

Van Audenhaege K., Van Holen R., Vandenberghe S., Vanhove C. (2015) Review of SPECT collimator selection, optimization, and fabrication for clinical and preclinical imaging, Med Phys, 42 (8): 4796–4813.

van Dalen J. A., Vogel W. V., Corstens F. H.M., Oyen W. J.G. (2007) Multi-modality nuclear medicine imaging: artefacts, pitfalls and recommendations, Cancer Imaging, 7 (1): 77–83.

Van den Wyngaert T., Strobel K., Kampen W. U., Kuwert T., van der Bruggen W., Mohan H. K., Gnanasegaran G., Delgado-Bolton R., Weber W. A., Beheshti M., Langsteger W., Giammarile F., Mottaghy F. M., Paycha F. (2016) The EANM practice guidelines for bone scintigraphy, European J Nucl Med Mol Imaging, 43 (9): 1723.

Vaquero J. J., Kinahan P. (2015) Positron emission tomography: current challenges and opportunities for technological advances in clinical and preclinical imaging systems, Annu Rev Biomed Eng, 17: 385–414.

Vicente E. M., Lodge M. A., Rowe S. P., Wahl R. L., Frey E. C. (2017) Simplifying volumes-of-interest (VOIs) definition in quantitative SPECT: beyond manual definition of 3D whole-organ VOIs, Med Phys, 44 (5): 1707–1717.

Wagenknecht G., Kaiser H. J., Mottaghy F. M., Herzog H. (2013) MRI for attenuation correction in PET: methods and challenges, MAGMA, 26 (1): 99–113.

Wang H. (2012) An MR image-guided, voxel-based partial volume correction method for PET images, Med Phys, 39 (1): 179–194.

Wang Y. Y., Wang K., Xu Z. Y., Song Y., Wang C. N., Zhang C. Q., Sun X. L., Shen B. Z. (2017) High-resolution dynamic imaging and quantitative analysis of lung cancer xenografts in nude mice using clinical PET/CT, Oncotarget, 8 (32): 52802–52812.

Wu G., Inscoe C. R., Calliste J., Shan J., Lee Y. Z., Zhou O., Lua J. (2017) Estimating scatter from sparsely measured primary signal, J Med Imaging (Bellingham), 4 (1): 013508.

Xiong G., Kola D., Heo R., Elmore K., Cho I., Min J. K. (2015) Myocardial perfusion analysis in cardiac computed tomography angiographic images at rest, Med Image Anal, 24 (1): 77–89.

Yan J., Schaefferkoette J., Conti M., Townsend D. (2016) A method to assess image quality for low-dose PET: analysis of SNR, CNR, bias and image noise, Cancer Imaging, 16 (1): 26.

Yan Y., Zeng G. L. (2009) Attenuation map estimation with SPECT emission data only, Int J Imaging Syst Technol, 19 (3): 271–276.

Yu H., Chen Z., Zhang H., Wong K. K. L., Chen Y., Liu H. (2015) Reconstruction for 3D PET based on total variation constrained direct Fourier method, PLOS ONE, 10 (9): e0138483.

Zaidi H. (2006) Quantitative Analysis in Nuclear Medicine Imaging, Springer, Germany.

Zaidi H., Hasegawa B. (2003) Determination of the attenuation map in emission tomography, J Nucl Med, 44 (2): 291–315.

Zarinabad N., Chiribiri A., Hautvast G. L. T. F., Breeuwer M., Nagel E. (2015) Influence of spatial resolution on the accuracy of quantitative myocardial perfusion in first pass stress perfusion CMR, Magn Reson Med, 73 (4): 1623–1631.

Zhao W., Esquinas P. L., Hou X., Uribe C.F., Gonzalez M., Beauregard J.-M., Dewaraja Y. K., Celler A. (2018) Determination of gamma camera calibration factors for quantitation of therapeutic radioisotopes, EJNMMI Phys, 5: 8.

Zhu A., Lee D., Shim H. (2011) Metabolic PET imaging in cancer detection and therapy response, Semin Oncol, 38 (1): 55–69.

Zimmerman B. E., Grošev D., Buvat I., Coca Pérez M. A., Frey E. C., Green A., Krisanachinda A., Lassmann M., Ljungberg M., Pozzo L., Afroj Quadir K., Terán Gretter M. A., Van Staden J., Poli G.L. (2017) Multi-centre evaluation of accuracy and reproducibility of planar and SPECT image quantification: an IAEA phantom study, Z Med Phys, 27 (2): 98–112.

8 Quality Control of Nuclear Medicine Equipment

Maria Argyrou

CONTENTS

8.1 INTRODUCTION

The basic components of a γ-camera are a collimator, a scintillation crystal, photomultiplier tubes, and electronic circuitries to sort out and localize the signal for display. A computer interface and an image display system are typically associated with the γ-camera. Scintillation cameras are used as single-detector or multi-detector

223

systems as well as whole body systems. In response to the demands for better sensitivity and high-energy resolution, a new generation of solid state imaging has been developed, by replacing photomultiplier tubes with semiconductors.

PET camera systems consist of several full-ring detectors crystals organized in blocks with a reduced number of photomultiplier tubes in order to detect annihilation photons. This block design reduces the dead time compared with that of γ-cameras.

The material that detectors in γ-camera or PET systems are made off is of great importance. There is a growing interest in the development of new scintillator materials because of the increment in the number of medical or industrial applications (Khoshakhlagh M et al. 2015).

The most commonly used scintillator in nuclear medicine is sodium iodide activated with thallium, NaI(Tl). Pure NaI crystals are scintillators only at liquid nitrogen temperature. They become efficient scintillators at room temperatures with the addition of small amounts of thallium. NaI crystal is a good absorber, cost effective, and an efficient detector of photon energy in the range of 50–250 keV. However, it has some disadvantages: it is fragile, hygroscopic, and large volumes are required for photon energies greater than 250 keV.

At higher energies, in PET systems, denser scintillators generally are preferred. Bismuth germanate oxide, BGO, ($Bi_4Ge_3O_{12}$), a common scintillator in PET imaging, has a very good efficiency at 511 keV. Lutetium oxyorthosilicate, LSO, [$Lu_2SiO_5(Ce)$] has similar density and effective atomic number to BGO, but it is brighter and faster and thus preferred in the case of high count rates. However, it is rather expensive to grow, because of its high melting point and its raw material costs. Lanthanum bromide (LaBr) has been proposed for time-of-flight (TOF) clinical PET, because of its excellent energy resolution and very fast light decay.

Concerning solid state imaging systems the most recently proposed material is Cadmium Zinc Telluride (CZT). These detectors have the ability of detecting different photon energies, high spatial and energy resolution and very low decay time (Abbaspour S et al. 2017).

In general, cost issues and availability remain limiting factors to the widespread adoption of new scintillating materials.

SPECT and PET systems are frequently integrated with a Computed Tomography (CT) scanner so that SPECT or PET images are acquired along with spatially registered CT images, in quick succession. Both SPECT and PET provide functional information for imaged tissues whereas for clinical evaluation anatomical details are also required. For that reason CT and/or MRI scans are used to complement with emission tomography imaging. CT scan also provides a map of tissue attenuation values to compute the corrections for photon attenuation and scatter for PET and SPECT studies (Kinahan PE et al. 2003; Patton JA et al. 2008). Attenuation correction is critical for reconstructing quantitatively accurate SPECT or PET images. There are many clinical situations in which MRI, rather than CT, is the anatomic imaging technique of choice; therefore, hybrid systems PET/MRI and SPECT/MRI have been developed.

In hybrid systems, additional CT quality control (QC) procedures must be followed. Their description is beyond the scope of this chapter.

The limits presented for the described QC procedures are, in general, larger as compared with those routinely used because of the existence of different detector types and manufacturers specifications.

8.2   PLANAR γ-CAMERA QC

The performance of γ-camera is affected by several parameters including uniformity, sensitivity, and spatial resolution. QC procedures are described for these parameters and they are mainly based on (NEMA (2001), EANM Physics Committee (2010), and Zanzonico (2008).

8.2.1   VISUAL INSPECTION

Before any QC procedure, a visual inspection is required as it may reveal obvious defects which can be degrade the imaging efficacy. A visual inspection of collimators may reveal obvious defects, which may compromise the safety or the imaging efficacy of the system. Signs of denting or scratching may indicate mechanical damage to the collimator, and surface stains may be a sign of possible contamination. Both of these may produce artefacts such as cold or hot spots on planar images and rings on SPECT images. Inspection concerning other possible defects which compromise patient or staff safety (e.g. frayed or damaged electrical cables, mechanical faults in the camera, or scanning table) should be carried out before patient examination. If any other defects are detected, the equipment should not be used until it is established that it is safe to do so.

8.2.2   BACKGROUND LEVEL MEASUREMENTS

High or even moderate intensity background radiation is possible to compromise any type of imaging because it may seriously degrade intrinsic uniformity or other intrinsic measurements.

The background level measurement should be carried out daily with the same acquisition conditions (collimator, camera position, energy window) which are most frequently used for imaging but in the absence of radioactive sources.

The count rate must be compared with the reference value. The reference value has to be determined at installation as the mean value of multiple background measurements over some days.

Excess of background measurements may be indicative of a contamination or a presence of unsealed radiation source which should be eliminated before patient image acquisition. The actual background value should read between 50% and 200% of the reference value.

8.2.3   ENERGY RESOLUTION

Energy resolution refers to the ability of the detector to accurately determine the energy of the incoming radiation. The system can only determine within a range of values (energy window), what energy radiation it is detecting. Thus, the energy resolution is expressed as a percent of the energy of the incoming photons.

The energy window setting affects uniformity, sensitivity, and scatter contribution to the image. In particular, a large energy window degrades uniformity and sensitivity, whereas increases scattered radiation. In older γ-camera systems, the photopeak can change due to slight variations in high voltage, photomultiplier drift, changes in temperature, and other factors.

An important measurement to assess the efficiency of the scintillation γ-camera is the Full-Width at Half-Maximum (FWHM), which should typically be less than 10%. Energy resolution is expressed as a percent of the FWHM of a specific energy E_γ according to Equation (8.1):

$$\Delta E(\%) = \frac{FWHM}{E_\gamma} \times 100\% \qquad (8.1)$$

Peak settings should be checked daily, adjusted in a consistent manner and should be recorder to detect any long-term drift. Sudden changes in peak setting indicate a possible fault in the camera and should be fully investigated before clinical examinations. The energy window setting should be checked for all radionuclides used in a daily routine. If a change in the peak setting for one radionuclide is detected, it is likely that the settings for other radionuclides also need to be adjusted.

Peaking should be checked without collimator to reduce scatter and to ensure that an average peak for the whole field of view (FOV) is obtained. A point radioactive source is used giving a count rate not exceeding 20 kcps.

Acceptable discrepancies should be less than ±3% of window center; otherwise the camera must be recalibrated before imaging can proceed.

8.2.4 Sensitivity

The sensitivity of a γ-camera system depends on a variety of factors, such as the geometric efficiency of the collimator, the detection efficiency of the detector, PHA discriminator settings, and the dead time. A deviation of the sensitivity from the reference value indicates faults in the crystal, in the electrical camera units, or a degradation of the energy resolution.

To measure the sensitivity, a source with a γ-energy <200 keV should be used.

The collimator, the source-collimator distance, the geometrical conditions, and the energy window must be always identical.

If a Tc-99m source is used, the decay between the timings of the dose calibrator and camera measurements must be considered.

Using high-energy collimators, sensitivity of a point source depends on the position of the source at the collimator because of the septal thickness. Therefore, if sensitivity measurement is carried out using high-energy collimators, a surface source should be used instead of a point source.

The measured values of the sensitivity should not exceed ±5% of the reference values determined at installation.

8.2.5 Uniformity

An adequate image interpretation of nuclear medicine modality rely on the assumption that differences seen are due to differences in the patient only and not differences introduced by the γ-camera.

Uniformity is the ability of a γ-camera to accurately produce an image of the data it detects. It can be degraded by several factors, such as a failure of a multiplier tube, an inappropriate spatial linearity or energy correction.

Therefore, a thorough QC should be scheduled in a weekly basis to assure that nonuniformity issues on the γ-camera system are absent.

Intrinsic uniformity (that is when collimator is removed) can be checked simply by the use of point radioactive source suspended at a distance equal to five times the FOV (Figure 8.1). To detect gradual deterioration in uniformity, it is important that uniformity measurements are carried out in a consistent manner (e.g. same orientation, same number of counts, etc.) and records are kept to allow comparisons over periods of weeks or even months.

Regular analysis of uniformity by a computer can facilitate detection of gradual deterioration prior to any visible change. Two different uniformity parameters can be quantified: integral uniformity (IU) and differential uniformity (DU).

IU is a measure of the maximum pixel count deviation in the FOV and is given by Equation (8.2):

$$IU(\%) = 100\% \times \frac{Max - Min}{Max + Min} \tag{8.2}$$

where Max and Min denote the maximum and minimum counts per pixel in the FOV.

DU is a measure of the maximum deviation of a limited area designed to approximate the size of a photomultiplier tube, usually 5×5 pixels (see Equation 8.3):

$$DU(\%) = 100\% \times \frac{Max(area) - Min(area)}{Max(area) + Min(area)} \tag{8.3}$$

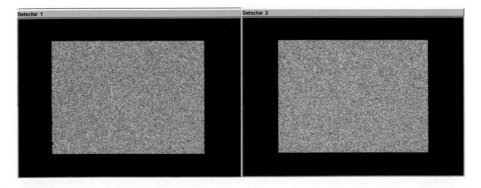

FIGURE 8.1 An intrinsic uniformity image of a dual-head γ-camera, using a Tc-99m point source. This is an example of a good uniformity, since a uniform distribution across the entire field of view, is shown. (From Athens Medical Center, Nuclear Medicine Department.)

The calculation should be done for the X- and the Y-directions independently and the maximum change represents the DU.

The uniformity check might be evaluated extrinsically using a flood source (usually Co^{57}) placed on the detector with a collimator attached.

The flood source must exceed the FOV on all sides by a minimum of 20 mm. The overall total thickness must be at least 80 mm.

The count rate should not exceed 20 kcps. A measuring time should be selected that guarantees a count density >5000 counts per cm^2. The camera should be peaked and a consistent set-up has to be used (e.g. same distance of source, same orientation, same radionuclide, same collimator).

- Significant background radiation from other sources should be eliminated.
- Pronounced nonuniformity in the image should be checked visually.
- Windowing may be used to highlight nonuniform areas if the study is collected on computer.
- The nonuniformity should not differ from that in the reference image, which is prepared at installation. An example of a good uniformity is given in Figure 8.1.
- If a quantitative analysis is used, intrinsic IU should be <8%.

8.2.6 SPATIAL RESOLUTION

Spatial resolution refers to the ability of the camera to image two separate objects in small distance. The images are expected to blur at a separation approximately equal to the Full Width Half Maximum (FWHM); therefore, one method of estimating γ-camera spatial resolution is the determination of FWHM of a source.

Spatial resolution is evaluated every six months by the use of a bar phantom. Bar phantoms consist of four sets of parallel lead bar strips arranged perpendicular to each other in four quadrants. The widths and spacing of the strips are the same in each quadrant but differ in different quadrants. The thickness of lead should be sufficient to stop photons of a given energy for which spatial resolution is being estimated. The bar phantom is placed over the γ-camera detector. A flood source of equivalent dimension containing sufficient activity (usually, 200–400 MBq) is placed on the top of it and an image is taken (Figure 8.2). The image is visually inspected and spatial resolution is estimated from the smallest strips or spacing distinguishable on the image (B_{min}). Then the FWHM is given by Equation (8.4):

$$FWHM = 1,75 \times B_{min} \tag{8.4}$$

If the image is taken extrinsically, the spatial resolution depends primarily on the collimator resolution. An improved method is based on the use of a line source. The line source is placed in the FOV and an image is obtained. The FWHM value of the resulting line in the image gives the spatial resolution. For routine quality checks, the use of a bar phantom is sufficient.

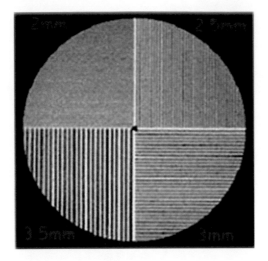

FIGURE 8.2 Four quadrant bar phantom quality control image (www.medimaging.gr). The smallest space distinguishable is 2.5 mm, thus the FWHM is 4.3 mm.

The smallest width and spacing of the strips that is distinguishable on the image should be less than 4 mm for intrinsic measurement and less than 6 mm for extrinsic measurement.

8.2.7 LINEARITY

Linearity, or spatial distortion, is the ability of a γ-camera to reproduce straight-line sources of radioactivity as straight lines. Linearity check is performed half-yearly using the procedure of uniformity determination and a bar phantom that consists of one set of parallel lead bar strips. The phantom should be rotated by 45° to the X- and Y-direction. Using proper software deviation of the strips from a straight line can be determined. The presence of nonlinearity diminishes the spatial resolution and leads to artefacts in whole body scans and SPECT (Figure 8.3). Nonlinearity can be inspected visually by comparing the actual image with the reference image at the installation. Up to now, there are no general limits for the quantitative nonlinearity estimation.

8.2.8 WHOLE BODY RESOLUTION

During whole body scans the relative physical position between bed and detector has to be accurately synchronized with the electronic offset applied to the image data to form the whole body image in order to preserve resolution along the scanning direction. Any mechanical problem or inappropriate adjustment can cause a loss of resolution.

The whole body resolution is assessed every six months, using two sources, equally separated and placed on the bed.

The system resolution as expressed by the FWHM of the line spread function is evaluated both for the parallel and perpendicular to the direction of the γ-camera

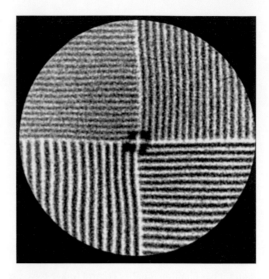

FIGURE 8.3 An example of bad γ-camera linearity: the four-quadrant bar phantom quality control image demonstrates wavy lines. (From Mettler FA et al, 2012.)

system motion, in whole body scanning acquisition mode, with no scatter medium. Minimizing the distance to the detector, whole body scan is acquired for all available velocities for patient scanning (Figure 8.4). At the final image, the distance between the sources is measured and compared to the real distance on the bed.

8.3   SPECT CAMERA QC

In addition to the QCs procedures for γ-camera imaging discussed earlier, rotating γ-camera SPECT requires specific controls (Murphy 2009; Halama JR et al. 2019). Among the components of a routine SPECT QC program are the center of rotation (COR) alignment, tomographic uniformity and an overall system performance. These tests apply to both SPECT-only scanners and the SPECT subsystem of SPECT/CT scanners.

8.3.1   DETECTOR HEAD TILT

In order to achieve accurate tomographic imaging, the detector should be aligned correctly, that is parallel to the rotation axis and that the angular readouts displayed on the gantry truly represent the detector's head position. Otherwise, the errors occur in the reconstruction process due to an incorrect geometrical configuration.

Therefore, the position of each detector as defined by the gantry angular readouts should be checked along the X- and Y-directions using a spirit level. Any deviation in detector position implies that there is either a detector misalignment or an error in the angular readout which should be investigated.

A visual check should be performed to assure that the detector position corresponds to the required angle prior to each acquisition.

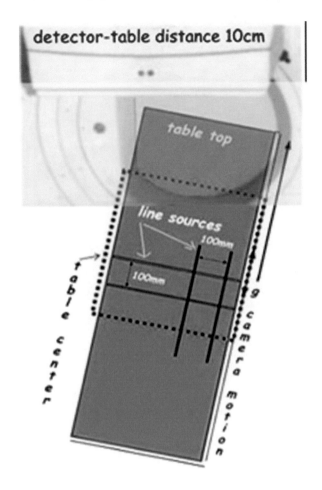

FIGURE 8.4   Whole body scan spatial resolution setup.

8.3.2   CENTER OF ROTATION (COR)

During the reconstruction of tomographic projection data, it is assumed that the center of each image corresponds to a single fixed point in space. Any movement in the detector head position (e.g. caused by gantry sag) can induce positioning errors of the center of the image and consequently artefacts in the reconstructed data.

It is expected that the mechanical rotation of such heavy detectors cannot be accurate about a single point and as a consequence, corrections must be applied to the acquired data to minimize the effects of any lateral shift.

There are standard protocols in γ-camera systems which incorporate routine protocols for the evaluation of COR offsets. The most common setup for testing the COR is the use of a point source, placed near the center of the field and off the central axis. Tomographic data are acquired over an orbit of 360° and a radius of 20 cm.

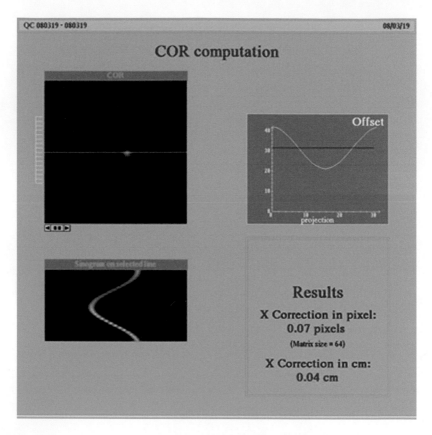

FIGURE 8.5 An example of COR quality control on a single head γ-camera (Athens Medical Center, Nuclear Medicine Department). Both the continuity of the sinogram and the calculated offset indicate a well-fixed COR.

The qualitatively assessment of the COR can be performed by creating a sinogram and looking for discontinuities in the data (Figure 8.5). In addition, quantitative assessment involves the calculation of the center of gravity for each frame and the detection of variations during the acquisition. The Y-axis position should be constant, whereas the position of X-axis should follow a smooth sinusoidal motion.

Concerning dual head γ-cameras, the COR must be checked for different gantry set-up (e.g. "H-mode" and "L-mode") and rotation and for each set of collimators used for SPECT. Variations should be less than 2 mm. It is advisable to check the offsets weekly and incorporate corrections into a monthly QA program.

8.3.3 Overall System Performance

One method to ensure that the images are of a consistent quality is to acquire suitable phantom data, incorporating objects for the qualitative assessment of resolution and uniformity. The most widely used is the Jaszczak ECT phantom (see Figure 8.6).

FIGURE 8.6 The Jaszczak phantom provides consistent performance information for any SPECT or PET system. Multiple performance characteristics of camera-based SPECT systems are evaluated by a single scan of the phantom. (https://www.biodex.com.)

The phantom is imaged under idealized conditions with signal-to-noise ratios much higher than those found with patient studies. While these studies do not reflect true patient data, the improved quality of the images allows a more effective qualitative assessment and the identification of any changes in performance before they are likely to impact on routine clinical investigations.

The phantom can be used for overall system performance including tomographic uniformity and resolution, collimator, attenuation correction (Saha K et al. 2016), (Toyama H 2015) calibration, and reconstruction parameters assessment.

Tomographic image reconstruction propagates and may amplify the effects of nonuniformities. Therefore, uniformity of γ-camera response is another consideration in obtaining high-quality SPECT images. For instance, a nonuniformity (integral or differential of 5%) image may produce significant ring, or bull's-eye, artifacts in tomographic images.

Tomographic uniformity should be evaluated by visually inspecting the resulting reconstructed images for the absence of artifacts.

The resolution of a SPECT imaging system is worse than that attainable for a comparable system performing planar imaging due to the influence of additional factors that do not impact on static imaging, including the COR corrections. As such, a measurement of the resolution represents an overall test of SPECT image quality rather than an investigation of an individual system parameter.

The Jaszczak phantom allows clear identification of multiple objects in the image and thus spatial resolution can be evaluated. There should also be a uniform transaxial area to allow better identification of ring artefacts caused by COR misalignment (Figure 8.7). Data should be reconstructed using commonly used protocols to

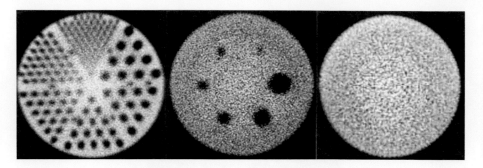

FIGURE 8.7 Jaszczak phantom quality control images: different sections are shown e.g. cold rods, inserted cold spheres, and uniformity section allowing for spatial resolution evaluation as well as for the impact of reconstruction process, attenuation effects, and COR errors on image quality. (From Zanzonico 2008.)

ensure that the reconstruction process is not subject to change. Any variation in the reconstructed image quality from reference data should be investigated. It can also allow evaluation of attenuation and scatter compensation as well as single slice and total volume sensitivity calculation.

The overall performance check should be repeated every six months.

There are also phantoms dedicated for specific SPECT studies like myocardial perfusion (see Figures 8.8 and 8.9) or lung emission acquisition data (Figure 8.9). Details can be found in Imbert et al. (2012), Pourmoghaddas et al. (2016), and Kroiss et al. (2017).

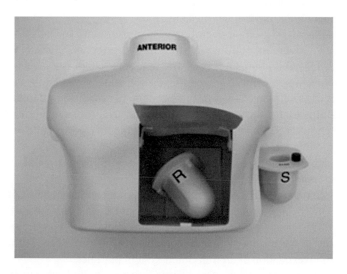

FIGURE 8.8 SNMMI 2012 Cardiac SPECT phantom is used to check the ability to acquire and process stress and rest myocardial perfusion studies, to identify and quantify areas of perfusion abnormality, and to determine the diagnostic and the prognostic significance of the findings.

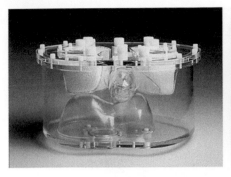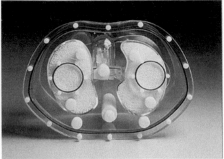

FIGURE 8.9 Evaluation of cardiac and lung emission computed tomography data acquisition and reconstruction methods. Evaluation of nonuniform attenuation and scatter compensation methods. (From http://www.spect.com/.)

8.4 PET CAMERA QC

PET camera QCs aim to ensure continued optimal operation of PET equipment in terms of image quality and accuracy, as well as the safety of both patients and operators.

Routine QC is necessary to detect issues that affect image quality and PET/CT co-registration accuracy (IAEA 2009; Demir M 2018; NEMA 2 2007).

8.4.1 PET Detector Stability Test

The constancy of the detector performance should be checked in a daily basis, prior to clinical examination, to allow early detection of a possible module failure.

Depending on the system, this test can be accomplished using different test sources, e.g. a rotating Ge^{68} line source, a cylindrical phantom filled uniformly with Ge^{68} centered properly in the FOV of the PET, or a Na^{22} point source placed at the center of the FOV.

The test is carried out using the system's daily PET QC acquisition protocol. Sinograms should be subjected to a visual inspection for the presence of streak artefacts and then compared to previews acquired reference sinograms (Figure 8.10). Most manufactures supply quantitative software for these daily QC procedures.

The difference between the two sinograms is characterized by the value of the average variance, which is a sensitive indicator of various detector problems. It is expressed by the square sum of the differences of the relative crystal efficiencies between the two scans weighted by the inverse variances of the differences. The sum divided by the total number of crystals is the average variance. A chi-square (χ^2) calculation is performed between the daily QC and the reference QC volume phantom scans:

If $\chi^2 < 2.5$ patient scanning may be performed normally.

If $2.5 < \chi^2 < 7$ a PET calibration should be performed as soon as it is convenient. Patient scans can still be performed; however, the test scans' images

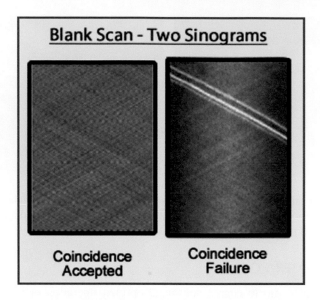

FIGURE 8.10   Typical sinograms of a daily blank scan: on the left image, a normal sinogram is presented whereas on the right, diagonal strike artefacts appear. (From www.people.vcu.edu/~mhcrosthwait/PETW.)

must be checked for obvious artifacts. A single detector block among other well-tuned blocks may not result in a $\chi^2>7$.
In the case of $\chi^2>7$ normalization or service on the scanner is required. Patient scanning is not advisable.

The final report is digitally stored and charts are produced to detect any trends in the values of quantitative parameters.

The tolerances are provided by the manufacturers as a part of the daily QC software protocol. If the results are consistent with a sudden change in uniformity or other parameters a warning message is displayed.

8.4.2   Uniformity of the Reconstructed Image

The rational of this test is similar to that performed on γ-cameras. There is no general consensus on the methodology or analysis parameters such as quantitative indices of residual nonuniformity. The usual procedure is based on the NEMA 1994 standard and it is included to the software of most scanners. This test should be performed quarterly.

A phantom filled with a uniform solution of F^{18} or cylindrical uniform source of Ge^{68} are used. The distributed radioactivity should be calculated at the time of measurement. Reconstruction should be performed with all the corrections applied (for normalization, dead time, decay, randoms, scattering, and attenuation) and according to the standard parameters (e.g. matrix size, pixel size, slice thickness, algorithm, and filters) used in the clinical practice. Reconstructed images should be carefully inspected visually for artifacts.

To obtain a quantitative estimation of nonuniformity, similar mathematical expressions to the case of γ-camera integral and DU QC are applied.

A suggested tolerance is that the calculated nonuniformity percentage should be less than 1.05% of the reference nonuniformity.

8.4.3   PET Normalization

The aim of this procedure is to acquire crystal efficiency data for use in correcting acquired sinograms for detector nonuniformities. The use of incorrect normalization data will compromise image quality. It is recommended that normalization should be performed quarterly and additionally whenever the results of daily OET QC indicate the need for renormalization or a service is carried out on the PET detector.

This procedure can be accomplished usually by means of a uniform cylindrical Ge^{68} phantom centered properly at the FOV of the PET. Normalization data should be acquired following the instructions of the manufacturer.

A visual inspection of the normalization sinograms should be made. If no problems are observed, the new normalization data should be stored in a file, according to the flowchart established by the manufacturer.

8.4.4   XYZ-Offset Calibration

In order to determine the offsets along the X-, Y-, and Z-axes, all PET/CT manufacturers have a standard procedure for the co-registration of PET and CT data. The use of incorrect offsets will result in erroneous registration in fused images and attenuation artefacts. This test should be performed quarterly and whenever the PET and CT gantries are separated for service. An alignment phantom comprising sources extending over the FOV of PET is used. A PET/CT scan is acquired according to the manufacturer's instructions.

The x, y, and z offsets required to register PET and CT images are computed and stored in a file from where they can be accessed by PET/CT fusion software.

8.4.5   Routine Image Quality Test for PET/CT

The aim of this test is to monitor the consistency of image quality parameters using a widely available nonuniform phantom (e.g. a Jasczsak phantom). The suggested test frequency is quarterly. After centering the phantom properly in the FOV of the PET/CT, a standard brain PET/CT scan is acquired. Data are reconstructed taking into account all correction normally applied in clinical scans.

At first, the reconstructed image is investigated for any visible artefacts. Then, the following analyses should be performed.

8.4.5.1   Uniformity

Within the uniform section of the phantom, a slice is selected and a circular ROI smaller than the internal diameter of the phantom is drawn. The IU is given by Equation (8.5):

$$U(\%) = 100 \frac{C_{max} - C_{min}}{C_{max} + C_{min}} \tag{8.5}$$

where C_{max} and C_{min} are the maximum and minimum voxel values within the ROI, respectively.

8.4.5.2 Radioactivity Concentration

Within the same slice and ROI, the mean voxel value in Bq/cm^3 is computed and compared with the actual concentration at the time of the scan. The difference is expressed as a percentage of the known resolution.

8.4.5.3 Spatial Resolution

The smallest diameter rods that can be distinguished in the reconstructed images are visually determined and recorded.

The existence of an apparent significant change in the aforementioned parameters should be investigated.

8.5 SYSTEM PARAMETER DETERMINATION BY USING MATLAB

New models of γ-cameras are frequently equipped with QC software tools developed by the company. The use of this facility in managing the periodical QC shows the advantage to perform the tests directly by the equipment console, with optimization in saving working time and results record. Nevertheless, in some cases it is not so clear the processing steps and algorithm followed by the company to analyze the QC images and one could wonder if the evaluation method follows or not the analysis recommendations.

It is advisable to perform the tests by the company QC tools as well as by independent measurements in such way to find an agreement in the results. Additionally, manufacturer-provided QC software is usually camera-specific; this means that often data from two different cameras cannot be analyzed by the same program, even if both cameras were built by the same manufacturer.

To overcome these problems, a software application should be developed that implements the basic scintillation camera QC analyses, as described in the most recent QCs guidelines. It must also have the ability to support DICOM format files.

Toward this direction, an application based on MATLAB, Nuclear Medicine-QC (NMQC) was proposed. MATLAB is a programming language that is widely used in medical image processing due to its reliability and cross platform compatibility. It offers functions that can read various image formats and tools to perform image registration, filtering, corrections, segmentation, image enhancement, and background subtraction.

Another feature of MATLAB is the Graphical User Interface (GUI) Design Environment that allows quick development of algorithmic code. Additionally, anyone can run this software without MATLAB or Image Processing Toolbox, just by installing the free MATLAB Common Runtime.

NMQC follows NEMA NU 1-2001 (NEMA 2001) in image processing and analysis including basic QC tests for γ-camera i.e. intrinsic uniformity, intrinsic resolution, COR, and tomographic uniformity (Rova A et al. 2008). In Figure 8.11, a typical example of intrinsic nonuniformity is presented.

NM Toolkit is also a free software package supporting NEMA-based acceptance testing procedures and routine SPECT QC testing (intrinsic resolution, uniformity, Jaszczak phantom measurements, and Specphan phantom shown in Figure 8.12, measurements).

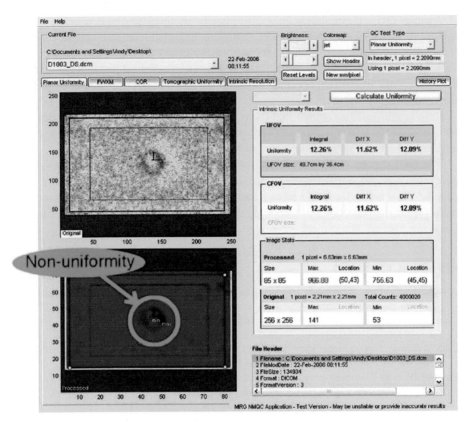

FIGURE 8.11   Example of planar uniformity screen with a nonuniformity highlighted. (From Rova A et al. 2008.)

FIGURE 8.12   A new design phantom, the Specphan™ phantom for quality assurance tests for SPECT and coincidence systems. (http://phantomlab.client-proof.com/products/specphan.php.)

Another GUI (Parimalah V et al. 2017) has been developed to determine spatial resolution based on IAEA protocols, that is to calculate the line-spread function (LSF) and modulation transfer function from the count profile of an acquired line source image. The reliability of the GUI program was assessed through comparison between results of FWHM obtained from the developed GUI program and the standard processing program.

Results analysis has shown that there is no significant deviation (approximately 1%) between the two software. Consequently, the GUI program can be considered a reliable and advanced program compared to the standard program.

The GUI can be run in any operating system that has MATLAB installed and purchase of additional software to analyze spatial resolution is not required. The developed program is considered as a cost-effective alternative and provides an easy platform for the physicist and technologist to analyze the spatial resolution after the line source image is acquired.

REFERENCES

Abbaspour S. et al. "Cadmium telluride semiconductor detector for improved spatial and energy resolution radioisotopic imaging." World J Nucl Med. 16, no. 2 (2017): 101–107.

Demir M, et al. "Evaluation of PET scanner performance in PET/MR and PET/CT systems: NEMA tests." Mol Imaging Radionucl Ther. 27, no. 1 (2018): 10–18.

EANM Physics Committee. "Routine quality control recommendations for nuclear medicine instrumentation." Eur J Nucl Med Mol Imaging. 37, no. 3 (2010): 662–671.

Halama JR et al. Gamma Camera, SPECT, and SPECT/CT Acceptance Testing and Annual Physics Surveys (TG177). AAPM, Task Group No. 177, 2019.

International Atomic Energy Agency. Quality Assurance for PET and PET/CT Systems, Human Health Series No. 1, IAEA, Vienna (2009). https://www.iaea.org/publications/8002/quality-assurance-for-pet-and-pet/ct-systems

Imbert L et al. "Compared performance of high-sensitivity cameras dedicated to myocardial perfusion SPECT: a comprehensive analysis of phantom and human images." J Nucl Med. 53, no. 12 (2012): 1897–1903.

Khoshakhlagh M et al. "Development of scintillators in nuclear medicine." World J Nucl Med. 14, no. 3 (2015): 156–159.

Kinahan PE et al. "X-ray-based attenuation correction for positron emission tomography/computed tomography scanners." Semin Nucl Med. 33, no. 3 (2003): 166–179.

Kroiss AS et al. "CT-based SPECT attenuation correction and assessment of infarct size: results from a cardiac phantom study." Ann Nucl Med. 31, no. 10 (2017): 764–772.

Mettler, FA, & Guiberteau, MJ. Essentials of Nuclear Medicine Imaging. Philadelphia, PA: Elsevier/Saunders. 2012.

Murphy PH. "Acceptance testing and quality control of gamma cameras, including SPECT." J Nucl Med. 28 (2009): 1221–1227.

NEMA 2. "Performance measurements of positron emission tomographs." National Electrical Manufacturers Association. 2007: 26–33.

NEMA, NU. Performance measurements of scintillation cameras. http://www.nema.org/stds/nu1.cfm. Rosslyn: National Electrical Manufacturers Association, 2001.

Parimalah V et al. "Determining spatial resolution of gamma cameras using MATLAB." J Med Imaging Radiat Sci. 48, no. 1 (2017): 39–42.

Patton JA et al. "SPECT/CT physical principles and attenuation correction." J Nucl Med Technol. 36 (2008): 1–10.

Pourmoghaddas A et al. "Quantitatively accurate activity measurements with a dedicated cardiac SPECT camera: physical phantom experiments." Med Phys. 43 (Jan 2016): 44.

Rova A et al. "Development of NEMA-based software for gamma camera quality control." J Digit Imaging. 21, no. 2 (2008): 243–255.

Saha K et al. "Application of Chang's attenuation correction technique for single-photon emission computed tomography partial angle acquisition of Jaszczak phantom." J Med Phys. 41, no. 1 (2016): 29–33.

Toyama H et al. "Evaluation of scatter and 2 attenuation correction methods using a originally designed 3D striatal phantom with 3D printer for quantitative brain SPECT." J Nucl Med. 56 (2015): 1813.

Zanzonico P. "Routine quality control of clinical nuclear medicine instrumentation: a brief review." J Nucl Med. 49, no. 7 (2008): 1114–1131.

9 Introduction to MATLAB and Basic MATLAB Processes for Nuclear Medicine

Marios Sotiropoulos

CONTENTS

9.1 PREFACE

This chapter intends to introduce the very basic usage of the MATLAB environment. It targets the reader who is not familiar with MATLAB, but has a very basic understanding of programming concepts. We will start with the basic concepts and gradually introduce more complicated techniques. At this stage every mistake is acceptable. In fact, we endorse you to challenge your curiosity and try things. Later, mistakes are going to be much more unwanted.

Rather than providing exhausting description of the different syntaxes that an object might have, we choose to introduce every object with an example of its use. We can always look into MATLAB's documentation, as described later, for more details.

9.2 INTRODUCTION TO MATLAB

MATLAB is a proprietary numerical computing environment and programming language, developed by MathWorks. It offers a strong computing framework with strong visualization tools, in an easy to use programming environment. For those reasons, it has been widely used by the image processing community.

Free compatibles to MATLAB are available, with the most dominant being GNU Octave. In fact, Octave syntax has a very high compatibility with MATLAB, allowing an almost seamless translation between the two software packages. However, the majority of the advanced features available in MATLAB are not available in Octave.

Underneath MATLAB's interface is a very strong matrix manipulation system. In principle all the main computations can be realized as operations between matrices. MATLAB offer a great variety of appropriate tools to manipulate our data arranged into matrices. Not surprisingly, in MATLAB even an image is a matrix!

9.2.1 INTERFACE

The tools, mainly function, are arranged into groups for similar application that are called ToolBox. In this book we are going to heavily use the Image Processing Toolbox and some other Toolboxes. Not all MATLAB installations come with all the ToolBoxes installed, as separate license might be needed for some Toolboxes.

But before we get into the computing procedures let's see the interface.

MATLAB has a fully customizable user interface. When it starts for the first time is divided into five main sections as in Figure 9.1 (MATLAB, 2018a):

- The Toolstrip is a series of tabs that organize MATLAB actions into sections. It was introduced in MATLAB (2012b).
- The File Browser helps to identify the current directory and its containers.
- The Command Window is where the actual work is being done. Here is where we can type our commands and execute existing scripts.
- The Workspace shows all the variables that are currently active and in our computer's memory.
- The Command History stores the past commands that have been issued.

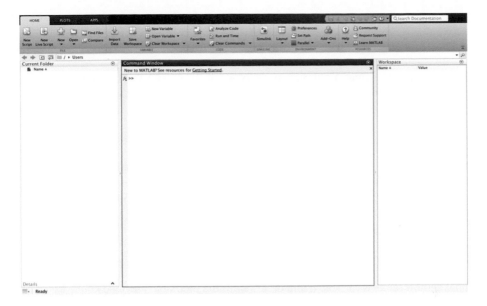

FIGURE 9.1   MATLAB interface when it opens for the first time.

Although not visible in the current arrangement, MATLAB also provides a script editor, but we will discuss this further later in this chapter.

9.2.2   MATLAB AS A CALCULATOR

Now it is time to run our first command. It will be a simple calculation. For doing this, in the Command Window and after the >> type:

```
>>5 + 5
```

Press Enter. We should be able to see the result.

Obviously, MATLAB can do much better than a simple calculation. Let's create a variable and assign a value.

```
>>a = 5
>>b = 5
```

In MATLAB we don't have to define the type of the variable. Instead, MATLAB will select the most appropriate type for the input provided.

We can also perform an operation, such as a summation:

```
>>a + b
```

MATLAB automatically stores the last result in a variable called ans. But it is always better to use our own variables – ans is going to be the last result every time.

In that case, we could use another variable:

```
>>c = a + b
```

Notice that every time a variable is assigned a value, the value is automatically printed in the command line. To suppress the automatic printing we can use a semicolon after the expression.

```
>>c = a + b;
```

That is extremely important when working with large files or images.

We can now look at the workspace. We should be able to see the variables that we created. However, quite often we want to remove variables that we are not going to use anymore, so that we free some memory and keep everything tidy.

We can remove a variable as follows:

```
>>clear a
```

That should remove variable *a* and de-allocate the memory. Then we could try:

```
>>clear b c
```

to remove the *b* and *c*. Also, if clear is used without any variable name, it deletes all the workspace's variables. However, use it with caution; we do not want to delete a variable that stores the result from a computationally intensive process. It is recommended to have a clean workspace every time we start something new, for example when we are going to try a new example in this book.

9.2.3 MATRICES

MATLAB exceeds in the manipulation of matrices. Let's see how we can do that.

```
>>matrix1D = [1 2 3];
```

This will create a one-dimensional matrix. We have used a new symbol, the square brackets, []. Everything inside the square brackets separated by space or comma is considered the elements of the matrix. However, all the elements of the matrix have to be of the same type.

To create a two-dimensional matrix, separate the matrix's lines with a semicolon

```
>>matrix2D = [1 2 3; 4 5 6];
```

We can avoid suppressing the output (no semicolon at the end of the expression) to see the matrix properly arranged.

To access an element of a matrix simply use the name of the matrix with the name. For example, to get the very first element of the two-dimensional matrix and store it to a variable write:

```
>>first = matrix2D(1, 1);
```

Next is to try to do some basic matrix calculations.

```
>>clear
>>matrix1 = [1 2 3; 4 5 6];
>>matrix2 = [9 8 7; 6 5 4];
```

We can actually do any calculation with the matrices, such as addition (+), subtraction (−), multiplication (.*), division (./), etc. It is important to note that we have used a dot before the actual operator. This is because we want an elemental-wise calculation (i.e. first element of the first matrix with the first element of the second matrix etc.), otherwise MATLAB will perform the operation according to the linear algebra rules. Of course we can add, subtract, multiply, or divide a matrix by any (scalar) value.

```
>>matrix3 = 7*matrix1 - 1;
```

It is very common to need a matrix starting from a value vmin, up to a maximum value, vmax, with a specific step. MATLAB has a specific syntax to produce such a matrix.

```
>>matrix = 0:10:100;
```

This will create a matrix from 0 to a 100 with the step of 10; we can always omit the semicolon to allow MATLAB to show us the input. If the step (and one semicolon) is omitted the default value of 1 is assumed for the step.

Occasionally it is useful to extract a part of an existing matrix.

```
>>matrix2D = [0:10; 10:20];
>>submatrix = matrix2D(:, 3:6);
```

Confusing? The semicolon without any numbers means all the elements in this dimension. The expression 3:6 means to get only the elements 3 to 6, i.e. 3, 4, 5, and 6. If needed, the keyword end can be used to denote the last element in the matrix.

There are also many other functions to create and populate a new matrix. For example:

```
>>matrix4 = ones(5, 5);
```

But wait, what is ones? Well, we have actually used our first (built-in) MATLAB function. All MATLAB functions are called by their name and if we have to provide some input should be inside the parenthesis. Different input arguments are separated by a comma. Of course we can use one or more variables to assign the output to, separated by comma (,).

The function ones takes as an input argument the number of elements per dimension separated with a comma and creates a matrix where all the elements are one. In the previous example, ones (5, 5) will create a two-dimensional matrix with 5 elements in each dimension, i.e. a 5×5 array.

9.2.4 Data Types

Although MATLAB will assign the correct type of data automatically to a variable, in many cases is useful to know what type our variable.

Generally in MATLAB we have:

- Logical. A data type that can hold only two values: 0 or 1.
- Numeric, which can further divided to:
  - Integers (int). An integer variable can further be divided based on the maximum number that can hold. For example we have int8, int16, int32, and int64. Also an unsigned version is available.
  - Floating point (single or double). We can have floating point numbers of single or double precision.
- Character type (char), which can hold exactly one character. However, a matrix of char can hold some text.
- String (string), which is in principle a type based on the character but has been optimized to hold text. A series of tool to manipulate a string type is also available.

MATLAB also has some advanced data types, such as structures, and cells, but for now we are going to work with the basic types.

If we would like to see what is the type of our variable, we can ask MATLAB using the whos command:

```
>>x=5;
>>whos x
```

The reason why the type of a variable is important is that each type needs a different amount of memory. For example, an int8 type needs 1 byte of memory and the largest number that can hold is 127. On the other hand, an int64 needs 8 bytes and can hold a number up to 2^{63}-1.

Usually in MATLAB we don't care about the type of a variable. MATLAB will take care of this. But in some occasions, for example when we work with very big matrices (and guess, a series of images is usually the case!), it is highly recommended to use the smallest type that can hold our data, saving precious memory. MATLAB by default will store numeric data in a double type, but that might not be the optimal choice. For this reason MATLAB offers all the tools needed to convert between compatible types. For example:

```
>>x = [1 2 3 4 5];
>>whos x
x holds 40 bytes of memory.

>>y = int8(x);
>>whos y
y holds 5 bytes of memory.
```

Also, sooner or later we will come across the two special data values in MATLAB:

- The representation of infinity, with the keyword Inf.
- An operation result that is not-a-number, represented by NaN.

The Inf and NaN are treated as any other numeric type in MATLAB. We can even use them into our calculations, or to check that a calculation has given an unwanted result.
We can produce these unusual values:

```
>>a = 1/0;
>>b = 0/0;
```

In the previous example a equals Inf and b NaN.

9.2.5 FILES AND DIRECTORIES

Equally important with data manipulation is the ability to work with files and directories. For example, we might want to group our data into different folders. So we need the ability to navigate into our system's directories, create new folders, or make changes to the file system (Table 9.1).
To check our current directory, we can look at our file browser or the pwd command.

```
>>pwd
```

Similarly, we can change the current directory either by using the user interface, or by the command

```
>>cd('C:/myWorkingFolder');
```

that will change the current folder to the myWorkingFolder located in the C: drive. Each operating system uses its own file system, so be careful to use the correct way to address a folder.
We can create a new folder named 'myFolder' inside the current folder:

```
>>mkdir('myFolder')
```

TABLE 9.1
Useful Directory Manipulation Commands

Command	Usage	Description
cd	cd('C:\myfolder')	Changes the current directory to "C:\myfolder"
mkdir	• mkdir('newFolder')	• Creates a new folder named "newFolder" in the current directory
	• mkdir('C:\myfolder)	• Creates a new folder named "myFolder" in C:\
pwd	pwd	Shows the current directory
ls	ls	Shows the contents of the current directory

9.3   MATLAB DOCUMENTATION

As we have already mentioned, MATLAB is a collection of Functions. It would be impossible to keep this book in a reasonable size if we try to describe every Function available in MATLAB. For this purpose MATLAB has an excellent Documentation module. Every Function included in the Toolboxes available in our installation is described in the Documentation. We can bring the Documentation module by clicking the HELP icon on the Home tab on the Toolstip or write directly the function or keyword we are looking for at the search panel. More importantly, the Documentation can be used to look for suitable functions and its use. In addition, all the main functions of a module are presented and relevant examples are given.

Generally, a short description of the function is presented, along with the different syntaxes that the function can be used. Usually, some basic examples of how to use the function are presented. We can have a look at the MATLAB documentation on the ones() function. Similarly, we can try to find out more about the zeros() and eye() functions.

9.4   PLOTS

MATLAB offers a great variety of data visualization tools. We will only focus to the most useful tools for image processing. The simplest way to visualize data is by a scatter plot. Let's create a very simple data set and visualize.

```
>>x = [0:10];
>>y = x^2;
>>plot(x, y);
```

Of course that is the $f(x) = x^2$ function. Let's see how we can annotate our plot.

```
>>title('f(x) = x^2');
>>xlabel('x');
>>ylabel('y');
```

This MATLAB code should be self-explanatory now. There are many more options to modify in order to annotate the figure to our personal preference. We can find more details in the documentation. Later we are going to see a few more options, but for now that is enough.

9.5   FLOW CONTROL

A program runs when a number of sequential commands are executed. However, in most of the cases, we need to execute different code depending on the input we have, and it is also important to repeat code without having to rewrite everything. So we need to have flow control methods. In MATLAB a number of common flow control methods are provided, so let's have a look of how we can use them to improve our programming skills.

9.5.1 FOR LOOP

The syntax of the standard 'for' loop in MATLAB is

```
for counter = elements in a matrix
    code to repeat
end
```

That will execute whatever code is in the 'code to repeat' section once for every element in the matrix defined in the counter. Here the counter is in principle a 1D matrix.

For example:

```
>>matrix = [];
>>for i = [1:10]
matrix = [matrix; i];
end
```

With the previous code we created a column matrix, called matrix. If we wanted a line matrix we have to change the code inside the loop as:

```
matrix = [matrix, i];
```

Now let's have a look at the loop code. After the execution of the second line, MATLAB waits for further input. The loop section will stop when there is the keyword 'end'. Of course we can give any name permitted to our counter variable. This loop will create a matrix containing the numbers 1 to 10. If the matrix hasn't been defined before the loop, we will get an error; so we define the matrix as an empty matrix before the loop. The drawback with this method is that the matrix will increase its size in every loop. Generally, this is not a good practice. It is much better if we could know the size of the matrix and we had defined the matrix with the correct size. Rewriting the previous example, defining the size of the matrix:

```
>>matrix = zeros(10, 1);
>>for i = [1:10]
matrix(i) = i;
end
```

In this example we used the matrix(i) command to access the ith element of the matrix and give it the value of i. That's the most common way to access the element of a matrix.

Of course we can have nested for loops, for example in the case we want to work with 2D matrices. For instance:

```
>>matrix = zeros(10, 10);
>>for i = [1:10]
for j = [1:10]
matrix(i, j) = i;
end
end
```

In some cases, due to a mistake or other reasons, a calculation might take longer than expected. For instance, there is always the possibility to accidentally create an infinite loop. In that case we can use the Ctrl+C to terminate the execution of the code.

9.5.2 Logical Expressions

Before we go any further, we should talk about logical expressions. A logical variable is quite important in programming, as in many cases we work with expressions that result in a logical value (Table 9.2). For example:

```
>>x=5;
>>y=6;
>>x>y
```

The final expression will give 1 if x>y and 0 if not. That 1 and 0 can be interpreted as true or false, respectively. Another very important situation is when we want two expressions to be true simultaneously:

```
>>x1=5;
>>y1=6;
>>x2=3;
>>y2=8;
>> (x1>y2) && (x2<y2)
```

The result is 1, meaning that both comparisons are true. The && is the logical AND and means that both expressions in the parenthesis have to be true for the expression to be true. In case that we want one of the two expressions to be true (but don't care which one) we can use ||.

As we will see in the next sections, these expressions are very useful.

TABLE 9.2

Logical and Comparison Operators

Operator	Description
&&	AND
||	OR
==	EQUAL
~=	NOT EQUAL
~	NOT
>	greater
>=	greater or equal
<	less
<=	less or equal

9.5.3 WHILE LOOP

In many cases, we do not know how many times we have to repeat the code in the loop, but it is defined by a logical expression. For this purpose, we can use the while loop, which will keep executing as long as the logical expression is true.

```
While logical expression
    code to repeat
end
```

For example, the following code will calculate 10 times the expression i=i+1, printing each time the result.

```
i=0;
while i<10
    i=i+1
end
```

Obviously there are many similarities with the for loop, although there are specific occasions where the one rather than the other might be needed.

9.5.4 IF STATEMENT

One other very important flow control method is the if statement.

```
if logical expression
    code to execute
elseif logical expression
    code to execute
else
    code to execute
end
```

In the if or elseif blocks the first expression that is true is executed. If none is true then the else block is executed. We can have as many as elseif blocks as we want, but only one else block is allowed. The elseif and else blocks are optional.

Let's try to use our knowledge in an example. For instance we would like to plot only the non-negative values of the $f(x) = x^3$ function, and have zero instead of a negative value.

We will calculate the function from -10 to 10, with a step of one.

```
>>x = [-10:1:10];
>>y1 = zeros(1, length(x));
>>y2 = zeros(1, length(x));
```

The function length(*matrix*) will calculate the largest dimension of the matrix. It would allow to input the size of the x matrix as a variable. So, if we later want to change the calculation range, we will have to do that only in the first line.

```
>>for i = [1:1:length(x)]
value = x(i)^3;
y1(i) = value;
if(value<0)
y2(i) = 0;
else
y2(i) = value;
end
end
```

And to plot the results we will use the subplot function, which allows to put more than one plot in a figure, arranged in a grid. The syntax is subplot(m, n, p), where m x n is the grid and p is the position index. We will actually need three plots, one below the other.

Top plot:

```
>>subplot(3,1,1)
>>plot(x, y1)
>>title('y1 = x^3');
>>xlabel('x');
>>ylabel('y1');
```

Middle plot:

```
>>subplot(3,1,2)
>>plot(x, y2)
>>title('y2 = x^3');
>>xlabel('x');
>>ylabel('y2');
```

Another way to get the non-positive values of the y1 matrix is through some matrix manipulation. We can create a mask matrix, which its elements will be 0 if the same element in the y1 is non-positive and 1 if the value is positive. Please remember how MATLAB treats matrices.

```
>>mask = ~(y1<0);
```

The tilde (~) inverts the values, as otherwise we will have the opposite, 0 for positive values and 1 otherwise.

```
>>y3 = mask.*y1;
```

Bottom plot:

```
>>subplot(3,1,3)
>>plot(x, y2)
>>title('y3 = x^3');
>>xlabel('x');
>>ylabel('y2');
```

9.5.5 SWITCH

Another useful flow control method is the switch, where one block of code is executed based on a set of cases. The value of the expression in the switch is compared with the expressions in the case, and when the comparison is true the code in the block is executed. We can have as many cases as we would like. Unlike other programming languages, where the switch expression is usually based on an integer value, a comparison for strings is also possible.

```
switch expression
      case expression1
            code to execute
      case expresion2
            code to execute
      otherwise
            code to execute
end
```

9.6 MATLAB SCRIPTS

It has been annoying that we have to write each command every time we want to have a calculation, without the possibility of saving our work. What has not been shown is that MATLAB offers a script editor that makes the organization of the code much easier.

We can open a new scripting document in the Home tabs, by pressing the New Script icon. It is in principle a text editor. However, we write on MATLAB's language, and MATLAB understands that, highlighting the text; even making suggestions for suitable functions or variables.

9.6.1 SECTIONS AND COMMENTS

In the editor, and generally in any text that will be read by MATLAB, we can add comments by using the % (percentage) symbol. Anything after the % will be ignored by MATLAB. That can be in the beginning of a line or after a command. More importantly, the editor allows dividing our work in sections. Each section starts with %% (double percentage) symbol and can be (optionally) followed by a title. It is also good programming practice to always comment on our code. It will make it easier to understand how our code is working after we haven't worked on it for a while, or more importantly, it will make it easier for a colleague to understand. However, avoid extensive descriptions and analysis within the comments sections.

Another very useful feature of the scripting editor is that we can select a portion of our code and execute it with F9 (for a MAC: Shift+F7). Or we can calculate a section by pressing Shift+ENTER (for a MAC: cmd+ENTER).

From now on, we CAN start a new script file for each chapter of the book, creating sections when necessary.

9.6.2 MATLAB as a Computing Environment

The most common way to use MATLAB is as a computing environment, where we have a long (or better a short) script file containing the code for calculations. Functions are used to separate the calculations into different parts and avoid unnecessary repeats. Alternatively, we can write a stand-alone program, with graphical user interface, but that is a far more advanced and less common approach.

What is usually recommended is to create sections representing each part of a calculation. At the end of this part we can use the clear command, to delete variables that are not useful anymore de-allocating some memory. It can be useful to save the contents of our workspace after a complicated and long calculation, so that we can have the data safe and ready any time we need them. The save and load commands are what we need in that case.

```
save('filename', 'variable1', 'variable2', …)
load('filename', 'variable1', 'variable2', …)
```

MATLAB will save the data in the file filename.mat. In the save function, if no variable name is given, all the variables in the workspace will be saved. Similarly, in the load function, if no variables are given will load all the variables from the file.

9.6.3 MATLAB Interface Arrangement

MATLAB's interface is fully customizable, so that every user can arrange it as would fit better. Regardless, a very useful arrangement would be the following (Figure 9.2):

- Divide vertically the window into three unequal parts, with the center being more than half of the window.

FIGURE 9.2 MATLAB interface with the arrangement suggested.

- On the left side put the directory browser and the history, and on the right side put the workspace.
- In the middle section, use about three to four fifths for the script editor, and below (dividing horizontal) place the command line. Add the variable viewer window as a tab in the same place as the script editor.

This way we have most of the space for the script editor, which is where we will spend most of our time. We also have a large window for the workspace where we can see all the variables.

9.7   FUNCTIONS

As we proceed more into programming and start making more complicated scripts, is becoming important to keep the code organized. A possible way to do that is to organize parts of the code that is often used to our own functions. This is also an efficient way to share our code.

To define a function we need to open a new script file. We have to name the function with a valid name, which means that has to start with an alphabetic character and can contain letters, numbers, and underscore. The name of the script file should be the same as the first function inside that file. More than one function is allowed per file, including nested functions.

Let's assume that we would like a function where we could give the X and Y data, the x and y axes labels, and the title, and create that plot. Let's see how we can make that function. Start a new script file and save it with the name myplot, in a directory named myFunctions, located in C:\. Of course we can change the name and save the function in a different location. Then we can write the code for the function.

```
Function myplot(X, Y, xlabel, ylabel, title)
% It is recommended to write a few sentences describing the
function
% In this case myplot is a function that simplifies plotting
% and annotating a figure

figure % this will open a new figure window,
    % we don't want to accidentally modify an existing figure

plot(X, Y)
xlabel(xlabel)
ylabel(ylabel)
title(title)

end
```

And we can try the function as follows. That can be done by writing to a new script, or we can type directly to the command line:

```
% Don't forget to change current directory
% to the one containing the myplot function
```

```
addpath('C:\myFunctions')
t = 0:100;
f_t = t^2;
myplot(t, f_t, 'time', 'position', 'f(t)=t^2')
```

Notice that in the title the square appears as superscript. In fact MATLAB offers many options for writing the text for a title or an axis label.

Here we can stop for a bit a look into the function declaration details. The function declaration has to be in the first (executable) line in the file. A function is declared as follows:

```
Function [out1, out2, …, outN] = functionName(in1, in2, …,
inM)
    code
    …
End
```

where out1 to outN are the output arguments and in1 to inM are the input arguments. Both the input and output arguments are optional.

A good practice when we create a function is to try to catch the cases where the user will give a wrong number of input or output arguments or wrong data type. This can be done by using the nargin and nargout function. When the nargin or nargout function is called inside a function, they will return the number of input and output argument(s) the function was called with.

In some cases, it is useful to design our function so that the input or output arguments are not known until the call of the function. For example, the built in plot() function can be used with an arbitrary number of x and y pairs. In that case we can use the varargin or varargout words at the end of the input or output list, respectively.

Another set of MATLAB functions that are useful to check for proper input is the is* detect state set. Useful function in this set is for example the isempty, which checks if an array is empty. For the full list please check MATLAB's documentation.

Now, we can make our function a bit better.

```
function myplot(X, Y, xlabl, ylabl, titl)

% Check the input arguments' number
if(nargin < 5)
    disp('ERROR: input arguments not enough.')
    % The function disp() will print its input to
    % the command line. Accepts only string data.
    return % return will terminate function
end

if( ~ismatrix(X) || ~ismatrix(Y) ) % check if X and Y are
matrices
    disp('ERROR: X and Y have to be matrices')
    return
end
```

```
figure % this will open a new window,
    % we don't want to accidentally modify an existing figure

plot(X, Y)
xlabel(xlabl)
ylabel(ylabl)
title(titl)

end
```

9.7.1   MATLAB Search Path

When a function is called, either build-in or user defined, MATLAB will look into a set of predefined folders to find it. That means that if our functions are not into one of the predefined folders, MATLAB will not find them. In that case we have in principle three options:

- Place our function files into a directory already monitored. For this purpose during the installation of MATLAB a folder called MATLAB has been automatically placed into our Documents folder. That's a very good place to store our functions. This folder might be different from system to system, so consult the documentation.
- Add the folder containing our functions files in the path. Let's assume that our functions are stored inside the 'C:\myFunctions'. To add them in the path for the current session:

```
>>addpath('C:\myFunctions')
```

- Set the current path with our working path containing the functions. This is very simple; just navigate to the folder containing the functions.

At any time we can run the PATH command to check if our folder is in MATLAB's path.

Be very careful when making changes to MATLAB's path, as this can create many problems. A large amount of folders in the path will slow down the search. For this reason it is highly advised to add new folders in the path temporarily at the begging of each session. That is also very useful to keep our projects separate.

9.7.2   Debugging

We all make mistakes. No matter how experienced we are in programming, it is certain that we will receive an error message. In MATLAB what we usually do is to run our code section by section, so that if there is a mistake we will figure it out. However, in case of a loop, or more importantly, a function, it is very helpful to be able to stop our code at a specific point and check the variables' contents. For this purpose we have the debugger. With the debugger we can stop the execution of our code at any line we want, and see what is happening at the variables loaded in the workspace.

A breakpoint denotes the point where the execution of the code will stop. To add a breakpoint, we can click on the minus sign next to the line numbering of the script editor. A red circle is placed to denote an active breakpoint. We can have as many as breakpoints we need, and we can remove a breakpoint that is not needed anymore by clicking it again.

The debugging options are available through the Editor tab. The most common options are:

- Breakpoints:
  - Clear all: remove all breakpoints.
  - Set/Clear breakpoint on current line.
  - Enable/Disable breakpoint.
  - Set condition. We can ask from the debugger to stop the execution at a specific line when the specified condition is met. Very useful to check the contents of a loop.
- Continue: (after stopping at a breakpoint) continues the execution of the code. It will stop at the next breakpoint or we can select to stop when (i) an error or (ii) a warning occurs, or (iii) a NaN or Inf value is returned.
- Step: execute one line.
- Step in: get inside the next function, if exists. Use this to enter and see the execution of a function.
- Step out: get out of the current function. Use this when done with the current function.
- Function call stack: Chose which function to see, in case we have many nested function calls.
- Quit debugging. Quit debugging, the code execution stops as well.

As an example we can consider the function myplot of the previous section. We can start a new script:

```
% Don't forget to set-up the path for myplot
addpath('C:\myFunctions')
% Or work from a folder that contains myplot; eg
% cd('C:\myFunctions')

t = 0:100;
f_t = t^2;
t = 5; % intentionally assign a scalar to t
myplot(t, f_t, 'time', 'position', 'Debugging', 'Test')
```

Run this example as normally, from the script editor. We get an empty plot. In this case we know that the problem is that t is a scalar, but what about if that value had been assigned by a wrong calculation inside a function?

Let's create a break point at the line where we use the build-in plot function inside the myplot function, and run the script again. The execution will stop at the plot(X, Y) command. Have a look at the workspace and the variable X, what is its value? Well, it should be 5!

9.8   APPLICATION: IMPORTING THE ICRP PHANTOM

Enough with the basic examples! In order to practice what we have learnt so far, we shall try a real world example. In particular we shall import the male ICRP 110 voxelized phantom (ICRP, 2009) and do some further processing, applying all our current knowledge and expanding our MATLAB techniques.

The ICRP phantoms represent a reference male or female human adult and were based on computed tomography images of real persons. A special care was given to adapt the phantom in consistency with the reference ICRP organ masses and height. The phantoms have been extensively used in numerous studies including organ absorbed dose coefficient estimation and effective dose calculation (Patni et al., 2011; Petoussi-Henss et al., 2014; Cros et al., 2017; Jansen et al., 2018), and organ S-value calculations for internal dosimetry (Lamart et al., 2016).

A new high-quality polygon-mesh series of phantoms, based on the anatomy of the ICRP 110 voxelized phantoms is underway (Kim et al., 2018), which mainly contribute to overcome limitations associated on dose calculations of small organs and organ wall tissue, especial on weakly penetrating radiation. However, the original voxelized phantoms remain still in use in research and, more importantly, are a great educational tool.

The necessary phantom data are available through the ICRP website[1], as supplementary information.

9.8.1   The Voxel Phantom Details: The Male Phantom

The phantom data are provided in a compressed. zip file. When uncompressed, we get two folders, one for the male (AM) and female (AF) phantom and a readme file describing the electronic contents. Inside the folders we can find the data (.dat) files of the phantom. For this example we selected the male phantom, with the process for the female phantom being identical.

What we need to know for importing the phantoms into the MATLAB environment is how the data is arranged. The phantom contained in the AM.dat file (AF. dat for the female), with an integer number representing the organ ID for each voxel, separated by space, and having 16 values per line. In separate files, the tissue represented by each organ ID and the material composition of each tissue are given.

The organ IDs will fill the voxel phantom matrix starting with the column numbers increase right to left, row number increase from front to back, and the slice number increase from the feet to the head. The male phantom is composed of 254 columns, 127 rows, and 222 slices.

9.8.2   Importing the Phantom

Let's start writing our ICRP phantom reading script. The comments provide the details for the code. We will use a MATLAB script as our notepad, where we are going to collect all the necessary code to import the phantom.

---

[1] ICRP 110, http://www.icrp.org/publication.asp?id=icrp%20publication%20110.

```
% Import ICRP phantom
% Reference phantom dimensions: Table 5.1
%                                         male      female
% Slice thickness (voxel height, mm)  8.0       4.84
% Voxel in-plane resolution (mm)      2.137     1.775
% Voxel volume (mm3)                  36.54     15.25
% Number of columns                   254       299
% Number of rows                      127       137
% Number of slices                    222       348


%% Male phantom details

% Set-up directories
% Note that the actual path depends on our file organization
cd('C:\icrp\malePhantom');
addpath('C:\readICRPphantom')

% Phantom size
cols    = 254;
rows    = 127;
slices  = 222;

% File to import
file = 'AM.dat';
```

Firstly, we will try to read the first slice of the phantom.

```
% Read data

%% First slice

fid = fopen(file,'r'); % open file to read

slice = zeros(rows, cols); % allocate matrix

for x = rows:-1:1
  for y = cols:-1:1

    % use fscanf() to read one number each line
    slice(x, y) = fscanf(fid, '%d', 1);

  end
end

fclose(fid); % close open file

% Tidy-up workspace
clear x y fid % clear unused variables

% Display image
```

```
imagesc(slice);
axis off % do not display the axis
```

In the previous code section, the only unknown functions should be the fopen, fclose, and fscanf. The fopen function opens the file 'AM.dat' for reading (denoted by the 'r'), and will provide a file id, fid to use every time we have to read from the file. Similarly the fclose 'closes' the file. In short, the fscanf function reads the data from the file with file id fid, specified by the format string, '%d', 1 value at a time. The format type '%d' denotes an integer.

It is interesting seeing how our matrix was treated as an image from MATLAB.

```
%% Whole phantom

fid = fopen(file,'r');

phantom = zeros(rows, cols,  slices);

for z = slices:-1:1
  for x = rows:-1:1
    for y = cols:-1:1

      phantom(x, y, z) = fscanf(fid, '%d', 1);

    end
  end
  disp(['slice no: ' int2str(z)]) % display the slice number
  % used int2str to convert the integer z to string,
  % and the [] to % merge the text, since disp
  % accepts only one string as input.
end

fclose(fid);

% Tidy-up workspace
clear x y z fid

% save the phantom, do that only once
save('male_ICRP_phantom', 'phantom', 'rows', 'cols', 'slices')
    % that will save the phantom, rows, cols, slices variables
    % to a file named male_ICRP_phantom.mat

% load the phantom, in case we have already read the data
load('male_ICRP_phantom')
```

We CAN use the imagesc function to display our phantom again, but it would be more interesting to have a dedicated function to display every slice that we would like. Let's try to write this function. First, we have to create a new script file named 'showPhantom':

```
function showPhantom(phantom, varargin)
%showPhantom displays cut views of the voxel phantom
% Input:
%       the 3D phantom matrix
%       the position of at least one plane to visualize
% It will display the axial, coronal, and sagital planes
% X increases back to front
% Y increases left (heart side) to right
% Z increases head to toes
% Use:
%       showPhantom(matrix3D, 100)
%           will display the axial plane at position 100
%       showPhantom(matrix3D, 100, 50)
%           will display the axial plane at position 100, and
%           coronal plane at 50
%       showPhantom(matrix3D, 100, 50, 45)
%           will display the axial plane at position 100,
%           coronal plane at 50 and sagittal plane at 45
%

    % Check for valid argument number
    if(nargin < 2 || nargin > 4) % too few input arguments

        disp('not enough arguments')
        return

    else
        % check input
        % phantom has to be a 3D matrix, i.e.
                % each dimension greater than 0
        dim = size(phantom);
        % dim is a vector with the x, y, z dimensions
        % of the phantom matrix
        if( length(dim) ~= 3 || …
                ~(dim(1)>0 && dim(2)>0 && dim(3)>0) )
                    % the ellipsis (three dots) means that the
                    % statement continues to next line
            disp('phantom is not a 3D matrix')
            return
        end

        % check for appropriate matrix indices
        for i=2:nargin
          value = varargin{i-1}; % varargin is a cell matrix
          % check if the value is:
          %       - not numeric
          %       - less than zero
          %       - greater than the corresponding dimension in
          %         the phantom dimension
          if(~isnumeric(value) || value<0 || value>dim(i-1))
            disp('matrix index error')
```

```
              return
          end
      end

      figure % open a new window for the images
      noSubplots = nargin - 1;
      xdim = 2.137; ydim = 2.137; zdim = 8;
                  % the relative size of each dimension of
                  the
          % phantom was taken from the ICRP report
      for j=2:nargin
        subplot(1,noSubplots, j-1)
        switch j

          case 2
              imagesc(rot90(squeeze(phantom(:, :,...
               varargin{j-1})), 2))
              title(['Axial plane, Z= '
              int2str(varargin{j-1})])
              daspect([xdim ydim 1])
              axis off
          case 3
              image = ... rot90(squeeze(phantom(varargin
              {j-1}, :, :)), 3);
              text = ['Coronal plane, X= '...
               int2str(varargin{j-1})];
               aspratio = [zdim ydim 1];
              myimage(image, text, aspratio)

            case 4
            myimage(rot90(squeeze(phantom(:, ... varargin
            {j-1}, :)), 3), ...
            ['Sagital plane, Y= ' int2str(varargin{j-1})], ...
            [zdim xdim 1])

          end
        end
    end

    % Nested function, cannot be accessed outside of
    % showPhantom function
    function myimage(matrix, txt, aspect)

        imagesc(matrix)
        title(txt);
        daspect(aspect)
        axis off

    end
end
```

Coronal plane, X= 70

Sagital plane, Y= 100

Axial plane, Z= 100

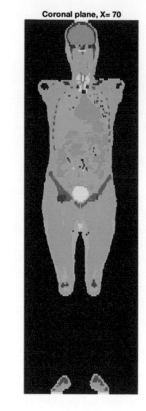

FIGURE 9.3   The ICRP male phantom as read and plotted by our function.

Now that we have our display function we can visualize our phantom (Figure 9.3).

```
% Display phantom
showPhantom(phantom, 100, 70)
showPhantom(phantom, 100, 70, 100)
showPhantom(phantom, 0)
```

The last one will give an error! Let's try to find what is happening. We always start from MATLAB's output:

```
Index in position 3 is invalid. Array indices must be positive
integers or logical values.
Error in showPhantom (line 58)
    Imagesc(rot90(squeeze(phantom(:, :, varargin{j-1}))), 2))
```

We can probably understand that the problem is because MATLAB uses matrix indexes from 1, rather than 0. But that might not be clear directly. So let's look into our function and see what is going on.

- Start with a breakpoint at line 58 (the line number shown in the error message).

- Run the code, and wait to pause at the breakpoint.
- Check the input arguments for the phantom matrix. It is (:, :, varargin{j-1}).
- What will be the value of varargin{j-1}?
- Looking at the workspace we get the value for j = 2, making the j-1 = 1 a valid index for the varargin cell array.
- Looking at the varargin{1}, we get the value of 0.
- We found it. Exactly as we have expected. One of the indices for the phantom array is 0, which is not allowed.

In that case it would be good to update our showPhantom function so that it does not accept a 0 value as an input for the phantom matrix indices. In fact we were checking the range of the indices at line 43. We could change the middle part to value<=0, so 0 value would not be accepted, and an error message will be displayed.

```
If( ~isnumeric(value) || value<=0 || value>dim(i-1) )
```

9.9   THE DICOM FORMAT

Nowadays, the international standard for storing and transferring medical images is the DICOM (Digital Imaging and Communications in Medicine) format. The standard is being managed by the Medical Imaging and Technology Alliance (MITA), a division of the National Electrical Manufacturers Association (NEMA). Before the establishment of the DICOM format, sharing medical images was a big problem. Each vendor had its own format rendering the transfer and read of the data by another machine almost impossible.

As the DICOM format is rather complicated, MATLAB offers the necessary function to read and write DICOM files.

9.9.1   READING DICOM FILES

In this example we will use the positron emission tomography (PET) phantom scans from the Quantitative Imaging Network (QIN) PET Segmentation Challenge to assess the variability of segmentations and subsequently derived quantitative analysis (Beichel et al., 2017). The scans (Beichel et al., 2015) can be retrieved free of charge from the Cancer Imaging Archive (Clark et al., 2013).

The phantom is based on the NEMA IEC Body Phantom Set (Model PET/IEC-BODY/P). However, the structures have been replaced with custom made spheres and ellipses.

Here, we will use the UW high statistics and high contrast image set and will try to segment the images and calculate the volume of each insert, implementing a simple segmentation method based on threshold identification. The method is very simple and is intended to demonstrate MATLAB's toolboxes, rather than demonstrating an extensive segmentation method.

Let's try to read the DICOM files:

```
%% Intro to DICOM format
```

```
% Change current Folder to the one containing the DICOM data.

% read file
I = dicomread('000000.dcm');
figure, imshow(I) % show image
```

This is the first image in the dataset. The DICOM file also contains all the details of the scan, which is a huge advantage, as the image information and the description cannot be separated. The details can be read as follows:

```
info = dicominfo('000000.dcm');
```

We can try to see the contents of the info structure. The info structure consists of data elements containing a tag and a value. In some cases a data element might contain a set of nested data elements. In MATLAB we can access the data of a tag as:

```
info.PatientAge
```

to get the data representing the age of the patient.

From the info structure all the information regarding the scan can be retrieved. As we shall see later, we can get the number of pixels in the image, the total slices of the study, and much other information.

A DICOM image is being represented by grayscale values. But, in principle, this should represent activity levels, right? Well, not exactly. For practical reasons the values are stored as integers. But in the DICOM file we can find the conversion scheme. A linear scheme is implemented:

$$f = m \times SV + b$$

The values for m and b are in the tags RescaleSlope and RescaleIntercept. The units are contained in the Units tag. Later we shall use these tags to normalize the images to the maximum activity. Note that the value RescaleSlope and RescaleIntercept can be different from slice to slice.

Next, we can read another slice of the phantom:

```
I = dicomread('000020.dcm');
figure, imshow(I)
```

A useful toolset for an initial assessment of an image is called as:

```
imtools(I)
```

This interface offers a set of tools, with the most useful:

- The adjust contrast
- Crop image
- Measure distance
- Zoom in and out
- Drag image to pan

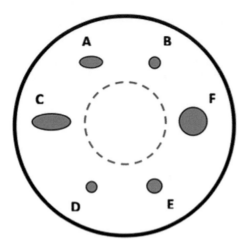

FIGURE 9.4 Arrangement of the inserts inside the phantom. B and F are spheres, A and C are horizontal ellipsoids, and D and E are axial ellipsoids. Figure not in scale.

With the contrast image we can see the histogram and make the necessary adjustments in the image. We can, for example, calculate the diameter of the big sphere (denoted as F in Figure 9.4). Here the distance is given in pixels, so we have to multiply with the size of the pixel. The values can be found in the DICOM file, under the tag PixelSpacing. The first value is the row and the second the column spacing in mm.

9.9.2 IMAGE SEGMENTATION

Since we have demonstrated the basic elements of a DICOM file, we will try to develop our segmentation code.

First we will read all the images, and create a matrix variable that holds all the images.

```
%% Read data

% Get info
info = dicominfo('000000.dcm');        % read DICOM info from
                                       % the first file
noSlices = info.NumberOfSlices;        % we can extract that
                                         information
                                       % from the DICOM file
rows = info.Rows;                      % Rows of the image
cols = info.Columns;                   % Columns of the image

image = zeros(rows, cols, noSlices);   % pre-allocate matrix

for i = 1:noSlices

    % files names are 000000.dcm, 000001.dcm to 000046.dcm;
    % use pad to
```

```
    % add the necessary zeros in front of the integer number
    fileName = [pad(num2str(i-1), 6, 'left', '0') '.dcm'];

    temp = dicomread(fileName); % read image
    info = dicominfo(fileName);    % read info

    % convert from image intensity to activity
    image(:, :, i) = info.RescaleSlope * double(temp) + ...
                    info.RescaleIntercept;

end

clear fileName i cols rows noSlices info temp % tidy up

% try to rescale the image values
Imax = max(max(max(image)));
Imin = min(min(min(image)));

scale = 2^16/Imax;

imageNorm = scale * image;
imageNorm = uint16(imageNorm);

% show all images
figure, montage(imageNorm, [])
colormap(hot)

clear Imax Imin scale
```

At this stage we should have the imageNorm, which should contain a matrix containing normalized activity of the phantom.

We will continue with calculating the threshold. For this we shall use the multithresh and imquantize MATLAB function. The first function requires as input the image and the number of different regions that we want to separate the image, based on its intensity. It will output the values of the threshold levels. The later function creates an image where the intensities are grouped into regions based on the levels calculated with the multithresh function.

```
%% Calculate threshold levels

I = imageNorm(:,:,19); % Select a slice with high activity

% Create a figure to compare our results
fig = figure;
subplot(1,3,1)
imshow(I,[]) % start with the unprocessed image
title('initial image')

% Thresholding: Create 7 regions of activity.
levels = multithresh(I, 6); % 6 thresholds to create 7 regions
```

```matlab
Imulti = imquantize(I, levels);

figure(fig)
subplot(1,3,2)
imshow(Imulti,[])
title('multithreshold image')

% Segment image
mask = Imulti>2;

Iseg = uint16(mask).*I;

figure(fig)
subplot(1,3,3)
imshow(Iseg,[])
title('segmented image')

clear I fig Imulti mask Iseg
```

At this stage we should have the threshold levels to segment the rest of our images.

```matlab
%% Use levels to segment all images

% Pixel dimensions
info = dicominfo('000000.dcm');
spacing = info.PixelSpacing;
dx = spacing(2); % in mm
dy = spacing(1);
dz = info.SliceThickness;
dV = dx*dy*dz; % in mm3
clear spacing info

% Initialize new matrices
dim = size(imageNorm);
noSlices = dim(3);

Iseg = uint16(zeros(dim));   % segmented image
BW = logical(zeros(dim));    % binary mask

% Do the segmentation
for i = 1:noSlices

    Itemp = imageNorm(:,:,i);

    Imulti = imquantize(Itemp, levels);

    mask = Imulti>2;

    BW(:,:,i) = mask; %  binary masks of images
    Iseg(:,:,i) = uint16(mask).*Itemp; % segmented image
```

```
end

clear dim i Imulti mask noSlices Itemp

% take the central slices
BWc = BW(:, :, 15:25);
BWc = padarray(BWc, [0 0 1], 0); % add an empty slice on both
% sides

% Visualize
figure, isosurface(BWc)
daspect([dx dy dz])  % use the pixel size to get correct
% proportions
axis( [100 160 100 160 0 15] )

% Create label for each volume
L = bwlabeln(BWc);

% Visualize the regions
s = regionprops(L(:,:,5), 'Centroid'); % calculate centroid
imshow(L(:,:,5))
hold on
for k = 1:numel(s)
    c = s(k).Centroid;
    text(c(1), c(2)-15, sprintf('%d', k), ...
        'HorizontalAlignment', 'center', ...
        'VerticalAlignment', 'middle', 'color', 'red');
end
hold off

clear k c

% Do the
s = regionprops3(L,"volume");

vol = s.Volume;

noVolumes = numel(vol);

volume = zeros(numel(s.Volume),1);

for i=1:noVolumes
    volume(i) = vol(i)*dV;
end

volume = volume/1000; % convert to ml

clear dx dy dz dV
clear i L s noVolumes vol
```

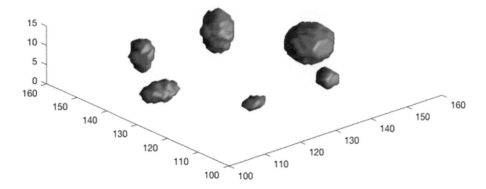

FIGURE 9.5 3D representation of the segmented image showing the spherical and ellipsoidal volumes.

Considering the simplicity of our method, the final volumes that we calculated are close enough to the nominal values. The shapes of the inserts can be clearly recognized in Figure 9.5.

9.10 CONCLUSION

In this chapter we have introduced the basic usage of MATLAB, infused with a few more advanced concepts, such as the debugging. We also demonstrated MATLAB's functionality through a simple segmentation problem. We hope that the reader will now be comfortable working with MATLAB and ready to continue with the more advanced examples in the following chapters. We have to stress that it is impossible to describe every function in MATLAB. Instead, searching in the documentation in the first instance when coming across an unknown function or method, or looking for advice on a function is recommended. Also don't forget that nowadays, the Internet is full of sites and forums with excellent advice on programming methods.

The examples here were mainly selected for demonstration purposes. In most of the cases, there might be a more efficient and fast method, but in this chapter we wanted the code to be simple, easy, and intuitive to understand and read.

REFERENCES

Beichel, R.R. et al., 2015. Data from QIN_PET_Phantom. *The Cancer Imaging Archive.* Available at: http://doi.org/10.7937/K9/TCIA.2015.ZPUKHCKB.

Beichel, R.R. et al., 2017. Multi,-site quality and variability analysis of 3D FDG PET segmentations based on phantom and clinical image data. *Medical Physics*, 44(2), pp. 479–496.

Clark, K. et al., 2013. The cancer imaging archive (TCIA): maintaining and operating a public information repository. *Journal of Digital Imaging*, 26(6), pp. 1045–1057.

Cros, M. et al., 2017. SimDoseCT: dose reporting software based on Monte Carlo simulation for a 320 detector-row cone-beam CT scanner and ICRP computational adult phantoms. *Physics in Medicine and Biology*, 62(15), pp. 6304–6321.

ICRP, 2009. Adult reference computational phantoms. ICRP publication 110. *Annals of the ICRP*, 39 (2).

Jansen, J.T.M. et al., 2018. Selection of bone dosimetry models for application in Monte Carlo simulations to provide CT scanner-specific organ dose coefficients. *Physics in Medicine and Biology*, 63(12), p. aac717.

Kim, C.H. et al., 2018. New mesh-type phantoms and their dosimetric applications, including emergencies. *Annals of the ICRP*, 47, pp. 45–62.

Lamart, S. et al., 2016. S values for131I based on the ICRP adult voxel phantoms. *Radiation Protection Dosimetry*, 168(1), pp. 92–110.

Patni, H.K. et al., 2011. Selected organ dose conversion coefficients for external photons calculated using ICRP adult voxel phantoms and Monte Carlo code FLUKA. *Radiation Protection Dosimetry*, 147(3), pp. 406–416.

Petoussi-Henss, N. et al., 2014. ICRP Publication 116 – the first ICRP/ICRU application of the male and female adult reference computational phantoms. *Physics in Medicine and Biology*, 59(18), pp. 5209–5224.

10 Morphology of Human Organs in Nuclear Medicine
MATLAB Commands

Elena Ttofi

CONTENTS

10.1 INTRODUCTION – USE OF MATLAB SOFTWARE

Advanced image processing and analysis technology are increasingly used in medicine. In medical applications, image data used to collect during the patient imaging process are obtained either by patients who suffer from a disease or by a normal screening processes (Li F. et al., 2005). Information from medical screening has become an integral part of patient care. The medical images give information about the structure of the organs and physiological functioning of an organism. In order to have high-quality diagnostic imaging and accurate quantitative information, processing and analysis of images is essential.

Matrix Laboratory – MATLAB is a high-performance interactive software developed by MathWorks reference and is an integral tool for image processing. MATLAB allows math operations, algorithms implementation, simulations, creation of functions and data, and signal-image processing, through various tools. It also provides the ability to analyze and visualize images taken during medical imaging examinations (SPECT, PET/CT, CT, MRI, etc.) (Fang Y.Y. et al., 2005). The MATLAB editing toolbox is an integrated standard algorithm and includes graphic image processing, visualization, and tools for algorithm development. It also provides the possibility of restoring degraded (noisy) images and improving

275

their clarity, shaping, and extracting the characteristics of an image. Therefore, with all these features, MATLAB can be used to perform any in-depth image analysis (Shirui G., 2012).

MATLAB and Image Processing Toolbox provide a wide range of advanced editing functions and interactive tools to enhance and analyze digital images. Interactive tools allow the user to perform spatial transformations, morphological operations such as edge detection and noise abstraction. At the same time, they provide the ability to process an area of interest, filtering, extract basic statistics; create curves, and various transformations (FFT – Fast Fourier Transform, DCT – Discrete Cosine Transform, and Radom Transform). MATLAB's ability to create translucent objects is a useful tool for 3D visualization, which provides information on spatial interfaces of different structures. MATLAB tools also give the user the opportunity to develop customized algorithms.

10.2 IMAGING AND QUANTITATIVE RESULTS FOR THE MORPHOLOGY OF THE ORGANS

The visualization and accurate measurement of organ size is important in estimating adequate growth of a healthy organ or patient point of the human body. Undoubtedly, the measurement of size of many organs plays a significant role to determine the growth and proper function. For example, decrease or increase in kidney size is an important indication of kidney disease. When evaluating a patient who presents for the first time an unexpected renal impairment, the size of the kidneys will help differentiate acute renal failure (where size is normal or greater) from acute exacerbation of chronic kidney disease (Matsaniotis N. et al., 2010; Kafetzis D.A. et al., 2011). The method is a key to avoid biopsy or immunosuppressive treatment for certain diseases. With MATLAB it is possible, easily, calculate measurements such as volume, in order to help physicians, diagnose more readily any abnormality.

10.3 OUTLINE OF THE HUMAN ORGAN BY MATLAB COMMAND 'CONTOUR'

In many Nuclear Medicine images, the organs' boundaries are presented unclear due to low resolution or presence of high percentage of noise. Also, the organ limits are not clearly defined, due to low resolution and existence of a high proportion of background. With the command contour in MATLAB, the delineating of organ boundaries in images is achieved. This variable is strongly related with the intensity of counts. The capabilities of the command are various; for example, it is possible to specify the contour lines to display as the last argument in any of the previous syntaxes, specify the style and color of the contour lines, specify additional options for the contour plot using one or more name-value pair arguments etc. In 2D isocontour surfaces, one can add labels using the *clabel* command, which indicates the level of elevation of counts and chooses the optimum colormap and font of the numbers for

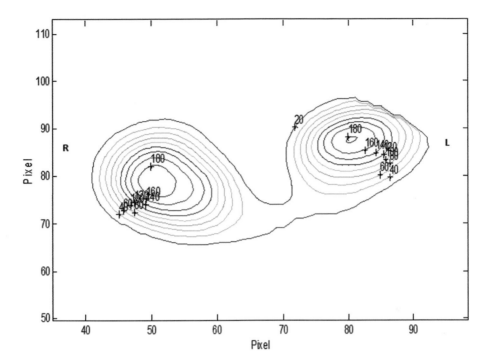

FIGURE 10.1   Examples of the command contour, clabel, and colormap tool, Renal Outline (one slice).

the presented figure (Figure 10.1). Another choice to represent a 2D slice of organ is *contourf*, which creates a filled contour plot containing the isolines. All these tools can give a valuable representation in 2D of a single organ.

10.4   MATLAB 2D ISOCONTOUR SLICES, LEVEL OF AN ELEVATION OF COUNTS, VOLUME AT VARIOUS ANGLES

Further than 2D representation in a slice, with MATLAB it is possible to obtain 3D imaging of organ's interior to illustrate the distribution of the radiopharmaceutical in the whole volume of the organ. To create this representation, there are a few options. Firstly, it must remove the background of the exam so that the organ is visualized clearly. The visualization can be achieved with the command *contourslice*, which is capable of drawing contours in volume slice planes. One of the capabilities of this command is that the end user can choose the level (z-axis) of the visualization of the organ and the color of the representation. Another option that gives this result is *contour3* which creates a 3D contour plot containing the isolines of a matrix as well as specifying the levels (heights) of counts in each slice, that can be achieved by setting contour3('ShowText','on').

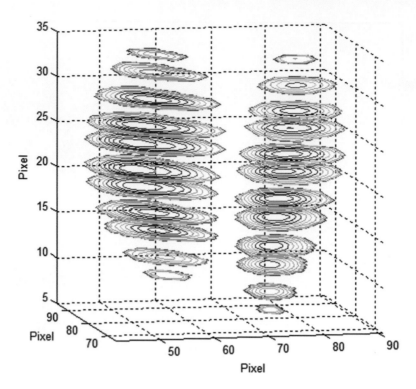

FIGURE 10.2   3D renal volume by the renal 2D isocontour slices (transaxial).

10.5   MATLAB 3D VOLUME RENDERING IMAGE – VOLUME OF THE ORGAN

MATLAB has a lot of options to visualize the volume of the human organs in Nuclear Medicine. Volume visualization in Nuclear Medicine consists of a method for extracting information from volumetric data utilizing and processing a Nuclear Medicine image (Lyra M. et al., 2010b). In MATLAB, this can be achieved by constructing a 3D surface plot which uses the pixel identities for (x, y, z) axes and the pixel value is transformed into surface plot height and, consequently, color (Lyra M. et al., 2010a). Volume rendering images mainly are used in 3D SPECT images for example in 3D myocardium, kidneys, thyroid, lungs, and liver studies. MATLAB has a lot of tools to display the volume rendering of an organ. For example, in Figure 10.3 the use of isosurface (V, isovalue) command which computes isosurface data V from the volume data at the isosurface value specified in isovalue. The isosurface connects points that have the specified value in the same way contour lines connect points of equal elevation.

There are several other commands in MATLAB's Image Processing Toolbox that give the volume rendering. The *setVolum(hVol,V)* updates the *volshow* object *hVol* with a new volume V. SetVolume preserves the current viewpoint and other

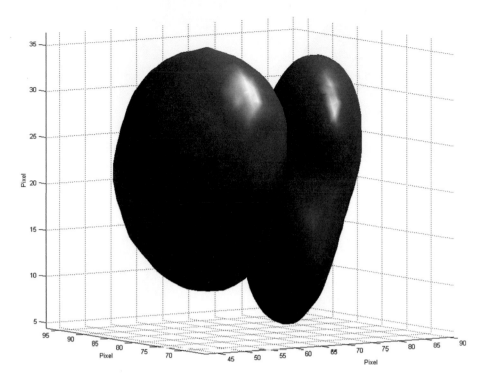

FIGURE 10.3 3D volume rendering by MATLAB of the renal surface with smoothing.

visualization settings remain unchanged. Another option to visualize Nuclear Medicine data is volshow; a volshow object with properties that control the appearance of the display is creating.

In addition to the above coding commands for 3D medical data visualization, MATLAB has toolbars that help the user to easily visualize 3D data such as 'Volume Viewer'.

The 'Volume Viewer' gives to the user the ability to view 3D volumetric data and 3D labeled volumetric data. It also provides the opportunity to view the data as a volume or as plane slices, view the data as a Maximum Intensity Projection (MIP) or an isosurface. An inside tool of this application, Rendering Editor, can manipulate opacity and illustrates wanted structures (MathWorks, 2019).

Nuclear Medicine exams such as renal DMSA can give a visualization of a single organ. Importing this data to the MATLAB in a loop that reads DICOM files it is easy to calculate the volume of an organ. In many studies, it is reported that the volume of an organ has a decisive role if the organ has an abnormality. MATLAB gives to the user the possibility to compile a code that can exclude or include data in many ways (see Chapter 14.2). Also, visualization and quantification of an organ volume can provide information of defects which indicate a possibility of pathogenesis.

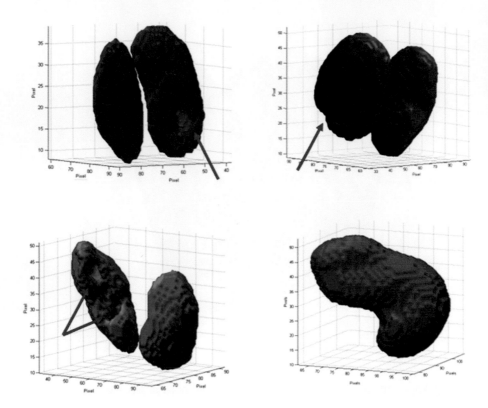

FIGURE 10.4   3D volume rendering of the kidney surface from different patients with renal abnormalities.

REFERENCES

Fang Y.Y., Sheng Y.Z., Shuang Z.Y., VB programming based on COM mixed with MATLAB. Comput Eng Des 01:61–65, (2005).

Kafetzis D.A., Koukoutsakis P.M., Miriokefalitakis N.E., Gouriotis D.I., Compendium pediatric, 2nd Pediatric Department, Athens University, Medical Versions Litsas, ch. 18, p. 469–524, (2011).

Li F., Ju L., Yun Hua D., V. Band MATLAB mixed programming method. Softw Technol 05:110–112, (2005).

MathWorks 2019, https://www.mathworks.com/Matsaniotis N., Karpathios Th., Nikolaidou-Karpathiou P., Compendium pediatric, Medical Versions Litsas, ch. 16, p. 477–506, (2010).

Shirui G. Research on medical image processing method based on the MATLAB, Proceedings of the International Conference on Information Engineering and Applications (IEA), Lecture Notes in Electrical Engineering 217, ch. 35, p. 269–276, (2012).

11 Internal Dosimetry by MATLAB in Therapeutic Nuclear Medicine

Nefeli Lagopati

CONTENTS

11.1 BASIC METHODOLOGY OF INTERNAL DOSIMETRY

Treatment planning requires the personalized dosimetry for the estimation of the absorbed dose in the target area as well as the other organs of each patient (Ljungberg M. et al., 2016). The appropriate calibration of the γ-camera is necessary in order to evaluate the biokinetics of the radiopharmaceutical (Zhao W. et al., 2018).

11.1.1 MIRD Scheme

MIRD scheme is a method which is employed to estimate the absorbed dose in the organs of interest if there is an administration of a specific dose of a radiopharmaceutical (Eberlein U. et al., 2011). The Society of Nuclear Medicine (SNM) is responsible for the invention and the documentation of the method (Donohoe K. et al., 2019). The fundamental principles and definitions are established and reported below.

If a radionuclide is administered in a patient's body, the initial radioactivity is considered as A_o and it is assumed that it finally fixed in a specific organ (Domínguez-Gadea L. et al., 2011). This organ can be considered as the source-organ, since it transmits radiation at the neighboring organs and toward itself (Rana S. et al., 2010). These organs which are burdened by this irradiation are the target-organs. During a radioactive decay, n particles are emitted and each of them is characterized by mean energy E. Thus, the total mean energy per decay is $n \times E$. The initial radioactivity fundamentally decreases over time, due to the physical decay of the radionuclide and the biological elimination of the radiopharmaceutical (Dash A. et al., 2015). Therefore, cumulative activity, $\tilde{A}$, is defined as the integral of the radioactivity as a function of time $(t_o \rightarrow t)$ (Equation 11.1) (Fathi F. et al., 2015):

$$\tilde{A} = \int_{t_o}^{t} A(t)\, dt. \tag{11.1}$$

The total energy which is emitted from the radionuclide of the source-organ for the same time is $\tilde{A} \times n \times E$ (Huang S.Y. et al., 2015). φ is a fraction of this energy, absorbed inside the target-organ, thus the absorbed energy is $\tilde{A} \times n \times E \times \varphi$ (Wayson M. et al., 2012). The absorbed dose is given by Equation 11.2:

$$D = \tilde{A} \times n \times E \times \varphi / m \tag{11.2}$$

where m is the mass of this organ. If $n \times E = \Delta$ and $\Phi = \phi/m$, then a factor S can be defined according to Equation 11.3:

$$S = \Delta \times \Phi \tag{11.3}$$

which is the absorbed energy per mass unit of the target-organ, during a radioactive decay inside a source-organ. The value of S factor depends on the radionuclide, the distribution of this radionuclide inside the source-organ, the shape and the morphology of the source-organ, and the target-organ and the distance between them (Divoli A. et al. 2009). The mean absorbed dose at the target-organ is given by Equation 11.4:

$$D = \tilde{A} \times S \tag{11.4}$$

Another important physical quantity in internal dosimetry in Nuclear Medicine, with MIRD, is the residence time of radionuclide, which is symbolized by τ. τ is the ratio of $\tilde{A}$ to A_o (Equation 11.5) (Ebrahimnejad Gorji K. et al., 2019):

$$\tau = \frac{\tilde{A}}{A_o} \tag{11.5}$$

where A_o is the administered radioactivity.

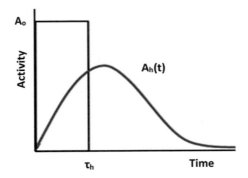

FIGURE 11.1   Distribution of cumulative radioactivity.

The residence time is estimated by diagram of radioactivity versus time from the equality of the areas of the rectangle $A_o - \tau h$ and the curve "$A_h(t) -$ time" (Figure 11.1). Thus, the mean absorbed dose is given by Equation 11.6 (Constantinescu C.C. et al., 2013):

$$D = A_o \times S \times \tau \qquad (11.6)$$

If the target-organ is irradiated by more than one source-organ, then the absorbed dose is given by Equation 11.7 (Momennezhad M. et al., 2016):

$$D = A_o \times \Sigma(S \times \tau) \qquad (11.7)$$

summing the corresponding products of S and τ for each combination of source and target organ (Figure 11.2). During each radioactive decay of a nuclide, a variety of

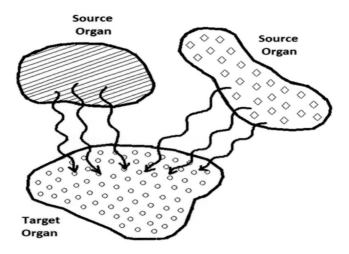

FIGURE 11.2   A target organ can be irradiated by many source-organs.

particles is emitted with different energy (Kassis A. I., 2008). The mean energy per category of particles is given by Equation 11.8:

$$\Delta_i = K \cdot n_i \cdot E_i \tag{11.8}$$

where K is a constant, which depends on the unit system.

The absorption fraction for a type of emitted particles i, which are emitted from the source-organ h, and is absorbed by the target-organ k, is defined as $\varphi_i(r_k \leftarrow r_h)$, where $0 \le \phi i \left(r_k \leftarrow r_h \right) \le 1$ (Hashempour M. et al., 2019). The specific absorption fraction (SAF) is given by Equation 11.9:

$$\Phi_i \left(r_k \leftarrow r_h \right) = \frac{\phi_i \left(r_k \leftarrow r_h \right)}{m} \tag{11.9}$$

where m is the mass of the source-organ. The SI values are available in tables, for emitted electron and photons, of a specific energy, for many different organs and for many different radionuclides, which are typically used in Nuclear Medicine either in diagnostic or in therapeutic applications (Yeong C. H. et al., 2014).

For the needs of MIRD scheme, the organs are modelized with mathematical and geometrical volumes, creating a model which simulates the human body (Sgouros G. et al., 2009). For this reason, the MIRD method cannot be used in individualized cases of internal dosimetry, if the patient has a body type, which is extremely different from the average body and thus from the model of MIRD (Ljungberg M. et al., 2016, Kuker R. et al., 2017). Another disadvantage of the method is that the radioactivity gathers in a non-homogenous way inside the source-organs, despite the fact that MIRD considers this concentration of the radionuclide homogenous (Vakili A. et al., 2012).

11.1.2 DOSIMETRY OF NON-HOMOGENOUS RADIOACTIVITY DISTRIBUTIONS

Inside a tumor, there are some areas with increased radionuclide absorption (hot spots) and others with low radionuclide absorption (cold spots) (Agrawal A. et al., 2016). The cold spots are suspicious for disease recurrence in the future, because the cancer cells which are located in these areas, receive low doses of radiopharmaceuticals and, therefore, it is possible to remain alive, capable of proliferation, and thus for disease progression (Buscombe J. R., 2007). In order to obtain a safe and accurate prognosis of the therapeutic result, it is important to study the histogram of dose versus tumor area instead of the simple calculation of the mean dose in the tumor (Miura H. et al., 2013). This histogram provides the opportunity for the estimation of the percentage of the tumor area which receives a low dose of radioactivity (Shah D. J. et al., 2012).

The limitation of the representation of non-homogenous radioactivity distribution inside the source-organs can be overcome with the utilization of voxels (Kost S. D. et al., 2015). Therefore, every source-organ is divided into voxels and each voxel corresponds to a value of cumulative radioactivity (Satterlee A. B. et al., 2017). In this way every abnormality in radioactivity absorption inside the source-organ can be represented (Rezaeejam H. et al., 2015). The smaller are the dimensions of the voxel, the more detailed is the obtained representation (Maret D. et al., 2012). The dosimetric calculations are achieved with the same way as with MIRD method (Ramos S. M. O. et al., 2017). In voxel dosimetry the source-organ is replaced by the source-voxel and the target-organ by the target-voxel (Lamart S. et al., 2011). Thus, Equations 11.10 and 11.11 are created:

$$\bar{D}\left(voxel_k\right) = \sum_{h=0}^{N} \tilde{A} \, voxel_h \cdot S\left(voxel_k \leftarrow voxel_h\right) \tag{11.10}$$

$$S\left(voxel_k \leftarrow voxel_h\right) = \sum_{i} \Delta_i \cdot \frac{\varphi_i\left(voxel_k \leftarrow voxel_h\right)}{m_{voxel_k}} \tag{11.11}$$

11.1.3 ESTIMATION OF RESIDENCE TIME

For the estimation of residence time of a radionuclide, the cumulative radioactivity $\tilde{A}_h$ and the initially administered radioactivity A_o are needed (Li T. et al., 2017). The residence time τ is given by Equation 11.5. The radioactivity decreases as time passes, due to physical and biological half-life, as will be thoroughly described in Chapter 12 (Sections 12.1 and 12.2) (Tanaka S. et al., 2018). Thus, the elimination of the radioactivity as a function of time is given by Equation 11.12:

$$A_t = A_o \cdot e^{-\lambda_e t} \tag{11.12}$$

The integration of Equation 11.1 from $t = 0$ to $t \rightarrow \infty$ leads to Equation 11.13:

$$\tilde{A}_t = \frac{A_t}{\lambda_e} \tag{11.13}$$

Since $\lambda_e = \lambda_p + \lambda_b$, $T_e = \frac{T_p \cdot T_b}{T_p + T_b}$, thus $T_e = \frac{ln2}{\lambda_e}$. Consequently, since $\tilde{A}_t = \frac{A_t}{\lambda_e} = A_t T_e 1.44$ and $\tau = \frac{\tilde{A}}{A_o}$, thus the residence time is given by Equation 11.14:

$$\tau = 1.44 T_e \frac{A_\tau}{A_o} \tag{11.14}$$

It is obvious that the residence time depends on the initially administered radioactivity A_o (von Gunten H. R., 1995). Another important factor is the chemical compound which is the tracer of the radiopharmaceutical (Kopka K., 2014). The composition of this compound can determine the percentage of the radiopharmaceutical absorption and the source-organs (Merrill J. R. et al., 2016). The tracer also defines the way of metabolic process and the residence time (Jang C. et al., 2018). Furthermore, the selection of the type of administration, the somatometric features, the age, the pathology, and the diet can affect the residence time. Pharmacokinetics of a pathologic area is significantly different compared to the normal one (Vamvakas I. C., 2016). Thus, it is clear that individualized dosimetry could take into account the personal characteristics of each patient (Ljungberg M. et al., 2016).

11.1.4 MONTE CARLO SIMULATION

Monte Carlo simulation is utilized in internal dosimetry as it is accurate approach of the dosimetric calculations (Momennezhad M. et al., 2016). It is a mathematical method which can simulate the energy deposition which is transferred by the particles that are emitted by the radioactive decays of the radionuclides, in a tissue level (AL Darwish R. et al., 2015). There are many software packages which are available for simulation in nuclear physics in general (MacLeod R. S. et al., 2009). Monte Carlo N-Particle eXtended (MCNPX) is among the most popular codes (Sadoughi H. R. et al. 2014). This software includes data about all the particles emitted by the radionuclides, the energy, and the possibility of interaction and energy deposition in the biological tissue (Ljungberg M. et al., 2016).

The most important step in Monte Carlo simulation is to build the geometrical characteristics of the system, which is about to be studied, and the general initial conditions (Harrison R. L., 2010). If the final challenge is the internal dosimetry, the design of geometries simulating the organs is the aim (Xu X. G., 2014). There are some mathematical equations to define, for example a cylindrical form like bones, or ellipsoids forms, like kidneys (Vamvakas J., 2016). The main problem is that this approach does not allow very good conformity to the real characteristics of a human body with its particularities. For the needs of individualized dosimetry, it is mandatory to create many mathematical models for many body types, for each gender, different height, for babies and children (Xu X. G., 2014). Recently some advanced codes use the data from CT and create a precise mathematical model of each patient (Krumholz H. M., 2014). This technique is an attempt to create a voxelized phantom in order to be used in Monte Carlo simulation (Bueno G. et al., 2009).

The following step includes the placement of the cumulative radioactivity of the radiopharmaceutical inside the body model (Ikuta I. et al., 2012). Then the simulation starts, and it represents the decays of the radionuclide and the way of energy deposition (Botta F. et al., 2013). The user selects the number of histories for the code to execute (Figure 11.3).

11.2 IMAGING-BASED AND PATIENT-SPECIFIC DOSIMETRIC MODELS IN TARGETED RADIONUCLIDE THERAPY

11.2.1 PLANAR IMAGE-BASED ESTIMATION OF RESIDENCE TIME AND INTERNAL DOSIMETRY

This method is simple and quick in acquisition and provides reliable dosimetric results. This process includes some distinct steps (Vamvakas I. C., 2016). First of all, the calibration of γ-camera is obligatory in order to convert the count rate (cpm) into radioactivity units (MBq). Rectangular water phantom is usually used for the calibration in order to simulate the patient's body (Khan F. M., 2010). The width of the phantom is similar to the anterior-posterior diameter of an average human body (Figures 11.4a and b) (Stepusin E. J. et al., 2017). A preselected administrative dose of a radiopharmaceutical is inserted in the phantom, in a close cavity, simulating the area of the tumor which is about to be displayed (Pereira J. M. et al., 2010). Then a

FIGURE 11.3 Results from a simulation of cumulative radioactivity 88.8 GBq·h ^{131}I in the thyroid gland. The simulation was undergone with VMC code Oak Ridge and the estimation of the dose considered only the dose which is related to photons emission. 1,000,000 histories were counted.

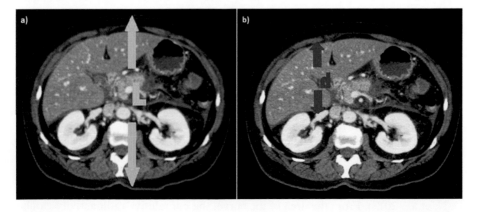

FIGURE 11.4 Measurement of the anterior-posterior diameter of: (a) the patient and (b) the source-organ through CT. (Revised figures from Vamvakas I. C., 2016.)

posterior and an anterior planar scintigraphic image is obtained and the count rates R_{post} and R_{ant} are reported, respectively (Middleton M. L. et al., 2012). The same process is repeated for a variety of administered doses, in order to create a graph of count rate versus radioactivity. The measured number of counts which is obtained by γ-camera through this procedure is less than the number of counts in the absence of the water phantom, due to water absorption of the radiation (Elschot M. et al., 2011). This phenomenon can be studied theoretically if the transmission coefficient T is estimated, which is given by Equation 11.15, where μ is the attenuation coefficient and L the width of the water phantom (Ji C.-Y. et al., 2016):

$$T = e^{-\mu L} \tag{11.15}$$

The value of the transmission coefficient T is used as a correction coefficient of the conversion function of the count rate into radioactivity (Frey E. C. et al., 2012).

There is an auto-absorption effect of the radiation inside the source-organ. The theoretical approach of this phenomenon is given by Equation 11.16, where μ is the attenuation coefficient of the water and d the width of target-organ:

$$f = \left[\frac{\frac{\mu d}{2}}{\sinh\left(\frac{\mu d}{2}\right)} \right] \tag{11.16}$$

The radioactivity of a source-organ is estimated by Equation 11.17, where E is the sensitivity of γ-camera:

$$A = \left(R_{ant} \cdot \frac{R_{post}}{T} \right)^{\frac{1}{2}} (f/E) \tag{11.17}$$

The anterior and posterior planar images are obtained half an hour, twenty-four hours, and forty-eight hours post injection of the radiopharmaceutical (Ebrahimnejad Gorji K. et al., 2019). It is possible to obtain images in subsequent moments for more accurate calculations (Veiga D. et al., 2014). It is important to consider the availability of γ-camera, due to the workload of a department of Nuclear Medicine, as well as the realism and practicality of the design of a dosimetric protocol (Zhao W. et al., 2018). In the scintigraphic images the boundaries of the tumor areas and the sensitive organs, namely the regions of interest (ROIs), must be defined with the appropriate software of γ-camera (Spanu A. et al., 2012). Afterward the count rate of each ROI in the posterior and the anterior image is reported, and the radioactivity of the ROI is estimated, based on the correlation curve, between count rate and radioactivity (Willegaignon J. et al., 2016). This procedure is repeated for every time point which is selected for data acquisition and in this way the curve of radioactivity as a function if time can be prepared (Shevtsova O. N. et al., 2017). From the area of the curve, the cumulative radioactivity can be estimated for the tumors and the sensitive organs (Schipper M. J. et al., 2012). The residence times can be obtained by dividing the cumulative radioactivity with the initially administered radioactivity (Thomas G. A. et al., 2016). The absorbed dose of every organ is estimated by MIRD scheme, based on the published S values or by employing MIRDOSE or OLINDA software (Ebrahimnejad Gorji K. et al., 2019). The accuracy of this approach is acceptable.

The overlays and overlaps of some organs in the anterior-posterior planar images do not allow the calculation of radioactivity of each ROI separately and this is among the main disadvantages of this method (Ljungberg M. et al., 2016, Vamvakas I. C., 2016). This phenomenon leads usually to overestimation of the dose received in the sensitive organs and underestimation of the dose in the tumor area (Garin E. et al., 2016). Additionally, the scintigraphic data do not provide anatomic information. So, the definition of ROIs requires experienced user and the use of CT images is considered as very important tool for fusion process and improved accuracy (Lyra M. et al., 2011).

11.2.1.1   MIRDOSE and OLINDA Software

MIRDOSE software was created by Oak Ridge Institute of Science and Education and provides the S values for many anthropomorphic phantoms. It has a collection of 28 available source-organs, 28 target-organs obtained by ten different phantoms (Lyra M. et al., 2012). The user has to put the residence time for each source-organ and the type of radionuclide (Baechler S. et al., 2008). Then the system calculates the absorbed dose in the target-organs and the effective dose (Fisher D. R. et al., 2017). OLINDA software is the transformation of MIRDOSE and is improved as it includes data from many more phantoms and from many more radionuclides (Kost S. D. et al., 2015). MIRDOSE and OLINDA software allow individualized calculations of internal dosimetry, giving the opportunity to choose the better phantom which simulates the characteristics of the patient (Stabin M. G. et al., 2005).

11.2.2   SPECT Image-Based Estimation of Residence Time and Internal Dosimetry

This method is time consuming during the data processing, but the results are more accurate, avoiding the effect of organ overlaps. The calibration of γ-camera is mandatory for the conversion of count rate per minutes into radioactivity units (Ljungberg M. et al., 2016). During data acquisition the γ-camera head turns around the patient's body in order to obtain many projections from many angles (Zanzonico P., 2008). The correct selection of the phantom is of crucial importance. Anthropomorphic or cylindrical phantoms with dimensions similar to a patient body are proposed as the more suitable (Johnson P. B. et al., 2011).

Firstly, the radioactivity is selected and injected inside a specific area of the phantom (Figure 11.5), simulating the tumor area and tomographic images are obtained (Frey E. C. et al., 2012). The protocol of the tomographic images acquisition must be similar during the whole process. The balance between the accuracy of the imaging and the convenience and affordability of the method must be considered during the design of the protocol. Three-dimensional reconstruction is following and afterward the count rate of the ROI in each section is measured (Dorbala S. et al., 2018). The counts are summed up and the total number of counts is divided with the total time

FIGURE 11.5   Tomographic representation of a preselected radioactivity of 74 MBq In-111-octreoscan. The radioactivity is included in a syringe of a total volume of 10 mL, placed at the center of a cylindrical water phantom (diameter 20 cm). Measurement of the anterior-posterior diameter of the source-organ through CT. (Revised figure from Vamvakas I. C., 2016.)

of tomographic data acquisition (Sitek A., 2012). Thus, the matching of count rates with the radioactivity is possible and if this process can be repeated for many values of radioactivity, then a correlation graph of count rate and radioactivity can be prepared (Dauer L. T. et al., 2014).

In clinical routine, the tomographic acquisitions are obtained in different time points post administration and the same imaging protocol with the calibration process of the cylindrical phantom is used (Li T. et al., 2017). According to the common protocol, the first tomographic data is obtained thirty minutes post injection and the other two after twenty-four hours and forty-four hours, respectively (Chen Q. et al., 2015). Three-dimensional reconstruction of the image in cross sections is following. ROIs are defined for the sensitive organs and tumor areas (Sgouros G. et al., 2008). For every ROI the count rate which is obtained by all the sections is reported. The correlation curve between count rate and radioactivity is estimated by the radioactivity of the volume of interest (VOI) for the specific time point (Frey E. C. et al., 2012). If this procedure can be repeated for all times which are selected for data acquisition, then a graph of radioactivity versus time can be prepared, for the tumors and sensitive organs which are defined (Lyra M. 2009; Vamvakas I. C., 2016).

From the estimation of the area of these graphs the cumulative radioactivity can be evaluated for each VOI. The residence time is obtained by dividing the cumulative radioactivity and the initial radioactivity (Li T. et al., 2017). If the residence times are estimated, then the absorbed doses for each organ or VOI can be calculated by the MIRD method and the previously mentioned software MIRDOSE or OLINDA (Momennezhad M. et al., 2016). CT and MRI data are used as an additional tool for fusion and accurate definition of ROIs and VOIs (Foster B. et al., 2014).

11.3 INTERNAL DOSIMETRY WITH MATLAB

MATLAB provides ready-made codes and routines for Nuclear Medicine dosimetry purposes (Botta F. et al., 2013). Image processing tool is a specialized package for digital image processing which allows accurate individualized dosimetric studies in radiopharmaceutical therapeutic applications (Vijayakumar C. et al., 2011).

11.3.1 VOXELIZED DOSIMETRY WITH MATLAB

This voxel-based method of dosimetry gathers many advantages compared to other methods. It allows imaging in full detail, using non-homogenous distributions from initial planar images and resulting in the final distribution of the absorbed dose (Mikell J. K. et al., 2015). The energy window for the calibration of γ-camera should be selected, considering the type of the radionuclide, the common administered dose, and estimated radioactivity that is supposed to be remained 24 post injection (Bhattacharyya S. et al., 2011). The radiopharmaceutical is inserted in the phantom (for example in a cylindrical phantom PMMA for thyroid gland studies) (Alqahtani M. S. et al., 2017). The tomographic images are obtained using the appropriate protocol. ROIs and VOIs are defined in order to mark the area of pre-known radioactivity and to correlate count rate and radioactivity, by making a graph (Ljungberg M. et al., 2016).

Usually, tomographic scintigraphic images in many different time points are obtained and these data are saved in DICOM and are inserted in the MATLAB environment (Vamvakas I. C., 2016). The first overview of these images is undergone with CERR program. The CERR program is an open-source program and it is supported by scientists in Memorial Sloan Kettering Cancer Center, New York (Vamvakas I. C., 2016). Then the pixels with the information of the total number of counts in each section are reported, and the ROI is defined for the area of pre-known radioactivity (Figure 11.6).

Since the images correspond to three-dimensional distribution, the total number of counts is included in every voxel of the distribution (Jobse B. N. et al., 2012). This is a kind of a three-dimensional phantom of the patient, obtained by his own scintigraphic images (Seo Y. et al., 2008). Since the time of acquisition is known for every tomographic image, the dose rate which corresponds to each voxel is estimated by dividing the total number of counts of each voxel with the total time of the acquisition. Based on the correlation curve of count rate and radioactivity, it is easy to estimate the radioactivity for each voxel of the tomographic data (Parodi K. et al., 2007). These calculations are automatically made by MATLAB routines if the appropriate mathematical fitting in the correlation curve of count rate and radioactivity is created.

From the data of the three-dimensional distribution of the radioactivity in every time point of tomographic acquisition, it is possible to create a three-dimensional

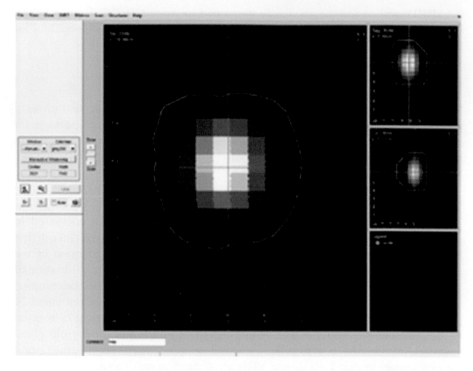

FIGURE 11.6 Definition of ROI to determine the number of counts which correspond to the pre-known radioactivity. (Revised figure from Vamvakas I. C., 2016.)

distribution of the cumulative radioactivity and the histogram of counts versus tumor area (Figure 11.7) (Cheng L. et al., 2013). The simple way is to create the graph of radioactivity versus time for each voxel. For example, if the tomographic data were obtained, the first one during the radiopharmaceutical administration and the others 24, 48, and 72 hours post injection and it is assumed that 150 hours post injection the radioactivity is practically zero then the graph of radioactivity versus time include five points (Ikuta I. et al., 2012). Linear interpolation is undergone, and the area of this curve is equal to cumulative activity of each voxel. These calculations can be programmed to execute by MATLAB (Kost S. D. et al., 2015).

Then from each transverse section of the scintigraphic tomographic images, the corresponding matrix (64 × 64) is created. Every point of the matrix is an index equal to the number of the counts at this position (Peterson T. E. et al., 2011). Sixty-four matrices are prepared. Every voxel which is obtained is a 6.1-mm cube. The matrices are formatted in grayscale (Vamvakas I. et al., 2015). The command *imshow3D* (filename) allows the imaging of the matrices, where the filename is the name of the file which includes the 64 matrices (Figure 11.8) (He L. et al., 2017).

These matrices are converted into matrices which represent cumulative activity based on the calibration of γ-camera and the assumption that the residence time is equal to an hour (3600 s) (Guan J. et al., 2015). Each matrix element of the matrix, which corresponds to counts, is multiplied with the factor 3600 and also with

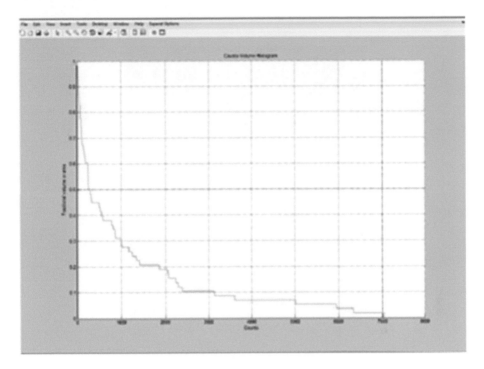

FIGURE 11.7   Histogram of counts versus tumor area for a specific radioactivity. (Revised figure from Vamvakas I. C., 2016.)

factor 10^{-4} for the conversion of counts into MBq. Thus, the matrices of cumulative radioactivity are obtained and each of their elements presents the cumulative radioactivity (MBq) of the specific position (Figure 11.9) (Vakili A. et al., 2012).

The next step includes the conversion of the three-dimensional distribution of radioactivity into a distribution of absorbed dose per voxel (Kost S. D. et al., 2015). Therefore, each voxel of a specific radioactivity is replaced with another of a specific absorbed dose (Sgouros G. et al., 2008). In fact, each voxel of a specific value of cumulative radioactivity can affect the neighboring voxels as well as itself (Botta F. et al., 2013). The dose of each voxel contributes to a total absorbed dose so these values per voxel must be summed up (Taschereau R. et al., 2007). For the study which is mentioned $64 \times 64 \times 64$ matrices are equal to 262,144 dose distributions. So, these 262,144 distributions are summed up.

Each voxel is relevant to a dose at its position as well as at the neighboring positions and these data are published at MIRD 17 for a variety of radionuclides. For example, the dose values for I-131 are presented in mGy/MBq for each 3 mm-cubic voxel. For the needs of the dose estimation, by MATLAB, a $5 \times 5 \times 5$ three-dimensional matrix with 6.1 mm-cubic voxel is created (Vamvakas I. C., 2016). Each voxel receives a coefficient for the conversion of the cumulative radioactivity into dose, using the values of MIRD 17 and considering the dimensional difference of the voxels between the published matrix and the prepared one (Dewaraja Y. K. et al., 2012).

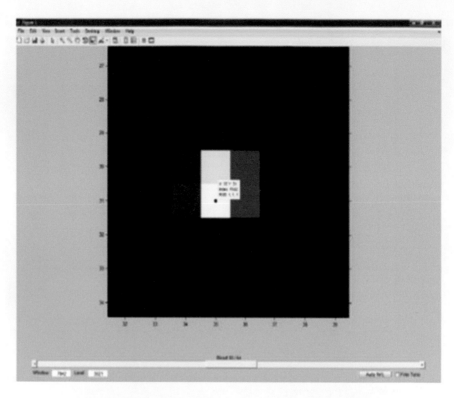

FIGURE 11.8 Image of the matrices which represent the counts of the tomographic scintigraphic image. (Revised figure from Vamvakas I. C., 2016.)

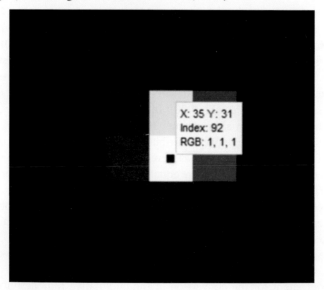

FIGURE 11.9 Image of the matrices which represent the cumulative radioactivity. (Revised figure from Vamvakas I. C., 2016.)

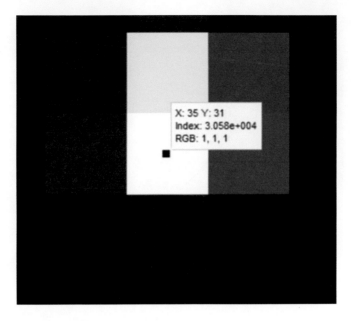

FIGURE 11.10   Image of the matrices which represent the absorbed dose. (Revised figure from Vamvakas I. C., 2016.)

The final calculation of the dose is achievable through MATLAB and particularly by the routine of convolution (*conv* routine) of the three-dimensional mathematical matrix of cumulative radioactivity and the three-dimensional matrix $5 \times 5 \times 5$.

Each voxel of the matrix contributes to the dose of itself and the dose due to the three neighboring voxels (distance is considered as $3·6.1 = 18.3$ mm) and the dose distributions are summed up. The required command for the convolution is *conv(matirx64, matrix5, 'Shape')*, where matrix64 is the name of the cumulative radioactivity matrix, matrix5 the name of the conversion matrix, and the parameter "shape" ensures that the result of the convolution will be a $64 \times 64 \times 64$ matrix. The elements of the obtained matrix correspond to dose (mGy) (Figure 11.10).

Conv routine of MATLAB can provide the three-dimensional distribution of the absorbed dose in every single voxel of a patient's body. Then VOIs can be defined and further statistical analysis of the data allows the creation of dose volume histograms (DVHs) (Figure 11.11), as well as the estimation of the minimum, the mean, and the maximum dose (Alfonso J. C. et al., 2015).

11.4   IMPACT OF PET AND SPECT ON TARGETED RADIONUCLIDE THERAPY

Targeted radionuclide therapy (TRT), also known as molecular radiotherapy (MRT) is a branch of radiotherapy which is related with the use of radioisotopes, radiolabeled molecules, or nanoparticles that controllably deliver radiation to diffuse primary or metastatic cancer cells in close vicinity to risk organs (Li T. et al., 2017,

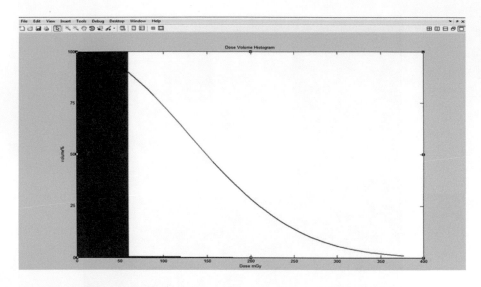

FIGURE 11.11   Dose volume histogram. (Revised figure from Vamvakas I. C., 2016.)

Chen H. et al., 2019). For this reason, TRT is the preferable therapeutic approach for a wide range of cancers. The importance of TRT is that it involves the use of dosimetry either at the phase of planning or retrospectively in order to ascertain the absorbed dose which was delivered during the treatment (Flux G. et al., 2006). The conventional planar images gather some pros, but tomographic methods are those which permit the determination of the activity volume.

The attenuation correction techniques, which are explained in previous chapter, require a patient-specific attenuation map. These corrections can be performed on the projection images or on the reconstructed images. Monte Carlo simulation is an additional tool that can help modeling interactions occurring in the patient or in the system of the detector (Li T. et al., 2017).

Planar image-based dosimetry is widely used, since the planar images are very common in clinical routine (Sapienza M. T. et al., 2019). Mostly the anterior-posterior method is used for the quantification (Flux G. et al., 2006). Planar imaging is less resource intensive than SPECT or PET-based imaging. For the anterior-posterior method, the matching of the two projections as well as the transmission image is crucial for accurate quantification (Seo Y. et al., 2008).

The main disadvantage of planar imaging is the lack of three-dimensional information (Ljungberg M. et al., 2018). The addition of classical three-dimensional anatomical images, like CT or MRI, can partly eliminate this limitation (Metcalfe P. et al., 2013). SPECT imaging can lead to a more accurate determination of the actual activity concentration in tissues (Seo Y. et al., 2008). It allows measuring organ activities in body structures with overlying structures. A time-sequential series of SPECT scans is usually employed, following the administration of a therapy or pre-therapy tracer radiopharmaceutical (Seo Y. et al., 2008). Post-acquisition image processing is also widely used. The potential of SPECT imaging to estimate the distribution of uptake within the target organ provides improved dosimetric accuracy. Tomographic image processing is usually performed by analytical methods (Frey E. C. et al., 2012). However, there are many reconstruction methods that work on an iterative basis in order to generate series of estimated projections from a first guess of the activity distribution (Tong S. et al., 2010). The comparison between the estimated projections and the measured ones leads to an update list of data, based on the differences (Flux G. et al., 2006). PET can be considered as the most accurate imaging method for the determination of the activity concentrations (Flux G. et al., 2006). The sensitivity and the spatial resolution of PET is greater than that of SPECT cameras and quantification techniques are well established (Khalil M. M. et al., 2011).

Clinical dosimetry requires standardized procedures for the estimation of the absorbed dose, but it is known that there is a heterogeneous activity distribution, depending on the organ or tumor (Sapienza M. T. et al., 2019). These calculations are based on the input of the spatial distribution of activity, which consequently relies on the imaging modality which is used (Ljungberg M. et al., 2016). Thus, it is reasonable that internal dosimetry for TRT is a research field with increasing interest (Li T. et al., 2017). The challenge of Nuclear Medicine is the improvement of these methodologies in order to be routinely available to the clinical practice (Williams E. D., 2016).

REFERENCES

Agrawal A., Purandare N., Shah S., Rangarajan V. (2016) Metastatic mimics on bone scan: All that glitters is not metastatic. Indian J. Nucl. Med. 31(3):185–90.

AL Darwish R., Staudacher A. H., Bezak E., Brown M. P. (2015) Autoradiography imaging in targeted alpha therapy with Timepix detector. Comput. Math. Methods Med. 2015:612580.

Alfonso J. C., Herrero M. A., Núñez L. (2015) A dose-volume histogram based decision-support system for dosimetric comparison of radiotherapy treatment plans. Radiat. Oncol. 10:263.

Alqahtani M. S., Lees J. E., Bugby S. L., Samara-Ratna P., Ng A. H., Perkins A. C. (2017) Design and implementation of a prototype head and neck phantom for the performance evaluation of gamma imaging systems. EJNMMI Phys. 4(1):19.

Baechler S., Hobbs R. F., Prideaux A. R., Wahl R. L., Sgouros G. (2008) Extension of the biological effective dose to the MIRD schema and possible implications in radionuclide therapy dosimetry. Med. Phys. 35(3):1123–34.

Bhattacharyya S., Dixit M. (2011) Metallic radionuclides in the development of diagnostic and therapeutic radiopharmaceuticals. Dalton Trans. 40(23):6112–28.

Botta F., Mairani A., Hobbs R. F., et al. (2013) Use of the FLUKA Monte Carlo code for 3D patient-specific dosimetry on PET-CT and SPECT-CT images. Phys. Med. Biol. 58(22):8099–120.

Bueno G., Déniz O., Carrascosa C. B., Delgado J. M., Brualla L. (2009) Fast Monte Carlo simulation on a voxelized human phantom deformed to a patient. Med. Phys. 36(11):5162–74.

Buscombe J. R. (2007) Radionuclides in the management of thyroid cancer. Cancer Imaging. 7(1):202–9.

Chen Q., Ma Q., Chen M., et al. (2015) An exploratory study on 99mTc-RGD-BBN peptide scintimammography in the assessment of breast malignant lesions compared to 99mTc-3P4-RGD2. PLOS ONE. 10(4):e0123401.

Chen H., Zhao L., Fu K., et al. (2019) Integrin $\alpha_v\beta_3$-targeted radionuclide therapy combined with immune checkpoint blockade immunotherapy synergistically enhances antitumor efficacy. Theranostics, 9(25):7948–60.

Cheng L., Hobbs R. F., Segars P. W., Sgouros G., Frey E. C. (2013) Improved dose-volume histogram estimates for radiopharmaceutical therapy by optimizing quantitative SPECT reconstruction parameters. Phys. Med. Biol. 58(11):3631–47.

Constantinescu C. C., Sevrioukov E., Garcia A., Pan M. L., Mukherjee J. (2013) Evaluation of [18F]Mefway biodistribution and dosimetry based on whole-body PET imaging of mice. Mol Imaging Biol. 15(2):222–9.

Dash A., Pillai M. R., Knapp F. F. Jr. (2015) Production of (177)Lu for targeted radionuclide therapy: available options. Nucl. Med. Mol. Imaging. 49(2):85–107.

Dauer L. T., Williamson M. J., Humm J., et al. (2014) Radiation safety considerations for the use of ^{223}RaCl$_2$ DE in men with castration-resistant prostate cancer. Health Phys. 106(4):494–504.

Dewaraja Y. K., Frey E. C., Sgouros G., Brill A. B., Roberson P., Zanzonico P. B., Ljungberg M. (2012) MIRD pamphlet No. 23: quantitative SPECT for patient-specific 3-dimensional dosimetry in internal radionuclide therapy. J. Nucl. Med. 53(8):1310–25.

Divoli A., Chiavassa S., Ferrer L., Barbet J., Flux G. D., Bardiés M. (2009) Effect of patient morphology on dosimetric calculations for internal irradiation as assessed by comparisons of Monte Carlo versus conventional methodologies. J. Nucl. Med. 50:316–23.

Domínguez-Gadea L., Cerezo L. (2011) Decontamination of radioisotopes. Rep. Pract. Oncol. Radiother. 16(4):147–52.

Donohoe K., Ahuja S. (2019) Society of Nuclear Medicine and molecular imaging efforts toward standardization: from procedure standards to appropriate use criteria. Semin. Nucl. Med. 49(2):153–60.

Dorbala S, Ananthasubramaniam K, Armstrong IS et al 2018, Single photon emission computed tomography (SPECT) myocardial perfusion imaging guidelines: instrumentation, acquisition, processing, and interpretation, J Nucl Cardiol, 25: 1784.

Eberlein U., Bröer J. H., Vandevoorde C., Santos P., Bardiès M., Bacher K., Nosske D., Lassmann M. (2011) Biokinetics and dosimetry of commonly used radiopharmaceuticals in diagnostic nuclear medicine – a review. Eur. J. Nucl. Med. Mol. Imaging. 38 (12):2269–81.

Ebrahimnejad Gorji K., Abedi Firouzjah R., Khanzadeh F., Abdi-Goushbolagh N., Banaei A., Ataei Gh. (2019) Estimating the absorbed dose of organs in pediatric imaging of 99mTc-DTPATc-DTPA radiopharmaceutical using MIRDOSE software. J. Biomed. Phys. Eng. (3): 285–94.

Elschot M., Nijsen J. F., Dam A. J., de Jong H. W. (2011) Quantitative evaluation of scintillation camera imaging characteristics of isotopes used in liver radioembolization. PLOS ONE. 6(11):e26174.

Fathi F., Moghaddam-Banaem L., Shamsaei M., Samani A., Maragheh M. G. (2015) Production, biodistribution, and dosimetry of (47)Sc-1,4,7,10-tetraazacyclododecane-1,4,7,10-tetramethylene phosphonic acid as a bone-seeking radiopharmaceutical. J. Med. Phys. 40(3):156–64.

Fisher DR, Fahey FH 2017, Appropriate use of effective dose in radiation protection and risk assessment, Health Phys, 113 (2): 102–109.

Flux G., Bardies M., Monsieurs M., Savolainen S., Strand S.-E., Lassmann M. (2006) The impact of PET and SPECT on dosimetry for targeted radionuclide therapy. Z. Med. Phys. 16:47–59.

Foster B., Bagci U., Mansoor A., Xu Z., Mollura D.J. (2014) A review on segmentation of positron emission tomography images. Comput. Biol. Med. 50:76–96.

Frey E. C., Humm J. L., Ljungberg M. (2012) Accuracy and precision of radioactivity quantification in nuclear medicine images. Semin. Nucl. Med. 42(3):208–18.

Garin E., Rolland Y., Laffont S., Edeline J. (2016) Clinical impact of (99m)Tc-MAA SPECT/CT-based dosimetry in the radioembolization of liver malignancies with (90)Y-loaded microspheres. Eur. J. Nucl. Med. Mol. Imaging. 43(3):559–75.

Guan J., Deboeverie F., Slembrouck M., van Haerenborgh D., van Cauwelaert D., Veelaert P., Philips W. (2015) Extrinsic calibration of camera networks using a sphere. Sensors (Basel). 15(8):18985–9005.

Harrison R. L. (2010) Introduction to Monte Carlo simulation. AIP Conf Proc. 1204:17–21.

Hashempour M., Ghorbani M., Amato E., Knaup C. (2019) Effect of beta particles spectrum on absorbed fraction in internal radiotherapy. Asia Ocean J. Nucl. Med. Biol. 7(1):71–83.

He L., Sneider A., Chen W., Karl M., Prasath V., Wu P. H., Mattson G., Wirtz D. (2017) Mammalian cell division in 3D matrices via quantitative confocal reflection microscopy. J. Vis. Exp. (129):56364.

Huang S. Y., Bolch W. E., Lee C., et al. (2015) Patient-specific dosimetry using pretherapy [^{124}I]m-iodobenzylguanidine ([^{124}I]mIBG) dynamic PET/CT imaging before [^{131}I]mIBG targeted radionuclide therapy for neuroblastoma. Mol. Imaging Biol. 17(2):284–94.

Ikuta I., Sodickson A., Wasser E. J., Warden G. I., Gerbaudo V. H., Khorasani R. (2012) Exposing exposure: enhancing patient safety through automated data mining of nuclear medicine reports for quality assurance and organ dose monitoring. Radiol. 264(2):406–13.

Jang C., Chen L., Rabinowitz J. D. (2018) Metabolomics and isotope tracing. Cell 173: 822–37.

Ji C.-Y., Guo Y.-C., Cui J., Yuan Z.-M., Ma X.-J. (2016) 3D experimental study on a cylindrical floating breakwater system. Ocean Eng. 125:38–50.

Jobse B. N., Rhem R. G., McCurry C. A., Wang I. Q., Labiris N. R. (2012) Imaging lung function in mice using SPECT/CT and per-voxel analysis. PLOS ONE. 7(8):e42187.

Johnson P. B., Geyer A., Borrego D., Ficarrotta K., Johnson K., Bolch W. E. (2011) The impact of anthropometric patient-phantom matching on organ dose: a hybrid phantom study for fluoroscopy guided interventions. Med. Phys. 38(2):1008–17.

Kassis A. I. (2008) Therapeutic radionuclides: biophysical and radiobiologic principles. Semin. Nucl. Med. 38(5):358–66.

Khalil M. M., Tremoleda J. L., Bayomy T. B., Gsell W. (2011) Molecular SPECT imaging: an overview. Int. J. Mol. Imaging. 2011:796025.

Khan F. M. (2010) The Physics of Radiation Therapy, 4th edition, Philadelphia, PA: Wolters Kluwer, Lippincott Williams & Wilkins.

Kopka K. (2014) Pharmaceuticals—Special issue on radiopharmaceutical chemistry between imaging and endoradiotherapy. Pharmaceuticals (Basel). 7(7):839–49.

Kost S. D., Dewaraja Y. K., Abramson R. G., Stabin M. G. (2015) VIDA: a voxel-based dosimetry method for targeted radionuclide therapy using Geant4. Cancer Biother. Radiopharm. 30(1):16–26.

Krumholz H. M. (2014) Big data and new knowledge in medicine: the thinking, training, and tools needed for a learning health system. Health Aff (Millwood). 33(7):1163–70.

Kuker R., Sztejnberg M., Gulec S. (2017) I-124 imaging and dosimetry. I-124 Görüntüleme ve Dozimetri. Mol. Imaging Radionucl. Ther. 26 (Suppl 1):66–73.

Lamart S., Bouville A., Simon S. L., Eckerman K. F., Melo D., Lee C. (2011) Comparison of internal dosimetry factors for three classes of adult computational phantoms with emphasis on I-131 in the thyroid. Phys. Med. Biol. 56(22):7317–35.

Ljungberg M., Sjögreen Gleisner K. (2016) Personalized dosimetry for radionuclide therapy using molecular imaging tools. Biomedicines. 4(4):25.

Ljungberg M., Pretorius P. H. (2018) SPECT/CT: an update on technological developments and clinical applications. Br. J. Radiol. 91(1081):20160402.

Li T., Ao E.CI, Lambert B., Brans B., Vandenberghe S., Mok G.S.P. (2017) Quantitative imaging for targeted radionuclide therapy dosimetry – technical review. Theranostics. 7(18):4551–65.

Lyra M. (2009) Single Photon Emission Tomography (SPECT) and 3D Images Evaluation in Nuclear Medicine, Image Processing, Yung-Sheng Chen, IntechOpen.

Lyra M., Phinou P. (2012) Internal dosimetry in nuclear medicine: a summary of its development, applications and current limitations. Radiation Safety Officer. 5(2):17–20.

Lyra M. E., Ploussi A., Georgantzoglou A. (2011) Chapter: 23. MATLAB as a tool in nuclear medicine image processing. MATLAB – A Ubiquitous Tool for the Practical Engineer. Ionescu CM (Editor). InTechEditors.

MacLeod R. S., Stinstra J. G., Lew S. et al. (2009) Subject-specific, multiscale simulation of electrophysiology: a software pipeline for image-based models and application examples. Philos. Trans. A Math. Phys. Eng. Sci. 367(1896):2293–310.

Maret D., Telmon N., Peters O. A. et al. (2012, Effect of voxel size on the accuracy of 3D reconstructions with cone beam CT. Dentomaxillofac. Radiol. 41(8):649–55.

Merrill J. R., Krajewski K., Yuan H., Frank J. E., Lalush D. S., Patterson C., Veleva A. N. (2016) Data on biodistribution and radiation absorbed dose profile of a novel (64) Cu-labeled high affinity cell-specific peptide for positron emission tomography imaging of tumor vasculature. Data Brief. 7:480–4.

Metcalfe P., Liney G. P., Holloway L., Walker A., Barton M., Delaney G. P., Vinod S., Tome, W. (2013) The potential for an enhanced role for MRI in radiation-therapy treatment planning. Technol. Cancer Res. T. 12(5):429–46.

Middleton M. L., Strober M. D. (2012) Planar scintigraphic imaging of the gastrointestinal tract in clinical practice. Semin. Nucl. Med. 42(1):33–40.

Mikell J. K., Mahvash A., Siman W., Mourtada F., Kappadath S. C. (2015) Comparing voxel-based absorbed dosimetry methods in tumors, liver, lung, and at the liver-lung interface for (90)Y microsphere selective internal radiation therapy. EJNMMI Phys. 2(1):16.

Miura H., Masai N., Oh R. J., Shiomi H., Sasaki J., Inoue T. (2013) Approach to dose definition to the gross tumor volume for lung cancer with respiratory tumor motion. J. Radiat. Res. 54(1):140–5.

Momennezhad M., Nasseri S., Zakavi S. R., Parach A. A., Ghorbani M., Asl R. G. (2016) A 3D Monte Carlo method for estimation of patient-specific internal organs absorbed dose for (99m)Tc-hynic-Tyr(3)-octreotide imaging. World J. Nucl. Med. 15(2):114–23.

Parodi K., Paganetti H., Cascio E., Flanz J. B., Bonab A. A., Alpert N. M., Lohmann K., Bortfeld T. (2007) PET/CT imaging for treatment verification after proton therapy: a study with plastic phantoms and metallic implants. Med. Phys. 34(2):419–35.

Pereira J. M., Stabin M. G., Lima F. R., Guimarães M. I., Forrester J. W. (2010) Image quantification for radiation dose calculations – limitations and uncertainties. Health Phys. 99(5):688–701.

Peterson T. E., Furenlid L. R. (2011) SPECT detectors: the anger camera and beyond. Phys. Med. Biol. 56(17):R145–R182.

Ramos S. M. O., Thomas S., Pinheiro M. A. et al. (2017) Internal radiation dose and modeling codes in nuclear medicine: a fresh look at old problems. Int. J. Radiol. Radiat. Ther. 4(5):00111.

Rana S., Kumar R., Sultana S., Sharma R. K. (2010) Radiation-induced biomarkers for the detection and assessment of absorbed radiation doses. J. Pharm. Bioallied. Sci. 2(3):189–96.

Rezaeejam H., Hakimi A., Jalilian A. R., Abbasian P., Shirvani-Aran S., Ghannadi-Maragheh M. (2015) Determination of human absorbed dose from [153Sm]-Samarium maltolate based on distribution data in rats. Int. J. Radiat. Res. 13(2):173–80.

Sadoughi H. R., Nasseri S., Momennezhad M., Sadeghi H. R., Bahreyni-Toosi M. H. (2014) A comparison between GATE and MCNPX Monte Carlo codes in simulation of medical linear accelerator. J. Med. Signals. Sens. 4(1):10–7.

Sapienza M. T., Willegaignon J. (2019) Radionuclide therapy: current status and prospects for internal dosimetry in individualized therapeutic planning. Clinics (Sao Paulo, Brazil), 74:e835.

Satterlee A. B., Attayek P., Midkiff B., Huang L. (2017) A dosimetric model for the heterogeneous delivery of radioactive nanoparticles In vivo: a feasibility study. Radiat. Oncol. 12(1):54.

Schipper M. J., Koral K. F., Avram A. M., Kaminski M. S., Dewaraja Y. K. (2012) Prediction of therapy tumor-absorbed dose estimates in I-131 radioimmunotherapy using tracer data via a mixed-model fit to time activity. Cancer Biother. Radiopharm. 27(7):403–11.

Seo Y., Mari C., Hasegawa B. H. (2008) Technological development and advances in single-photon emission computed tomography/computed tomography. Semin. Nucl. Med. 38(3):177–98.

Sgouros G., Frey E., Wahl R., He B., Prideaux A., Hobbs R. (2008) Three-dimensional imaging-based radiobiological dosimetry. Semin. Nucl. Med. 38(5):321–34.

Shah D. J., Sachs R. K., Wilson D. J. (2012) Radiation-induced cancer: a modern view. Br. J. Radiol. 85(1020):e1166–e1173.

Shevtsova O. N., Shevtsova V. K. (2017) Mathematical simulation of transport kinetics of tumor-imaging radiopharmaceutical 99mTc-MIBI. Comput. Math. Methods Med. 2017:2414878.

Sitek A. (2012) Data analysis in emission tomography using emission-count posteriors. Phys. Med. Biol. 57(21):6779–95.

Spanu A., Sanna D., Chessa F., Manca A., Cottu P., Fancellu A., Nuvoli S., Madeddu G. (2012) The clinical impact of breast scintigraphy acquired with a breast specific γ-camera (BSGC) in the diagnosis of breast cancer: incremental value versus mammography. Int. J. Oncol. 41(2):483–9.

Stabin M. G., Sparks R.B., Crowe E. (2005) OLINDA/EXM: the second-generation personal computer software for internal dose assessment in nuclear medicine. J. Nucl. Med. 46(6):1023–7.

Stepusin E. J., Long D. J., Marshall E. L., Bolch W. E. (2017) Assessment of different patient-to-phantom matching criteria applied in Monte Carlo-based computed tomography dosimetry. Med. Phys. 44(10):5498–508.

Tanaka S., Adati T., Takahashi T., Fujiwara K., Takahashi S. (2018) Concentrations and biological half-life of radioactive cesium in epigeic earthworms after the Fukushima Dai-ichi Nuclear Power Plant accident. J. Environ. Radioactiv. 192:227–32.

Taschereau R., Chatziioannou A. F. (2007) Monte Carlo simulations of absorbed dose in a mouse phantom from 18-fluorine compounds. Med. Phys. 34(3):1026–36.

Thomas G. A., Symonds P. (2016) Radiation exposure and health effects – is it time to reassess the real consequences?, Clin. Oncol. (R Coll Radiol). 28(4):231–6.

Tong S., Alessio A. M., Kinahan P. E. (2010) Image reconstruction for PET/CT scanners: past achievements and future challenges. Imaging Med. 2(5):529–45.

Vakili A., Jalilian A. R., Moghadam A. K., Ghazi-Zahedi M., Salimi B. (2012) Evaluation and comparison of human absorbed dose of (90)Y-DOTA-cetuximab in various age groups based on distribution data in rats. J. Med. Phys. 37(4):226–34.

Vamvakas I., Lyra M. (2015) Voxel based internal dosimetry during radionuclide therapy. Hell. J. Nucl. Med. 18(Suppl 1):76–80.

Vamvakas I. C. (2016) Patient specific dosimetry during radionuclide treatment, new techniques, National and Kapodistrian University of Athens, PhD Dissertation. Athens, Greece.

Veiga D., Pereira C., Ferreira M., Gonçalves L., Monteiro J. (2014) Quality evaluation of digital fundus images through combined measures. J. Med. Imaging (Bellingham). 1(1):014001.

Vijayakumar C., Gharpure D. C. (2011) Development of image-processing software for automatic segmentation of brain tumors in MR images. J. Med. Phys. 36(3):147–58.

von Gunten H. R. (1995) Radioactivity: a tool to explore the past. Radiochimica Acta. 70/71:305–16.

Wayson M., Lee C., Sgouros G., Treves S. T., Frey E., Bolch W. E. (2012) Internal photon and electron dosimetry of the newborn patient – a hybrid computational phantom study. Phys Med Biol. 57(5):1433–57.

Willegaignon J., Pelissoni R. A., Lima B. C., Sapienza M. T., Coura-Filho G. B., Queiroz M. A., Buchpiguel C. A. (2016) Estimating (131)I biokinetics and radiation doses to the red marrow and whole body in thyroid cancer patients: probe detection versus image quantification. Radiol. Bras. 49(3):150–7.

Williams E. D. (2016) Chapter 18. Development of Computers in Nuclear Medicine A History of Radionuclide Studies in the UK: 50th Anniversary of the British Nuclear Medicine Society [Internet]. McCready R, Gnanasegaran G, Bomanji JB, editors. Cham (CH): Springer.

Xu X. G. (2014) An exponential growth of computational phantom research in radiation protection, imaging, and radiotherapy: a review of the fifty-year history. Phys. Med. Biol. 59(18):R233–302.

Yeong C. H., Cheng M. H., Ng K. H. (2014) Therapeutic radionuclides in nuclear medicine: current and future prospects. J. Zhejiang Univ. Sci. B. 15(10):845–63.

Zanzonico P. (2008) Routine quality control of clinical nuclear medicine instrumentation: a brief review. J. Nucl. Med. 49(7):1114–31.

Zhao W., Esquinas P. L., Hou X., Uribe C. F., Gonzalez M., Beauregard J. M., Dewaraja Y. K., Celler A. (2018) Determination of gamma camera calibration factors for quantitation of therapeutic radioisotopes. EJNMMI Phys. 5(1):8.

12 Pharmacokinetics in Nuclear Medicine/ MATLAB Use

Nefeli Lagopati

CONTENTS

12.1 PHARMACOKINETICS IN NUCLEAR MEDICINE – INTRODUCTION

The possible path that a radiopharmaceutical follows inside the human body and its final biodistribution, post-administration, is governed by the pharmacokinetics as well as the physical-chemical properties of the selected radionuclide (Ding H. et al., 2012). These are very important factors for the determination of the efficacy and safety of a treatment in Nuclear Medicine.

12.2 RADIOACTIVE DECAY EQUATIONS

Radionuclides decay by spontaneous fission, α–, β–, and β+ particle emissions, electron capture, or isomeric transition (L'Annunziata M. F., 2016). Since the radioactive decay is a random process, only the average number of radionuclides disintegrating during a period of time can be estimated, giving the disintegration rate of a particular radionuclide. Thus, it gives the number of disintegrations per unit time and is proportional to the total number of radioactive atoms present at that time (Equation 12.1) (Saha G., 2013):

$$-\frac{dN}{dt} = \lambda_p N \qquad (12.1)$$

where N is the number of radioactive atoms present, and λ_p is referred to as the decay constant of the radionuclide (Groch M.W., 1998). The disintegration rate $-\frac{dN}{dt}$ is referred as the radioactivity or simply the activity of the radionuclide and denoted by A (Equation 12.2):

$$A = \lambda_p N \qquad (12.2)$$

Mathematically, solving the first-order differential Equation (12.1), another equation is generated (Equation 12.3):

$$N_t = N_o \cdot e^{-\lambda_p t} \qquad (12.3)$$

where N_0 and N_t are the number of radioactive atoms at $t = 0$ and time t, respectively. Therefore, the radioactivity decays exponentially. By multiplying both sides of Equation (12.3) by λ_p, one obtains Equation (12.4):

$$A_t = A_o \cdot e^{-\lambda_p t} \qquad (12.4)$$

Every radionuclide is characterized by a physical half-life, which is defined as the time required to reduce its initial activity to one half (Arthofer W. et al., 2016). It is usually denoted by $t_{1/2}$ and is unique for a radionuclide. It is related to the decay constant λ_p (Equation 12.5):

$$t_{1/2} = \frac{0,693}{\lambda_p} \qquad (12.5)$$

Another relevant quantity of a radionuclide is its mean life, which is the average lifetime of a group of radionuclides (Masok F. B. et al., 2016). It is denoted by τ and is related to the decay constant λ_p and half-life $t_{1/2}$ as follows (Equation 12.6):

$$\tau = \frac{1}{\lambda_p} = \frac{t_{1/2}}{0,693} = 1.44 t_{1/2} \qquad (12.6)$$

12.3   EFFECTIVE HALF-LIFE (T_e)

The physical half-life of a radionuclide is independent of its physicochemical conditions (Nayak T. K. et al., 2009). Analogous to physical decay, radiopharmaceuticals administered to humans disappear exponentially from the biological system through fecal excretion, urinary excretion, perspiration, or other routes (Lin C.-c. et al., 2006). As it is reasonable, after in vivo administration every radiopharmaceutical has a biological half-life (T_b), which is defined as the time needed for half of the radiopharmaceutical to disappear from the biologic system (Sharp P. F. et al., 2005). It is related to decay constant λ_b by $\lambda_b = 0.693/T_b$ (Leslie W. D. et al. 2003). In any biological system, the loss of a radiopharmaceutical is due to both the physical decay of the radionuclide and the biological elimination of the radiopharmaceutical (Thie J. A., 2012). Thus:

$$\lambda_e = \lambda_p + \lambda_b \qquad (12.7)$$

and sequentially

$$\frac{1}{T_e} = \frac{1}{T_p} + \frac{1}{T_b} \qquad (12.8)$$

or

$$T_e = \frac{T_p \cdot T_b}{T_p + T_b} \qquad (12.9)$$

which gives the equation for the effective half-time.[1]

12.4   THEORETICAL APPROACH OF THE KINETICS OF RADIOPHARMACEUTICALS

The kinetics of a substance in a biological system includes its spatial and temporal distribution in that system (Shargel L. et al., 2016). These are the results of several complex events, such as circulatory dynamics, transport into cells, and utilization (Cox P. H. et al., 1986). The last of them requires a series of biochemical transformations which are characteristics of the substance (Kowalsky R. J. et al., 2004). The substance can be a chemical element or a compound, such as amino acids, proteins and sugars, or a radionuclide bound to a pharmaceutical molecule (Knapp F. F., 2016). Some of them may exist normally in the body, and can be of endogenous or exogenous sources, or both. The challenge of the kinetic events is the rapid metabolism of a substance and the maintenance of specific levels of the substance in the various components of its systems (Committee on State of the Science of Nuclear Medicine, 2007). Internal control mechanisms are employed for this process. Understanding

---

[1] The effective half-life, T_e, is always less than the shorter of T_p or T_b. For a very long T_p and a short T_b, T_e is almost equal to T_b. Similarly, for a very long T_b and short T_p, T_e is almost equal to T_p.

the kinetics of a substance under normal circumstances is a very important issue, in order to better understand pathophysiological conditions, since these may be a result of abnormal kinetics (Biersack H.-J. et al., 2007).

The quantification of the kinetics of substances existing in the body is a fundamental problem. Tracers allow this approach. The term *tracer* is used to describe a substance, introduced externally into the system to provide useful data for quantitative estimations of events characterizing the kinetics of the substance (Cherry S. R. et al., 2012). Actually, tracers can be substances such as dyes or substances labeled with radioactive or stable isotopes. A naturally occurring substance is called a *tracee*. An ideal tracer is a substance which gathers some specific characteristics. First of all, it must be detectable by an observer and allow to be quantified. Secondly, its introduction into a system must not perturb the system being studied and having no effect on the ongoing metabolic processes which characterize the system under study (Bentourkia M. et al., 2006). This requirement is usually met by introducing an extremely small amount of tracer compared with the amount of tracee already existing (Yeong C.-H. et al., 2014). Last but not least, an ideal tracer has to be indistinguishable with respect to the properties of the tracee system being studied. This means that both tracer and tracee follow the same processes with equal probabilities (Pecile A. et al., 1988).

12.5 THE TRACER AND THE TRACEE

The kinetics of a substance is simply presented schematically in Figure 12.1, where the circles represent the masses of two interacting substances in specific forms at specific locations, and the arrows correspond to the flux of material and biochemical transformations existing (Lambrecht R. M., 1996). Two different substances, A and B, are

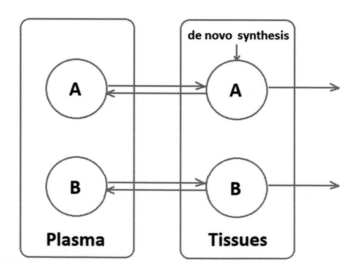

FIGURE 12.1 A schematic representation of the kinetics of a substance, with the circles representing masses and the arrows the fluxes of the substance. From this pool, it can (i) be irreversibly removed, (ii) exchange with a plasma pool, or (iii) be transformed into B. In turn, B can be exchange with a plasma pool, or irreversibly removed.

simulating the kinetics which include both transport between different locations, and biochemical transformation (Cobelli C. et al., 2002). An ideal tracer study must succeed in determination of the masses and fluxes in the selected system. Of course, it must be taken for granted that at least one component in the system is accessible for tracer administration, and tracer and tracee sampling and this special component is called the accessible pool and maybe the plasma or a tissue or a substance in expired air (Williams L. E., 1983). If it is necessary to describe this system with an emphasis in the accessibility of this component for tracee measurement, then the schematical representation should be slightly modified in order to include this parameter.

The aim of this approach is the characterization of the system by identifying the components and interconnections (Committee on the Mathematics and Physics of Emerging Dynamic Biomedical Imaging, 1996). Also, another point which needs clarification is the type of input, which can be exogenous or endogenous. Generally speaking, the tracee system (the tracer experiment), the tracer system (the relationship between the tracee and tracer systems), and the quantitation of the tracee system from the tracer data are the main subjects that must be discussed during the design of a model which focuses on a realistic representation of drug kinetics (Cobelli C. et al., 1992).

The tracee system is given in Figure 12.2. It is a single pool system which is accessible for measurement and in which it is assumed that the tracee is uniformly distributed, with the accessible pool and the system concurring in this particular situation (Moran N. E. et al., 2016).

With the assumption of that tracee system is in the steady-state case, de novo production I and disposal E are equal and constant (Marini J. C. et al., 2014). So, this means that the tracee mass M remains constant. In mathematical terms, a desired formalism could be expressed in Equation (12.10) as follows:

$$\frac{dM(t)}{dt} = I - E = 0 \tag{12.10}$$

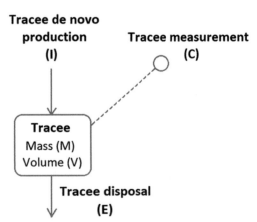

FIGURE 12.2   A schematic representation of the tracee system. The tracee system consists of a single pool of volume V, containing mass M. Tracee de novo production, I (mass/time), and disposal (utilization, elimination, or excretion), E (mass/time), occurs from this pool. The dotted line indicates tracee concentration measurement.

So, actually M is does not change with time, therefore,

$$M(t) = M = \text{constant} \tag{12.11}$$

The tracer system is also a single pool system which is accessible for measurement. In this system, the tracer is assumed to distribute uniformly (Vella A. et al., 2009). Because of tracer-tracee indistinguishability, the volume V is considered as equal to the volume of distribution of the tracee (Ramakrishnan R. et al., 2008). In this case, there is a dependence of some variables such as mass on time t ($i(t)$, $e(t)$, $m(t)$). Thus, this model can be described by Equation (12.12):

$$\frac{dm(t)}{dt} = i(t) - e(t) \quad \text{and} \quad m(0) = 0 \tag{12.12}$$

In the beginning of this hypothetic experiment, there is no tracer mass in the system, so $m(0)$ is a kind of initial condition. $m(t)$ changes with time, and hence is no longer equal to zero. The manner in which the amount of tracer is actually quantified depends upon the tracer chosen (Brooks F. J. et al., 2014). A radioactive tracer is usually quantified in terms of tracer concentration $c(t)\left(c(t) = \frac{m(t)}{v}\right)$. The most convenient way to express stable isotope measurements is the tracer mass per unit tracee mass (Equation 12.13) (Tuvdendorj D. et al., 2013):

$$z(t) = \frac{m(t)}{M} \tag{12.13}$$

As the volume V is the same for both the tracee and tracer, $z(t)$ represents the ratio between tracer and tracee concentrations (Equation 12.14):

$$z(t) = \frac{m(t)}{M} = \frac{c(t)}{C} \tag{12.14}$$

The link between the tracer and tracee system comes from the tracer-tracee indistinguishability assumption which implies that the probability that the tracer leaves the pool is equal to the probability that a particle in the pool is a tracer (Magkos F. et al., 2009). This can be written as:

$$\frac{e(t)}{E + e(t)} = \frac{m(t)}{M + m(t)} \tag{12.15}$$

or

$$e(t) = \frac{E}{M} m(t) \tag{12.16}$$

thus,

$$\frac{dm(t)}{d(t)} = i(t) - e(t) = i(t) - \frac{E}{M} m(t) = e(t) - km(t) \tag{12.17}$$

where $k = \frac{E}{M}$.

This means that a linear, constant coefficient differential equation provides the link between the tracer and tracee systems, while the tracer parameter k reflects tracee events.

In a single pool system, the unknown parameters of interest are usually E and M. If the experiment consists of injecting the tracer as a bolus of dose d at time zero, then the solution of Equation (12.17) is then converted into:

$$m(t) = de^{-kt} \tag{12.18}$$

For radioactive tracers, the concentration $c(t)$ is given by Equation (12.19):

$$c(t) = \frac{m(t)}{V} = \frac{d}{V}e^{-kt} \tag{12.19}$$

The ratio $\frac{d}{V}$ is equal with the tracer concentration at time zero, hence $V = \frac{d}{c(0)}$.

Briefly and generally, the conservation of mass principal applied to the tracer can be written as:

$$d = \int_{0}^{\infty} i(t)\,dt = \int_{0}^{\infty} e(t)\,dt \tag{12.20}$$

Since d, the total amount of tracer introduced into the system, one obtains Equation (12.21):

$$d = \int_{0}^{\infty} \frac{E}{M} m(t)\,dt \tag{12.21}$$

The recommended standard SI unit of radioactivity is undoubtedly the disintegration per second (dps) or bequerel (National Research Council [US] Committee on Evaluation of EPA Guidelines for Exposure to Naturally Occurring Radioactive Materials, 1999). Practically, the units of activity used in biomedical research are disintegrations per minute, dpm, or the curie which equals 3.7×10^{10} disintegrations per second.

For each radioactive isotope of an element, there is a proportional relationship between the mass of the isotope and the dpm emitted by that mass (Saha G., 2013). This can be written as:

$$dpm \ of \ m(t) = v \cdot m(t) \tag{12.22}$$

where v is the proportionality constant.

If $c(t)$ denotes the measurement of tracer concentration, in terms of dpm per unit volume, one obtains:

$$c(t) = \frac{vm(t)}{V} \tag{12.23}$$

While the variables for the general isotopic tracer are given in units of mass, the measurements of radioactive tracers are not in terms of mass, but energy. The mass balance (Equation 12.17) after multiplying both sides by the proportionality constant v to obtain

$$v\frac{dm(t)}{d(t)} = \frac{d(vm(t))}{d(t)} = -kvm(t) + vi(t) \tag{12.24}$$

12.6   IN VIVO KINETICS OF RADIOPHARMACEUTICALS

In vivo kinetics of radiopharmaceuticals is usually called ADME, from the initial letters of the words "Absorption", "Drug", "Elimination", and "Metabolism" (Bocci et al., 2017). The process of ADME is schematically presented in Figure 12.3.

There are many factors which influence ADME, such as the physical-chemical properties of the element and the radioactive isotope itself, the physical-chemical properties of the radiolabeled drug and the characteristics of the target – organ or the target – tissue (Bocci G. et al., 2017). The most common elements used in Nuclear Medicine, with their dominant localization and excretion, are gathered at the Table 12.1 Some physical-chemical properties of the radiolabeled drug are also very important, such as the ionization/oxidation state, the solubility at physiologic pH, and the protein/tissue/cell binding affinity (Alavijeh M. S. et al., 2005).

The rate of delivery and potential amount of drug distributed into tissues is determined by cardiac output, regional blood flow, capillary permeability, tissue volume, presence of characteristic receptors for the radiopharmaceutical administered (Jacob M. et al., 2016). The distribution to well perfused organs is faster (liver, kidney, and brain). The distribution to poorly perfused organs is slower (muscle, most visceral organs, skin, and fat).

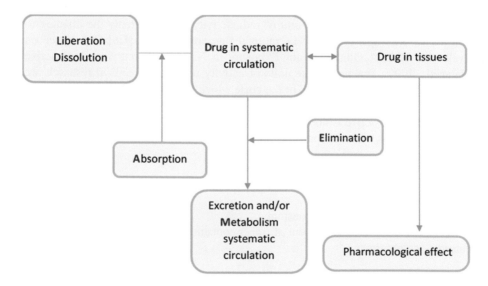

FIGURE 12.3   ADME – In vivo kinetics of radiopharmaceuticals.

TABLE 12.1

Common Elements Used in Nuclear Medicine/Their Dominant Localization and Excretion

Element	Dominant Localization	Dominant Excretion
Rb	Muscle	Urine
Cs	Muscle	Urine
Cu	Liver	Urine
Ag	Protein	Feces
Au	Liver	Urine
Sr	Bone	Feces
Ba	Bone	Urine + Feces
Ra	Bone	Urine + Feces
Ga	Bone, muscle	Feces
In	Bone, liver	Feces
Tl	Muscle	Feces
Sc	Bone, liver	Feces
Y	Bone	Urine
Sm	Liver, bone	Feces + Urine
Zr	Bone	Feces
P	Muscle, bone	Urine
Cr	Bone, blood	Urine
Mo	Liver	Urine
F	Bone	Urine
I	Thyroid	Urine
Tc	GIT	Urine

12.6.1 DRUG BIODISTRIBUTION MODELS

The simplest way to describe the process of drug distribution and elimination in the body is via the one-compartment open model (Figure 12.4a). According to this model, the drug can enter or leave the body and the body acts like a single, uniform compartment (Vauquelin G. et al., 2010). Also, the drug is injected all at once into a box or compartment and finally the drug distributes instantaneously and homogenously throughout the compartment (Zou P. et al., 2012).

The simplest route of drug administration from a modeling perspective is a rapid intravenous injection (IV bolus) (Turner P. V. et al., 2011). Drug elimination also occurs from the compartment immediately after injection. There are also more complicated modes to describe the excretion process (Figures 12.4b and c).

An important parameter which is usually utilized to characterize the distribution of a drug is the volume of distribution (V_d), which is determined as the ratio of the dose administered, divided by the plasma concentration (Equation 12.25):

$$V_d = \frac{D}{C_o} \qquad (12.25)$$

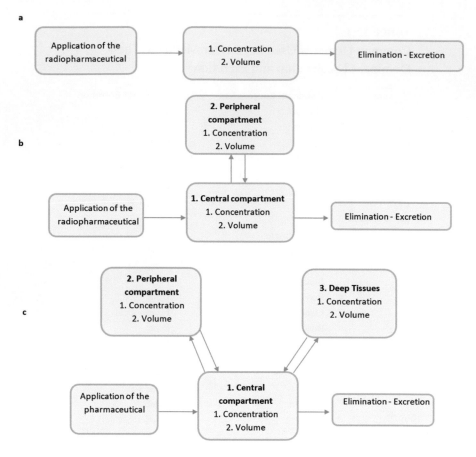

FIGURE 12.4   Compartmental distribution of radiopharmaceuticals. (a) Simple model, (b, c) complicated models.

12.6.2   ADMINISTRATION TYPES

There are many types of radiopharmaceutical administration, such as intravenously (IV), intramuscularly (IM), beneath the skin (subcutaneously, SC), oral (PO), and many others, with the first one to be the most reliable (Steele M. L. et al., 2014). The concentration of the radiopharmaceutical as a function to time is presented in Figure 12.5.

It seems that intravenous administration allows an exponential decrease of the radiopharmaceutical concentration in the bloodstream over time, while intramuscular and subcutaneous administrations show an initial increase of the radiopharmaceutical concentration and a gradual decrease (Shevtsova O.N. et al., 2017). The oral administration presents a different pattern, as there is an increase in the radiopharmaceutical concentration later, compared to the other types, as it needs more time until the digestive system through the small intestine allows the absorption of the pharmaceutical into the bloodstream (Vertzoni M. et al., 2019).

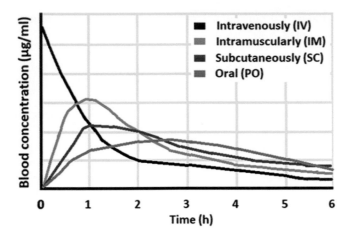

FIGURE 12.5 Radiopharmaceutical concentration according to type of administration. (Smilkov, 2017 – reprinted with further processing.)

12.6.3 BIODISTRIBUTION OF RADIOPHARMACEUTICALS

Biodistribution of radiopharmaceuticals is associated with many factors, such as the physical-chemical properties of the radiolabeled drug, the molecular weight, and the shape, especially for the protein radiopharmaceuticals, the protein binding, the lipid solubility, the partition coefficient, the presence or the absence of specific drug receptors, and the existence of specific transport mechanisms (Brandt M. et al., 2019). Moreover, among the important parameters, which have an impact on the biodistribution of radiopharmaceuticals is the stability of the radiolabeled compound, the purity of the radiopharmaceutical preparation, the pathophysiological state of the patient, and the presence or the absence of interfering drugs (Gutfilen B. et al., 2014).

Protein binding of radiopharmaceuticals is greatly influenced by a number of factors, such as the charge on the radiopharmaceutical molecule, the pH, the nature of the protein, and the concentration of anions in the plasma. Nonspecific binding to albumin and other plasma proteins is correlated positively and linearly with increasing lipophilicity (Ghafourian T. et al., 2013). Proteins contain hydroxyl, carboxyl, and amino groups, which determine their configuration and the extent and strength of protein binding to the radiopharmaceutical. Metal complexes can exchange the metal ions with proteins because of the stronger affinity of the metal for the protein. This process is called "trans-chelation" and leads to *in vivo* metabolism changes of the complex (Zeglis B. M. et al., 2014). For example, ^{67}Ga-citrate exchanges ^{67}Ga with transferrin to form ^{67}Ga-transferrin in the plasma.

The extent of protein binding of any new radiopharmaceutical should be determined before its clinical use (Matthews P. M. et al., 2012). Protein-binding of radiopharmaceuticals in plasma samples can be measured by several techniques including: Size-exclusion chromatography (SEC), Trichloroacetic acid (TCA), precipitation dialysis ultrafiltration, and the percentage of unbound proteins is given by Equation (12.26):

$$\% \text{ unbound} = \frac{C_u}{C_t} \cdot 100\% \qquad (12.26)$$

where C_u is concentration in ultrafine (unbound) and C_t – total drug concentration before the experiment.

The localization of radiopharmaceutical is related to the isotope dilution, and, via this process, the indirect estimation of the volume of RBC is feasible. Another reason of localization is the capillary blockage, through the physical trapping of particles inside capillaries and precapillary arterioles (Townsley M. I., 2012). The vascular permeability and the capillary leakage are also relevant mechanisms. Phenomena, such as the cellular migration and the cell sequestration also have an impact on the biodistribution of radiopharmaceuticals (Vallabhajosula S. et al., 2010). Furthermore, the passive diffusion is of crucial importance and follows the Fick's Law (Lecca P. et al., 2012). According to this law, the rate of diffusion is a function of the concentration gradient. It does not require the input of other external energy. There is no need of transporters, carriers, or other receptors to be involved; so passive diffusion is nonselective. It is not competitively inhibited by similar molecules and is a process which is not subject to saturation. The facilitated diffusion reflects glucose metabolism. It is a carrier-mediated and selective mechanism, which can be competitively inhibited (Stremmel W. et al., 1986). The active transport is carrier-mediated, selective, and relevant to Na$^+$/K$^+$ pump with great requirements for energy (ATP). Secretion is a special category of active transport out of glands and other tissues. Important examples are the gastric secretion, the kidney tubular secretion and the bile secretion (Yin J. et al., 2016).

Filtration or pore/convective transport is a kind of diffusion involving transit of molecules through pores or channels, due to hydrostatic or osmotic pressure gradient. No transporters, carriers, or other receptors are involved; thus, it is a nonselective process.

The phagocytosis also plays important role in localization of radiopharmaceuticals, as cells can engulf a particle and encapsulate it (Choudhury P. S. et al., 2012). The localization of radiolabeled metabolic precursors and substrates is based on the metabolic trapping (e.g. metabolic trapping of fludeoxyglucose [FDG]). The uptake of a radiolabeled amino acid within tumors may reflect the increased protein synthesis rate of proliferating tumor cells or simply an increased rate of amino acid transport across the tumor cell membrane.

Since some types of radiopharmaceuticals are transported into the cell cytoplasm, by diffusion, enzymatic reductions convert part of them into free radical anions (Maulucci G. et al., 2016). In hypoxic tissue, the free radical is further reduced to a series of reactive species, attached irreversibly to cellular macromolecules and retained within the cell. Normoxic cells can wash them out of their trap inside the hypoxic tissue. Regarding to cell proliferation, the growth rate of tumors correlates with their level of differentiation (Lee C.-T. et al., 2014). Thus, the most malignant tumors grow more rapidly than benign ones. This means that this increased mitotic activity in tumor tissue leads to an increased requirement of substrates (nucleotides) for DNA synthesis. The radiopharmaceuticals often have a nucleonic part for easier binding to cancer cells (Sharma P. et al., 2016).

The specific receptor binding is mainly based on somatostatin receptors (SSTRs). SSTRs have been identified on many cells of neuroendocrine origin as well as on lymphocytes. Most neuroendocrine tumors, small cell lung cancers, and medullary thyroid carcinomas express SSTRs in high density (Sharma P. et al., 2014). So, it is important to design imaging and therapy of SSTR-positive tumors. Moreover, increased Vasoactive Intestinal Peptide (VIP) receptor expression has been seen on adenocarcinomas, breast cancers, melanomas, neuroblastomas, and pancreatic carcinomas (Sharma P. et al., 2016).

The majority of breast cancers are hormone dependent, as indicated by increased expression of intracellular estrogen or progesterone receptors. The common radiopharmaceuticals, which are used in these cases, are transported into the cell by passive diffusion and bind to steroid receptors within the nucleus. Low-density lipoprotein (LDL) carries cholesterol to the adrenal glands as a substrate for synthesis of adrenal steroid hormones (cortisol and aldosterone) (Hu J. et al., 2010). So, specific radiopharmaceuticals, with a norcholesterol molecule, can be used for imaging of patients with adrenal cortical diseases, as they bind to LDL receptors. Furthermore, radiolabeled antibodies are monoclonal murine IgG antibodies directed to CD20 receptors on B-cells and non-Hodgkin's lymphoma (NHL) tumor cells (Mohammed R. et al., 2019).

12.6.4 Metabolism of Radiopharmaceuticals

The drug metabolism or, in other words, the biotransformation reactions are classified as: Phase I or functionalization reactions and Phase II or biosynthetic (conjugation) reactions. The functionalization reactions introduce or expose (unmask) a functional group (-OH, -NH$_2$, -SH, -COOH) of the active substance. These reactions are characterized by loss of pharmacological activity or formation of chemically reactive and more toxic, carcinogenic, or immunogenic metabolites. The biosynthetic reactions lead to the formation of a covalent linkage between a functional group on the active substance or Phase I metabolite with endogenously derived glucuronic acid, sulfate, glutathione, amino acids, or acetate (Lu S. C., 2013). These complexes are highly polar and are generally inactive and able to be excreted rapidly in the urine and feces.

The enzyme systems involved in Phase I reactions are located primarily in the endoplasmic reticulum. These biotransforming reactions are mainly carried out by Cytochrome P450s[2] isoforms (CYPs) and Flavin-containing monooxy-

---

[2] Cytochrome P450s (CYPs) are a superfamily of heme-thiolate containing enzymes, playing a major role in the metabolism of many drugs and other xenobiotics. A number of carcinogens are also metabolized by CYPs and it is often these metabolites which are the ultimate carcinogenic species. CYPs catalyze the oxidation of bound substrates through the redox action of the heme moiety and the activation of molecular oxygen and are able to carry out a variety of hydroxylations, dealkylations, and heteroatom oxidations. Substrate specificity is very low for the CYP enzyme complex. High-lipid solubility is the only common property that renders CYP substrates a wide variety of structurally unrelated drugs, ranging from a molecular weight of 28Da (ethylene) to 1203 Da (cyclosporine). Larger molecules, such as proteins, are not CYP substrates and little is known regarding their catabolism. It is believed that therapeutic proteins are metabolized by the same catabolic pathways as endogenous proteins and can be broken down into amino acid fragments. Generally, the metabolic products of proteins are not considered a safety risk and classical biotransformation studies as performed for small molecules are not needed (Saxena A. et al., 2008).

genase (FMO), esterases, and amidases. These reactions are catabolic, such as oxidation, reduction, or hydrolysis. The Phase II conjugation enzyme systems are mainly cytosolic.

In the process of designing new radiometal-labeled monoclonal antibodies (mAbs) and peptides for diagnostic imaging and targeted radiotherapy of cancer, the issue of metabolism of the radiopharmaceutical is often overlooked (Verel I. et al., 2003). When evaluating a Nuclear Medicine image of a radiolabeled mAb or peptide, a question arises related to the compound which is actually being observed in the image. The purity of the radiopharmaceutical is standard, but there is an uncertainty about the rate of metabolism of these compounds and also about the exact chemical form of the radiolabeled metabolite, presented at the various imaging times post-injection (Sharma R. et al., 2011). The fate of the radionuclide, meaning whether it remains bound to the mAb or peptide or whether it is metabolized, is of great significance, because it will ultimately determine the absorbed dose of the radiopharmaceutical to the tumor and normal tissues. The understanding of how radiometal-labeled proteins and peptides are metabolized requires taking several factors into consideration (Tibbitts J. et al., 2016). The most important of them is the enzymatic breakdown of a protein or peptide into smaller peptide fragments that may or may not be attached to the radiometal chelate, the enzymatic breakdown of the protein or peptide followed by acetylation or addition of another functional group and the dissociation of the radiometal from the chelator. According to recent studied, it seems that the metabolism of radiometal-labeled mAbs is highly dependent on the radiometal (DeNardo G. L. et al., 2001).

12.6.5 Routes of Drug Elimination

As it was previously mentioned, the possible routes of drug elimination include the urine (kidney), the secretion into the Gastrointestinal (GI) tract (feces), the liver, the lung, and other sites such as skin, milk, tears, saliva, and others. The mechanisms of renal excretion include the glomerular filtration, the active tubular secretion, and the passive and active tubular reabsorption. The glomerular filtration is nonselective. The unbound substances with molecular weight < 40,000 Da are eliminated by glomerular filtration. The substances tightly bound to plasma proteins are not filtered by the glomeruli (Kapusta D., 2007). The active tubular secretion is followed by many organic acids and some organic bases which are secreted by renal tubules. Lipid-soluble drugs tend to be reabsorbed by the renal tubules and are not excreted, exemplifying the passive and active tubular reabsorption.

The main pathways for excretion and clearance of radiopharmaceuticals are via urine, feces, and lungs (gases). Specifically, radiolabeled peptides due to small molecular weight are excreted rapidly in urine. The effective clearance of radionuclides is of crucial importance to avoid side effects (Gudkov S. V. et al., 2016). The general formula for calculating clearance (CL) in first-order kinetics for drugs is given by Equation (12.27):

$$CL = V_d \times k_e \qquad\qquad (12.27)$$

where V_d is the volume of distribution and k_e the clearance constant. In particular for radiopharmaceuticals, excretion is calculated easily by renal clearance of blood, with Equation (12.28):

$$CL = U \cdot V \cdot \frac{1}{B} \tag{12.28}$$

where U is the urinary concentration of the radiopharmaceutical, V the volume of urine per time, and B the blood concentration of the radiopharmaceutical.

12.7 MATLAB AS A TOOL FOR THE ESTIMATION OF THE KINETICS OF RADIOPHARMACEUTICALS

MATLAB can be used for the estimation of the chemical kinetics of the radiopharmaceuticals, through the interactive MATLAB document, allowing accuracy in the design of a diagnostic or a therapeutic scheme. For this reason, various models have been developed. Pharmacokinetic/Pharmacodynamic (PK/PD) model is thoroughly discussed below.

12.7.1 PK/PD Model

PK/PD models can describe the relation between drug dosing, concentration, and efficacy. Thus, PK/PD modeling, as an integral component of the drug development process, is an additional mathematical technique for predicting the effect and efficacy of drug dosing over time (Slater H. C. et al., 2017). Generally speaking, PK models describe how the body can react to a drug in terms of absorption, distribution, metabolism, and excretion. Actually, PD models describe how a drug affects the body by linking the drug concentration to an efficacy metric. If a PK/PD model is quite well-characterized, then it could be an important tool in guiding the design of future experiments and trials.

The PK/PD modeling process includes the following steps (MathWorks. https://la.mathworks.com):

- Import, process, and visualizing of time-course data
- Selection of a PK model from a library, or creation of mechanism-based PK/PD models using the interactive block-diagram editor
- Estimation of model parameters using nonlinear regression
- Exploration of system dynamics, using parameter sweeps and sensitivity analysis
- Simulation of dosing strategies and what-if scenarios

The most useful software to create a PK/PD modeling is *SimBiology* by MathWorks.

12.7.2 Creation of a PK/PD Model

For the creation of a PK model is of crucial importance to use a model construction wizard that let us to specify the number of compartments, the route of administration,

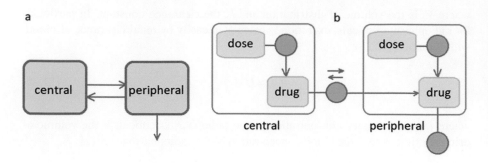

FIGURE 12.6 Comparison of two different views in PK. (a) Typical representation if one type of 2-compartment model, (b) Same 2-compertment model as represented in the *SimBiology* desktop.

and the type of elimination (Stroh M. et al., 2013). Otherwise, one can extend any model to build higher fidelity models or to build or load a new model, loading a *SimBiology* project or SBML model.

In Figure 12.6, there is a comparison between a model as typically represented in pharmacokinetics with the same model shown in the *SimBiology* model diagram. For this example, the assumption is that there is an administration of a drug using a two-compartment model with any dosing input and linear elimination kinetics (Peletier L. A. et al., 2017). The main model structure will remain the same with any dosing type.

SimBiology can represent the concentration or amount of a drug in a given compartment or volume by a species *object* contained within the compartment. It can also represent the exchange or flow of the drug between compartments and the elimination of the drug by reactions (Badhan R. K. S. et al., 2017). Moreover, *SimBiology* can give a sense of the intercompartmental clearance by a parameter (Q) which specifies the clearance between the compartments. *SimBiology* can drive the dosing schedule with a combination of species (Drug and/or Dose) and reactions (Dose -> Drug), depending on whether the administration into the compartment follows bolus, zero-order, infusion, or first-order dosing kinetics.

This model can also be considered as a regression function, $y = f(k,i)$, where y is the predicted value, given values of an input i, and parameter values k.

12.7.3 DOSING TYPES

SimBiology creates the following model components for each compartment in the model, regardless of the dosing type (MathWorks. https://la.mathworks.com):

1. Two species (Drug_CompartmentName and Dose_CompartmentName) for each compartment.
2. A reaction (Dose_CompartmentName -> Drug_CompartmentName) for each compartment, governed by mass action kinetics.

3. A parameter (ka_CompartmentName) for each compartment, representing the absorption rate of the drug when absorption follows first-order kinetics.

   ```
   This is the forward rate parameter for the Dose_
   CompartmentName -> Drug_CompartmentName reaction.
   ```

4. A parameter (Tk0_CompartmentName) for each compartment, representing the duration of drug absorption when absorption follows zero-order kinetics.
5. A parameter (TLag_CompartmentName) for each compartment, representing the time lag for any dose that targets that compartment and also that is specified as having a time lag.

For treatments that have a fixed infusion or absorption duration (infusion and zero-order), is also possible to use overlapping doses. The doses are additive (Mould D. R. et al., 2013).

12.7.4 CREATION OF A PHARMACOKINETIC MODEL USING THE COMMAND LINE

There are some specific steps in order to create a PK model with the specified number of compartments, dosing type, and method of elimination (Srimani J. et al., 2016).

1. Creation of a PKModelDesign object. The PKModelDesign object allows to specify the number of compartments, route of administration, and method of elimination, which *SimBiology* uses to construct the model object with the necessary compartments, species, reactions, and rules.

   ```
   pkm = PKModelDesign;
   ```

2. Addition of a compartment, specifying the compartment name, and optionally, the type of dosing, and the method of elimination. Also, it is important to specify whether the data contains a response variable measured in this compartment and whether the dose(s) have time lags (MathWorks. https://la.mathworks.com). For instance, for a tobramycin data set [1], one can specifies a compartment named Central, with Bolus for the DosingType property, linear-clearance for the EliminationType property, and true for the HasResponseVariable property.

   ```
   pkc1 = addCompartment(pkm, 'Central', 'DosingType',
   'Bolus', ...
               'EliminationType', 'linear-clearance', ...
               'HasResponseVariable', true);
   ```

3. Optionally, a second compartment named Peripheral, with no dosing, no elimination, and no time lag can be added. Set the HasResponseVariable property to true.

```
pkc2 = addCompartment(pkm, 'Peripheral',
'HasResponseVariable', true);
```

The model construction process adds the necessary parameters, including a parameter representing intercompartmental clearance Q. It is possible to add more compartments by repeating this step. The addition of each compartment creates a chain of compartments in the order of compartment addition, with a bidirectional flow of the drug between compartments in the model.

4. Construction of a *SimBiology* model object.

```
[modelObj, PKModelMapObj] = pkm.construct
```

The construct method returns a *SimBiology* model object (modelObj) and a PKModelMap object (PKModelMapObj) which contains the mapping of the model components to the elements of the regression function.

5. Performance of parameter fitting as shown in Perform Data Fitting with PKPD Models. The model object and the PKModelMap object are input arguments for the sbionlmefit, sbionlmefitsa, and sbionlinfit functions used in parameter fitting.

REFERENCES

Alavijeh M. S., Chishty M., Qaiser M. Z., Palmer A. M. (2005) Drug metabolism and pharmacokinetics, the blood-brain barrier, and central nervous system drug discovery, NeuroRx. 2(4): 554–571.

Arthofer W., Decristoforo C., Schlick-Steiner B. C., Steiner F. M. (2016) Ultra-low activities of a common radioisotope for permission-free tracking of a drosophilid fly in its natural habitat, Sci Rep. 6: 36506.

Badhan R. K. S., Khadke S., Perrie Y. (2017) Application of pharmacokinetics modelling to predict human exposure of a cationic liposomal subunit antigen vaccine system, Pharmaceutics. 9(4): 57.

Bentourkia M., Zaidi H. (2006) Tracer Kinetic Modeling in Nuclear Medicine: Theory and Applications, Quantitative Analysis in Nuclear Medicine Imaging, Springer, Germany.

Biersack H.-J., Freeman L.M. (2007) Clinical Nuclear Medicine, Springer, Germany.

Bocci G., Carosati E., Vayer P., Arrault A., Lozano S., Cruciani G. (2017) ADME-space: a new tool for medicinal chemists to explore ADME properties, Sci Rep. 7: 6359.

Brandt M., Cardinale J., Giammei C., Guarrochena X., Happl B., Jouini N., Mindt T. L. (2019) Mini-review: targeted radiopharmaceuticals incorporating reversible, low molecular weight albumin binders, Nucl Med Biol. 26(70): 46–52.

Brooks F. J., Grigsby P. W. (2014) The effect of small tumor volumes upon intra-tumoral tracer uptake heterogeneity studies, J Nucl Med. 55(1): 37–42.

Cherry S. R., Sorenson J. A., Phelps M. E. (2012) Physics in Nuclear Medicine, Fourth edition, Saunders, Elsevier.

Choudhury P. S., Savio E., Solanki K. K., Alonso O., Gupta A., Gambini J. P., Doval D., Sharma P., Dondi M. (2012) 99mTc glucarate as a potential radiopharmaceutical agent for assessment of tumor viability: from bench to the bed side, World J Nucl Med. 11(2): 47–56.

Cobelli C., Foster D., Toffolo G. (2002) Tracer Kinetics in Biomedical Research from Data to Model, Kluwer Academic Publishers, The Netherlands.

Cobelli C., Toffolo G., Foster D. (1992) Tracer-to-tracee ratio for analysis of stable isotope tracer data: link with radioactive kinetic formalism, Am J Physiol. 262(6 Pt 1): E968–E975.

(1996) Committee on the Mathematics and Physics of Emerging Dynamic Biomedical Imaging, National Academic Press, Washington D.C.

Committee on State of the Science of Nuclear Medicine (2007) Advancing Nuclear Medicine through Innovation, National Academy of Sciences, Washington, D.C., U.S.A.

Cox P. H., Mather S. J., Sampson C. B., Lazarus C. R. (1986) Progress in Radiopharmacy, Martinus Nijhoff Publishers, Belgium.

DeNardo G. L., DeNardo S. J., Kukis D. L., O'Donnell R. T., Shen S., Mirick G. R., Meares C. F. (2001) Metabolite production in patients with lymphoma after radiometal-labeled antibody administration, J Nucl Med. 42 (9): 1324–1333.

Ding H., Wu F. (2012) Image guided biodistribution and pharmacokinetic studies of theranostics, Theranostics. 2(11): 1040–1053.

Ghafourian T., Amin Z. (2013) QSAR models for the prediction of plasma protein binding, Bioimpacts. 3(1): 21–27.

Groch M. W. (1998) Radioactive decay, Radiographics. 18(5):1247–1256.

Gutfilen B., Valentini G. (2014) Radiopharmaceuticals in nuclear medicine: recent developments for SPECT and PET studies, Biomed Res Int. 2014: 426892.

Gudkov S. V., Shilyagina N. Yu., Vodeneev V. A., Zvyagin A. V. (2016) Targeted radionuclide therapy of human tumors, Int J Mol Sci. 17(1): 33.

Hu J., Zhang Z., Shen W.-J., Azhar S. (2010) Cellular cholesterol delivery, intracellular processing and utilization for biosynthesis of steroid hormones, Nutr Metab (Lond). 7: 47.

Jacob M., Chappell D., Becker B. F. (2016) Regulation of blood flow and volume exchange across the microcirculation, Crit Care. 20: 319.

Kapusta D. (2007) Drug Excretion, xPharm: The Comprehensive Pharmacology Reference, 1–2.

Knapp F. F. (Russ), Dash A. (2016) Radiopharmaceuticals for Therapy, Springer, India.

Kowalsky R. J., Falen S. W. (2004) Radiopharmaceuticals in Nuclear Pharmacy and Nuclear Medicine, Second Edition, American Pharmacists Association, Washington D.C.

Lambrecht R. M. (1996) Biological models in radiopharmaceutical development, In: Developments in Nuclear Medicine. Kluwer Academic Publishers, The Netherlands.

L'Annunziata M. F. (2016) Chapter 18 – Radionuclide decay, radioactivity units, and radionuclide mass, In: Radioactivity: Introduction and History, from the Quantum to Quarks, 2nd edition, pp. 621–638.

Lecca P., Morpurgo D. (2012) Modelling non-homogeneous stochastic reaction-diffusion systems: the case study of gemcitabine-treated non-small cell lung cancer growth, BMC Bioinformatics. 13(Suppl 14): S14.

Lee C.-T., Boss M.-K., Dewhirst M. W. (2014) Imaging tumor hypoxia to advance radiation oncology, Antioxid Redox Signal. 21(2): 313–337.

Leslie W. D., Greenberg I. D. (2003) Nuclear Medicine, Landes Bioscience, Texas, U.S.A.

Lin C.-c., Xu C., Zhu N., Yeh L.-T. (2006) Absorption, metabolism, and excretion of [^{14}C] viramidine in humans, Antimicrob Agents Chemother. 50(7): 2368–2373.

Lu S. C. (2013) Glutathione synthesis, Biochim Biophys Acta. 1830(5): 3143–3153.

Magkos F., Mittendorfer B. (2009) Stable isotope-labeled tracers for the investigation of fatty acid and triglyceride metabolism in humans in vivo, Clin Lipidol. 4(2): 215–230.

Marini J. C., Didelija I. C., Fiorotto M. L. (2014) Extrarenal citrulline disposal in mice with impaired renal function, Am J Physiol Renal Physiol. 307(6): F660–F665.

Masok F. B., Masiteng P. L., Mavunda R. D., Maleka P. P. (2016) Health effects due to radionuclides content of solid minerals within port of richards bay, South Africa, Int J Environ Res Public Health. 13(12): 1180.

MathWorks. https://la.mathworks.com

Matthews P. M., Rabiner E. A., Passchier J., Gunn R. N. (2012) Positron emission tomography molecular imaging for drug development, Br J Clin Pharmacol. 73(2): 175–186.

Maulucci G., Bačić G., Bridal L. et al. (2016) Imaging reactive oxygen species-induced modifications in living systems, Antioxid Redox Signal. 24(16): 939–958.

Mohammed R., Milne A., Kayani K., Ojha U. (2019) How the discovery of rituximab impacted the treatment of B-cell non-Hodgkin's lymphomas, J Blood Med. 10: 71–84.

Moran N.E., Novotny J. A., Cichon M. J. et al. (2016) Absorption and distribution kinetics of the ^{13}C-labeled tomato carotenoid phytoene in healthy adults, J Nutr. 146 (2): 368–376.

Mould D. R., Upton R. N. (2013) Basic concepts in population modeling, simulation, and model-based drug development – Part 2: introduction to pharmacokinetic modeling methods, CPT Pharmacometrics Syst Pharmacol. 2(4): e38.

National Research Council (US), Committee on Evaluation of EPA Guidelines for Exposure to Naturally Occurring Radioactive Materials (1999). Evaluation of Guidelines for Exposures to Technologically Enhanced Naturally Occurring Radioactive Materials, National Academies Press, Washington D.C.

Nayak T. K., Brechbiel M. W. (2009) Radioimmunoimaging with longer-lived positron-emitting radionuclides: potentials and challenges, Bioconjug Chem. 20(5): 825–841.

Pecile A., Rescigno A. (1988) Pharmacokinetics Mathematical and Statistical Approaches to Metabolism and Distribution of Chemicals and Drugs, Springer, Germany.

Peletier L. A., de Winter W. (2017) Impact of saturable distribution in compartmental PK models: dynamics and practical use, J Pharmacokinet Pharmacodyn. 44(1): 1–16.

Ramakrishnan R., Ramakrishnan J. D. (2008) Utilizing mass measurements in tracer studies – a systematic approach to efficient modeling, Metabolism. 57(8): 1078–1087.

Saha G. (2013) Physics and radiobiology of nuclear medicine, Fourth Edition, Springer, New York.

Saxena A., Parijat Tripathi K., Roy S., Khan F., Sharma A. (2008) Pharmacovigilance: effects of herbal components on human drugs interactions involving cytochrome P450, Bioinformation. 3(5): 198–204.

Shargel L, Andrew B.C. (2016) Applied biopharmaceutics & pharmacokinetics, Seventh Edition, McGraw Hill Education, USA.

Sharma R., Aboagye E. (2011) Development of radiotracers for oncology – the interface with pharmacology, Br J Pharmacol. 163(8): 1565–1585.

Sharma P., Singh H., Bal C., Kumar R. (2014) PET/CT imaging of neuroendocrine tumors with 68Gallium-labeled somatostatin analogues: an overview and single institutional experience from India, Indian J Nucl Med. 29(1): 2–12.

Sharma P., Mukherjee A. (2016) Newer positron emission tomography radiopharmaceuticals for radiotherapy planning: an overview, Ann Transl Med. 4(3): 53.

Sharp P. F., Gemmell H. G., Murray A. D. (2005) Practical Nuclear Medicine, Third Edition, Springer, London.

Shevtsova O. N., Shevtsova V. K. (2017) Mathematical simulation of transport kinetics of tumor-imaging radiopharmaceutical 99mTc-MIBI, Comput Math Methods Med. 2017: 2414878.

Slater H. C., Okell L. C., Ghani A. C. (2017) Mathematical modelling to guide drug development for malaria elimination, Trends Parasitol. 33(3): 175–184.

Smilkov K. (2017), In Vivo Kinetics of Radiopharmaceuticals – ADME, IAEE.

Srimani J., Moffitt R. A., Wang M. D. (2016) WebPK, a web-based tool for custom pharmacokinetic simulation, Conf Proc IEEE Eng Med Biol Soc. 2010: 1494–1497.

Steele M. L., Axtner J., Happe A., Kröz M., Matthes H., Schad F. (2014) Safety of intravenous application of mistletoe (Viscum album L.) preparations in oncology: an observational study, Evid Based Complement Alternat Med. 2014: 236310.

Stremmel W., Berk P. D. (1986) Hepatocellular uptake of sulfobromophthalein and bilirubin is selectively inhibited by an antibody to the liver plasma membrane sulfobromophthalein/bilirubin binding protein, J Clin Invest. 78(3): 822–826.

Stroh M., Hutmacher M. M., Pang J., Lutz R., Magara H., Stone J. (2013) Simultaneous pharmacokinetic model for rolofylline and both M1-trans and M1-cis metabolites, AAPS J. 15(2): 498–504.

Thie J. A. (2012) Nuclear Medicine Imaging: An Encyclopedic Dictionary, Springer: Springer-Verlag Berlin Heidelberg.

Tibbitts J., Canter D., Graff R., Smith A., Khawli L. A. (2016) Key factors influencing ADME properties of therapeutic proteins: a need for ADME characterization in drug discovery and development, MAbs. 8(2): 229–245.

Townsley M. I. (2012) Structure and composition of pulmonary arteries, capillaries and veins, Compr Physiol. 2: 675–709.

Turner P. V., Brabb T., Pekow C., Vasbinder M.A. (2011) Administration of substances to laboratory animals: routes of administration and factors to consider, J Am Assoc Lab Anim Sci. 50(5): 600–613.

Tuvdendorj D., Chinkes D. L., Herndon D. N., Zhang X.-J., Wolfe R. R. (2013) A novel stable isotope tracer method to measure muscle protein fractional breakdown rate during a physiological non-steady-state condition, Am J Physiol Endocrinol Metab. 304 (6): E623–E630.

Vallabhajosula S., Killeen R. P., Osborne J. R. (2010) Altered biodistribution of radiopharmaceuticals: role of radiochemical/pharmaceutical purity, physiological, and pharmacologic factors, Semin Nucl Med. 40(4): 220–241.

Vauquelin G., Charlton S. J. (2010) Long-lasting target binding and rebinding as mechanisms to prolong in vivo drug action, Br J Pharmacol. 161(3): 488–508.

Vella A., Rizza R. A. (2009) Application of isotopic techniques using constant specific activity or enrichment to the study of carbohydrate metabolism, Diabetes. 58(10): 2168–2174.

Verel I., Visser G. W., Boellaard R., Stigter-van Walsum M., Snow G. B., van Dongen G. A. (2003) 89Zr immuno-PET: comprehensive procedures for the production of 89Zr-labeled monoclonal antibodies, J Nucl Med. 44(8):1271–1281.

Vertzoni M., Augustijns P., Grimm M. et al. (2019) Impact of regional differences along the gastrointestinal tract of healthyadults on oral drug absorption: an UNGAP review, Eur J Pharmaceutical Sci. 134: 153–175.

Williams L. E. (1983) Radiopharmaceuticals. Introduction to Drug Evaluation and Dose Estimation, CRC Press, USA.

Yeong C.-H., Cheng M.-h, NG K.-H. (2014) Therapeutic radionuclides in nuclear medicine: current and future prospects, J Zhejiang Univ Sci B (Biomedicine & Biotechnology). 15(10): 845–863.

Yin J., Wang J. (2016) Renal drug transporters and their significance in drug–drug interactions, Acta Pharm Sin B. 6(5): 363–373.

Zeglis B. M., Houghton J. L., Evans M. J., Viola-Villegas N., Lewis J. S. (2014) Underscoring the influence of inorganic chemistry on nuclear imaging with radiometals, Inorg Chem. 53(4): 1880–1899.

Zou P., Yu Y., Zheng N., Yang Y., Paholak H. J., Yu L. X., Sun D. (2012) Applications of human pharmacokinetic prediction in first-in-human dose estimation, AAPS J. 14(2): 262–281.

13 Nanotechnology in Nuclear Medicine/ MATLAB Use

Nefeli Lagopati

CONTENTS

13.1 NANOPARTICLES: THERANOSTICS AND DRUG DELIVERY SYSTEMS IN NUCLEAR MEDICINE

The term and the concept of "theranostics" (also known as "theragnostics") refers to an integrated approach to diagnosis and therapy using suitable combinations of molecular targeting vectors and radionuclides (Yeong C.-H. et al., 2014). This can be achieved by using therapeutic radionuclides that also emit radiation for imaging, such as ^{111}In, ^{131}I, ^{177}Lu, and ^{166}Ho, for both therapy and diagnostic purposes (Li T. et al., 2017).

Theranostics is usually performed using molecular targeting vectors, such as peptides, labeled with either diagnostic or therapeutic radionuclides, targeted specifically by the vector at its molecular level (Bavelaar B. M. et al., 2018). The use of ^{68}Ga-labeled tracers is quite standard, and allows the diagnosis using ^{68}Ga to be effectively followed by therapy using therapeutic radionuclides such as ^{90}Y and ^{177}Lu labeled with the same tracer for personalized radionuclide therapy. The therapeutic radionuclides, which are most commonly used, are mainly nuclear reactor-produced radionuclides, including ^{47}Sc, ^{90}Y, ^{131}I, ^{166}Ho, ^{177}Lu, ^{188}Re, and ^{213}Bi (Müller C. et al., 2017).

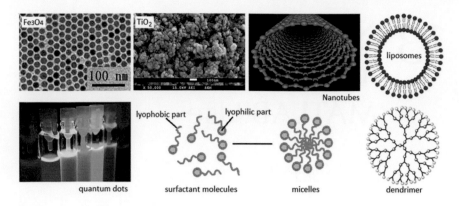

FIGURE 13.1 Various types of nanoparticles that can be radiolabeled for molecular imaging and targeted radionuclide therapy.

13.2 NANOTECHNOLOGY IN THERANOSTICS

Radiolabeled nanoparticles that can be used as platforms for attaching different functionalities for the purposes of multimodality molecular imaging and multivalent targeted therapy have been proven to be promising diagnostic and therapeutic tools (Ferro-Flores G. et al., 2014).

Recently, nanotechnology has shown a great potential for the early detection, accurate diagnosis, and personalized treatment of various diseases, especially in cancer therapy (Huang Q. et al., 2018). Nanoparticles have a size comparable to biological molecules such as antibodies, and they are about 100–10,000 times smaller than human cells (Hoshyar N. et al., 2016). It is well known that nanoparticles can interact with the biomolecules both on the surface and inside the cells (Xing Z.-C. et al., 2010). Nanoparticles can be very useful for internal radiation therapy through passive targeting and/or active targeting (Ventola C. L., 2017). Liposomes, spherical vesicles of lipid bilayers ranging from 100 to 800 nm diameter, are currently widely used nanoparticles for cancer therapy (Yingchoncharoen P. et al., 2016). There is great variety of nanomaterials, which have been approved for human use or are currently in clinical trials, nanoparticles (including iron oxide, titanium dioxide, etc.), nanotubes, quantum dots, micelles, and dendrimers (Figure 13.1) (Lagopati N. et al., 2014; Digesu C. S. et al., 2016).

The ultimate goal of nanoparticle-based radionuclide therapy is the achievement of an efficient and specific *in vivo* delivery of therapeutic radionuclides without systemic toxicity (Boschi A. et al., 2017). It will also facilitate imaging and evaluation of dose delivery and therapeutic efficacy. With the capacity to provide enormous sensitivity and flexibility, nanoparticle-based radionuclide therapy has great potential to improve cancer therapies in the near future (Yeong C.-H. et al., 2014).

13.3 DRUG DELIVERY SYSTEMS IN NUCLEAR MEDICINE

Drug delivery systems provide new perspectives in medicine. Recently, regarding the possible decrease in the number of new pharmacological agents, registered with the Food and Drug Administration (FDA), a significant number of tested entities' development is abandoned due to low activity at the target site, toxicity, unsuitable pharmacokinetic profiles, and the lack of absorption across biological membranes (Greig N. H. et al., 2013). Drug delivery systems can promote the drug concentration reaching the target site and decrease the pharmacokinetic profiles which are non-favorable (Wen H. et al., 2015). Efficacy and safety must be considered, as well as the possibility to administer a lower dose reducing the risk of adverse effects (Daughton C. G. et al., 2013).

Since there is an evaluating shortage of medical isotopes, an effective delivery of radiopharmaceuticals will ensure a decrease in the demand for these agents (Ballinger J. R., 2010). Drug delivery systems could provide a dose reduction without the need of any investment in new hardware by medical facilities (Coelho J. F. et al., 2010). Selective targeting to the regions of interest (ROIs) is also important for enhancing image quality and decreasing the presence of artefacts (Panjnoush M. et al., 2016). As Low As Reasonably Achievable (ALARA) principle will always be considered in Nuclear Medicine and drug delivery systems seem to contribute further to the safety of medical personnel and patients (Strauss K. J. et al., 2006).

The development of new theranostic radiotracers allows the targeting, increasing the amount of irradiation during the treatment phase, and decreasing the possible side effects in unspecific delivery areas (Yeong C.-H. et al., 2014; Zhu L. et al., 2017). With regard to radiopharmaceuticals, a theranostic agent would combine a positron emission tomography (PET) or single-photon emission computed tomography (SPECT) isotope to act as a biomarker for imaging and simultaneously incorporate a therapeutic entity that will be delivered at the same area of biodistribution (Agdeppa E. D. et al., 2009). Actually, radiopharmaceutical theranostics focus on cancer treatment, but other applications are currently under investigation in order to combat infectious diseases (Yordanova A., et al., 2017).

Diagnosis is recently based on the information received from more than one modality to increase the accuracy (Maas M. et al., 2011). The concurrent measurement of both anatomic and physiological data allows the accurate localization of nuclear signals. Hybrid imaging with PET/CT and SPECT/CT has become standard practice, and individual scanners are rarely employed (van Dalen J. A. et al., 2007, Kleynhans J. et al., 2018). Some innovative systems, such as PET/magnetic resonance imaging (MRI), sometimes require the combination of more than one tracer, perhaps a radiopharmaceutical in parallel with a contrast agent (Delso G. et al., 2011). These individual entities could be bound in one drug delivery system to ensure that all of the administered tracers behave in the same pharmacokinetic pattern (Tiwari G. et al., 2012). Radiopharmaceuticals in Nuclear Medicine are routinely labeled with an isotope, following their movement in living organisms without affecting the functionality of the biological system (Arthofer W. et al., 2016). So, from this point of view, radiopharmaceuticals can already be viewed as a carrier system locating the selected isotope at a specific target site (Kleynhans J. et al., 2018).

13.3.1 LIPOSOMES AND MICELLES IN DRUG DELIVERY SYSTEMS

Liposomes act as carrier of drug systems that incorporate bilayers of lipids encapsulating a hydrophilic interior where the active pharmaceutical ingredient accumulates (Figure 13.2) (Bozzuto G. et al., 2015, Daraee H. et al., 2016). Liposomes can be

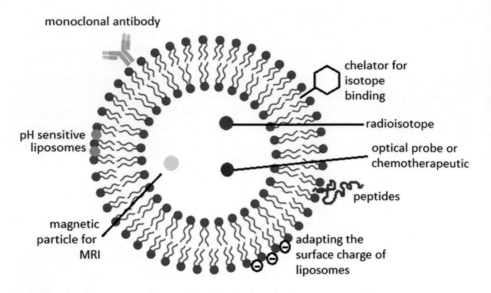

FIGURE 13.2   A liposome with the different adaptations.

tailor-made to suit the entity to be delivered and considered the characteristics of the cellular target to be reached (Kang J. H. et al., 2017, Kleynhans J. et al., 2018). The hydrophilic pharmaceutical active ingredient can become entrapped in the system which allows the transportation through hydrophobic biological barriers to release the load at the region of interest (Singh R. et al., 2009). An interesting characteristic of liposomes is the accumulation by enhanced permeation and retention (EPR) (Wong A. D. et al., 2015, Golombek S. K. et al., 2018). The modification of the surface of the liposome could include the encapsulation of monoclonal antibodies for immunoradiotherapy or targeting ligands and could ensure the distribution of the liposome to the target tissue (Liu D. et al., 2015). Also, another approach focuses on the development of pH sensitive liposomes, which could be destabilized in an acidic environment, such as the tumor issue, and thus could release the radiopharmaceutical. It is possible for lipids to be incorporated in the liposome and be adapted with chelators to allow the direct labeling of these drug carriers (Nisini R. et al., 2018). Various isotopes have been bound in liposomes, either alone or with other tracers or therapeutic agents (Zhang L. et al., 2010).

Recently, liposomes have been proposed as lipid theranostic agents as a part of drug delivery systems (Kleynhans J. et al., 2018). There are reported tests with liposome systems containing rhenium-188 as a dual imaging (gamma emission, 155 keV) and therapeutic isotope (beta emission, 2.12 MeV) bound to BMEDA (N,N-bis [2-mercaptoethyl]-N′,N′-diethylethylenediamine), acting as chelator (Goins B. et al., 2017). This system is proven to have the ability of the liposomes to passively target rhenium-188 with EPR to various neoplasms in the gastrointestinal tract (Li S. et al., 2012). The enhancement of the efficacy of this system is possible with concomitant administration of doxorubicin liposomes as well as external beam radiotherapy (Kouloulias V. E. et al., 2002).

Furthermore, another achievement includes the multimodal imaging liposomes, which are systems that contain the paramagnetic radiopharmaceutical zirconium-89 labeled octreotide as well as gadolinium (Glaus C. et al., 2010, Abou D. S et al., 2013). This system is proven to provide high-resolution anatomical information through MRI as well as functional information offered by the PET component (Fendler W. P. et al., 2016). Copper-64 labeled liposomes constitute another interesting category, manufactured by covering the superparamagnetic iron core for MRI contrast with phospholipids which allows chelating with copper-64 as PET imaging agents (Lee S. G. et al., 2016). An important characteristic of liposomes is that they have the ability to entrap radiotracers without having an effect on *in vivo* distribution characteristics of the drug carrier (Hoshyar N. et al., 2016). For a radioimmunotherapy (RIT), an isotope providing therapeutic radiation can be linked to a monoclonal antibody for targeting the malignancy. Also, labeling liposomes with technetium-99m can lead to a system for blood-pool imaging, providing a system independent from the injection of radiolabeled red blood cells (Liu M. et al., 2017).

Rhenium isotopes have shown intrinsic theranostic capabilities due the release of energy, capable of being both cytotoxic and useful for imaging (Phillips W. T. et al., 2014). The combination with a chemotherapeutic agent like doxorubicin can lead to substantial benefits (Ma J. et al., 2008). This approach of theranostic radiochemotherapeutics cannot be limited only to cancer therapy but also in anti-infective therapy (Shahbazi R. et al., 2016). Moreover, a companion diagnostic is a diagnostic agent used as pretreatment to determine the applicability of the therapeutic drug as treatment, based on a biomarker which predicts the efficacy or toxicity of the planned treatment (Olsen D. et al., 2014). In the case of the packaging of radioisotopes in one drug carrier system, with other pharmaceuticals, it is important that the dosing characteristics of these agents to be considered to avoid overexposure to a particular entity, according to ALARA principles and the safety of the patient (Strauss K. J. et al., 2006).

Multimodal imaging is of high interest, allowing combinations between dyes for near infrared fluorescence, a contrast factor for MRI, and a radiotracer for either SPECT or PET, followed by a theranostic approach with doxorubicin administration targeting tumor cells (Pratt E. C. et al., 2016).

Furthermore, the development of new agents which might proof to be more accurate and useful in neurological imaging is on-going (Chen S. et al., 2012). The goal is to achieve image neurodegeneration and evaluate cognitive function during therapy (Kabanov A. V. et al., 2007).

Alpha-particle emitters, such as actinium-225, are potent cytotoxic agents with a great DNA damage capacity, due to high linear energy transfer (LET) and short penetration ranges (~30 µm) (Scheinberg D. A. et al., 2011). Side effects can be eliminated if the alpha emitter is successfully targeted to tumor tissue (Sofou S., 2008). Encapsulation of alpha-emitters in drug carriers, such as polymersomes, has been shown to be very efficient (Anajafi T. et al., 2015). These carriers can be passively targeted to tumor tissue through the EPR effect.

Liposomal SPECT imaging agents had developed in the 1980s for the imaging of tumors (Sun X. et al., 2017). Long circulating pegylated liposomes containing In-111 had shown positive results indicating accumulation at tumor sites and reduced scavenging by the reticuloendothelial system (Jokerst J. V. et al., 2011). Other nano-systems allow the incorporation of technetium-99m, folated-γ-glutamic acid and chitosan and indicate positive effects on intratumoral applications *in vivo* (Peng S. F. et al., 2009).

Distinguishing between infection and sterile inflammation sometimes is a difficult approach but it is very important for the selection of the most appropriate therapy (Shen H. et al., 2013). There are some recent studies, which demonstrate that Technetium-99m labeled liposomes with a negatively charged surface tend to be localized at inflammation caused by rheumatoid arthritis (Ulbrich W. et al., 2010). The fundamental of this technique is based on the fact that this system uses the normally unwanted effect of the reticuloendothelial system to accumulate liposomes in the inflamed synovial tissue. It is not yet considered as a very accurate technique for distinguishing infection from sterile inflammation (Signore A. et al., 2010).

Micelles consist of another important category of possible carriers for dual nuclear-optical imaging probes (Fang C. et al., 2010). Micelles have been developed in order to harness the ability of quantum dots to generate light for the activation of a cisplatin prodrug, resulting in targeted chemotherapy (Glasgow M. D. K. et al., 2015). The quantum dots can add heat induction for photodynamic cancer therapy, generating reactive singlet oxygen, acting as an additional cancer treating mechanism (Sneider A. et al., 2017). Technetium-99m is incorporated as SPECT component to provide nuclear imaging additionally to the optical imaging provided by the quantum dots (Kleynhans J. et al., 2018).

13.3.2 MICROSPHERES IN DRUG DELIVERY SYSTEMS

Microsphere technologies are still effective delivery systems which can change the behavior of the materials they carry (Coelho J. F. et al., 2010). These are small spherical structures with dimensions between 10 nm and 1000 μm, manufactured from a wide range of materials (Champion J. A. et al., 2007). Bioactive glass made of samarium-153 with doxorubicin can be used for bone cancer palliative treatment (Sartor O., 2004). There are many studied about the development of microsphere system for selective internal radiation therapy (SIRT) in hepatocellular carcinoma (Kleynhans J. et al., 2018).

13.3.3 NANOPARTICLES IN DRUG DELIVERY SYSTEMS

Nanotechnology is a rapidly expanding frontier of modern research. Nanomaterials can be manufactured from a great variety of materials, including metals in different states, semiconductors, polymers, lipids, and larger biological molecules; as long as their sizes are restricted to the nanometer range and they exhibit novel properties (Khan I. et al., 2017).

Nanoparticles present a great capacity to incorporate different types of imaging and therapeutic applications (Baetke S. C. et al., 2015). The development of these agents is hampered by additional intensive investigations required to determine pharmacokinetics, biocompatibility, and nanotoxicology issues, due to the undefined characteristics of these structures (Bramini M. et al., 2018).

The selection of the core of a nanoparticle depends on the possible application. For example, a multimodal imaging agent will include a magnetic core for MRI contrast, such as iron oxide (Estelrich J. et al., 2015). Regarding to the production process, it is possible to either target the nanoparticle or to allow for the addition of a nuclear isotope or tracer (Pratt E. C. et al., 2016). Nanoparticles have been applied to nuclear medicine for multimodality imaging, theranostics as well as to target the radioisotope for enhancement of localization at the area of interest (Kleynhans J. et al., 2018).

Protein-based nanoparticles can be employed to develop dual-modality imaging agents (Glaus C. et al., 2010). Md-doped magnetism engineered iron oxide (MnMEIO) probes are functionalized with serum albumin, providing nanoparticles that can be used as probes for MRI. Iodo-beads (iodine-124) can be linked to provide with dual PET imaging capabilities (Choi J.-S. et al., 2008). Polyaspartic acid coated iron oxide nanoparticles conjugated with the tripeptide arginine-glycine-aspartic acid (RGD) are also synthesized, providing a substrate for binding with copper-64 for PET and MRI imaging (Lee H.-Y. et al., 2008). Smaller bisphosphonate iron oxide nanoparticles for functionalizing with SPECT agents have also been developed (Ai F. et al., 2016). Unfortunately, all these nanoparticles have not yet been modified to target specific disease mechanisms and need further development to allow for clinical use (Gobbo O. L. et al., 2015).

Quantum dots can be activated by internal radiation to produce fluorescence for the imaging of biological systems and can therefore be utilized as a carrier system for multimodal imaging (Key J. et al., 2014). Moreover, a trimodal PET/MR/optical nanoprobe system has developed, based on iron oxide nanoparticles functionalized by dopamine to afford the incorporation of other probes (Xing Y. et al., 2014). Furthermore, porphyrins are utilized to manufacture copper-64 containing nanoparticles (porphyrosomes) result in a dual-modality probe since porphyrin exhibits intrinsic fluorescence with *in vivo* efficacy for applications in prostate cancer and imaging of micrometastases (Ding Y. et al., 2017).

The development of gold nanoparticles has also been implemented to stabilize and allow the targeting of various isotopes including copper-64 and iodine-125 (Bao G. et al., 2013). Gold nanoparticles can also be excited through the surface plasmon resonance phenomenon, producing photothermal heating locally thereby destroying the cancer cells (Norouzi H. et al., 2018). The theranostic capabilities of a PET agent combined with gold nanoparticles could allow the imaging of the accumulation of these particles at the tumor area (Baetke S. C. et al., 2015). Additionally, gold nanorods labeled with iodine-125 have been developed for the imaging of infection and inflammation.

Lutetium-177 labeled cerasome nanoparticles with entrapped indocyanine green to enable optical imaging (NIRF), nuclear imaging as well as photothermal ablation of tumors (Hill T. K. et al., 2016). Magnetic nanoparticles can be combined with radioimmunoconjugates and then localized to tumor sites with external magnetic fields during clinical use (Zhang L. et al., 2010).

Photothermal agents have the ability to convert applied electromagnetic radiation to heat, damaging target cells in the vicinity of the molecule (Riahi R. et al., 2014). A possible combination of such a nanoparticle with a radiopharmaceutical provides the ability to monitor the distribution inside the patient and determine the optimal time of application of electromagnetic energy (Phillips W. T. et al., 2014). Copper-64-containing copper monosulfide nanoparticles are synthesized for a combination of PET with photothermal ablation therapy (Zhou M. et al., 2010). This category of nanoparticles has the potential to distribute, by passive accumulation to the tumor site and absorb near-infrared light.

Carbon nanotubes (CNTs) are very interesting nanostructures, due to their properties, which allow their use in a variety of applications (Lacerda L. et al., 2006). With high electrical and thermal conductivity, highly elasticity and flexibility and at the same time exhibiting considerable strength and a large surface area are the material of choice for many applications. Two main types of structures of CNTs are used, namely single walled (SWNT), with one rolled up sheet of carbon and multi-walled (MWNT) with multiple rolled up sheets (Ruggiero A. et al., 2010). The contribution of CNTs in drug delivery is important because this system does not illicit an immune reaction when introduced into biological systems (Kleynhans J. et al., 2018). CNTs are also very adaptable to incorporate various entities but the main disadvantage is their hydrophobic nature which commands the need for surface modification to allow for biological applications. In order to be used for PET imaging zirconium-89 can be linked to the SWNT and for therapy following imaging the zirconium-89 can be substituted with the alpha particle-emitting actinium-225 to deliver highly concentrated ionizing radiation. Generally, CNTs clear rapidly from the blood pool to the target area, having the potential to be used as radiotheranostic agents (Singh R. et al., 2006).

Systems under evaluation include hydroxyapatite particles, iron oxide nanoparticles, titanium dioxide nanoparticle, nanozeolite bioconjugates, and lanthanum phosphate nanoparticles (Wang J. et al., 2016).

13.4 LIMITATIONS IN THE USE OF NANOTECHNOLOGY IN NUCLEAR MEDICINE

It is very important to consider both the limitations dictated by the radiotracer and those contributed by the characteristics of the drug delivery system itself, in order to define the ideal characteristics of a drug carrier system, which would be successful in contributing positively to nuclear medicine (Coelho J. F. et al., 2010). There are some limitations for a drug carrier system if it is to be utilized to deliverer radiopharmaceuticals. Among them, we have to refer to the particle size, the stability of the system, the surface charge, and the physiological acceptability (Masarudin M. J. et al., 2015, Danaei M. et al., 2018).

With regard to particle size, it is important to limit the drug carrier system size below of 400 nm (Singh R. et al., 2009). Otherwise, the system will be removed from the blood circulation before it can accumulate at the target site. Recent studies support that this size should be even smaller to mimic the size of antibodies 10–35 nm.

Regarding to the system stability, it is well established that the system must be stable and should only release its load at the target site (Bozzuto G. et al., 2015). So, the carrier system should be coordinated with the radiopharmaceutical in a way that the radiotracer is delivered at the target site, in a manner providing accumulation which can allow imaging or therapeutic activity, considering the constraints of the decay half-life of the isotope.

With respect to surface charge, negatively charged particles reduce removal by the reticular endothelial system, so it is something that it should be taken into account during the development of the drug carrier system (Pisal D. S. et al., 2010). If the surface charge is negative there is a tendency of accumulation of the nanoparticles in the liver (Hoshyar N. et al., 2016). Thus, it is important in order to make a balance between the surface charge and the excretion functions of the body.

Referring to physiological acceptability, in active targeting, the ligands usually target the drug delivery system and the entrapped radiotracer should not illicit adverse events in the biological system (Muro S., 2012). The delivery system must not increase the toxicity of the radiotracer due to changes in biodistribution. The combination of two entities, namely the radiotracer and the chemotherapeutic agent, should be compatible in both the accumulation and dosing processes (Kleynhans J. et al., 2018).

The biodistribution, the toxicity, and the pharmacokinetics of a drug delivery system have to be totally clarified for an ideal system development.

13.5 MATLAB CODES FOR DOSE CALCULATIONS AND DISTRIBUTION IN NANOPARTICLE DELIVERY

Nanoparticle PBPK model is a package which contains a *Physiologically-Based PharmacoKinetic* (PBPK) model (Lin Z. et al., 2017). The accompanying code can be used to calculate confidence intervals related to parameter estimates and model predictions. The package *SimBiology* includes the appropriate functions "sbioparameterci" and "sbiopredictionci". Noncompartmental analysis (NCA) on PK datasets as well as model predictions to obtain clinically relevant parameters are also possible with the same software and the function "sbionca". Introducing of MATLAB functions in order to be performed in parallel with *SimBiology* is also possible, with MATLAB functions "datastore", "parfor", "mapreduce" (Dong D. et al., 2015).

During the development of code in *SimBiology*, it is crucial to make the definition of the unit and the general list of units, which are used. For example:

```
<unit kind="mole">
</unitDefinition>
<unitDefinition id="MWDERIVEDUNIT_1__molecule_1__hour"
name="(1/molecule)*(1/hour)">
<listOfUnits>
<unit kind="mole" exponent="-1"/>
<unit kind="second" exponent="-1"/>
```

The next step includes the definition if the compartments.

```
<listOfCompartments>
<compartment id="code number"
name="Extracellular" spatialDimensions="number-size"
size="number"
name="Intracellular" spatialDimensions=" number-size " size="
number-size "
</listOfCompartments>.
```

Afterward, it is important to definite the list of species, the list of substance units, the boundary conditions, and the list of predicted reactions. Thus, the reactants and the products, and the kinetic law have to be defined (Lin Z. et al., 2017).

Simbiology allows a number of commands which can model differential equations which describe many biological responses in the presence of radio-labeled nanoparticles. There is the possibility of an estimation of nanoparticle concentration and accumulation in the cytoplasm over time. The balance between the intracellular uptake (described usually by the parameter k1) and the degradation terms (described by a number of kinetic parameters, k2, k3, …), according to the proposed model is an important factor. The rated are also calculated as well as the output over time, depending on the effective half time (Dong D. et al., 2015).

The potential of *Simbiology* of MathWorks is unlimited and when it is used in combination with MATLAB functions, the results are desirable.

REFERENCES

Abou D. S., Thorek D. L. J., Ramos N. N., Pinkse M. W. H., Wolterbeek H. T., Carlin S. D., Beattie B. J., Lewis J. S. (2013) [89]Zr-labeled paramagnetic octreotide-liposomes for PET-MR imaging of cancer, Pharm Res. 30(3): 878–888.

Agdeppa E. D., Spilker M. E. (2009) A review of imaging agent development, AAPS J. 11(2): 286–299.

Ai F., Ferreira C. A., Chen F., Cai W. (2016) Engineering of radiolabeled iron oxide nanoparticles for dual-modality imaging, Wiley Interdiscip Rev Nanomed Nanobiotechnol. 8(4): 619–630.

Anajafi T., Mallik S. (2015) Polymersome-based drug-delivery strategies for cancer therapeutics, Ther Deliv. 6(4): 521–534.

Arthofer W., Decristoforo C., Schlick-Steiner B. C., Steiner F. M. (2016) Ultra-low activities of a common radioisotope for permission-free tracking of a drosophilid fly in its natural habitat, Sci Rep. 6: 36506.

Baetke S. C., Lammers T., Kiessling F. (2015) Applications of nanoparticles for diagnosis and therapy of cancer, Br J Radiol. 88(1054): 20150207.

Ballinger J. R. (2010) Short- and long-term responses to molybdenum-99 shortages in nuclear medicine, Br J Radiol. 83(995): 899–901.

Bao G., Mitragotri S., Tong S. (2013) Multifunctional nanoparticles for drug delivery and molecular imaging, Annu Rev Biomed Eng. 15: 253–282.

Bavelaar B. M., Lee B. Q., Gill M. R., Falzone N., Vallis K. A. (2018) Subcellular targeting of theranostic radionuclides, Front Pharmacol. 9: 996.

Boschi A., Martini P., Uccelli L. (2017) [188]Re(V) nitrido radiopharmaceuticals for radionuclide therapy, Pharmaceuticals (Basel). 10(1): 12.

Bozzuto G., Molinari A. (2015) Liposomes as nanomedical devices, Int J Nanomedicine. 10: 975–999.

Bramini M., Alberini G., Colombo E., et al. (2018) Interfacing graphene-based materials with neural cells, Front Syst Neurosci. 12: 12.

Champion J. A., Katare Y. K., Mitragotri S. (2007) Particle shape: a new design parameter for micro- and nanoscale drug delivery carriers, J Control Release. 121(0): 3–9.

Chen S., Li X. (2012) Functional magnetic resonance imaging for imaging neural activity in the human brain: the annual progress, Comput Math Methods Med. 2012: 613465.

Choi J.-S., Park J. C., Nah H., et al. (2008) A hybrid nanoparticle probe for dual-modality positron emission tomography and magnetic resonance imaging, Angew Chem. 47(33):6259–6262.

Coelho J. F., Ferreira P. C., Alves P., Cordeiro R., Fonseca A. C., Góis J. R., Gil M. H. (2010) Drug delivery systems: advanced technologies potentially applicable in personalized treatments, EPMA J. 1(1): 164–209.

Danaei M., Dehghankhold M., Ataei S., Hasanzadeh Davarani F., Javanmard R., Dokhani A., Khorasani S., Mozafari M. R. (2018) Impact of particle size and polydispersity index on the clinical applications of lipidic nanocarrier systems, Pharmaceutics. 10(2): 57.

Daraee H., Etemadi A., Kouhi M., Alimirzalu S., Akbarzadeh A. (2016) Application of liposomes in medicine and drug delivery, Artif Cells Nanomed Biotechnol. 44: 381–391.

Daughton C. G., Ruhoy I. S. (2013) Lower-dose prescribing: minimizing "side effects" of pharmaceuticals on society and the environment, Sci Total Environ. 443: 324–337.

Delso G., Füst S., Jakoby B., Ladebeck R., Ganter C., Nekolla S. G., Schwaiger M., Ziegler S. I. (2011) Performance measurements of the Siemens mMR integrated whole-body PET/MR scanner, J Nucl Med. 52: 1914–1922.

Digesu C. S., Hofferberth S. C., Grinstaff M. W., Colson Y. L. (2016) From diagnosis to treatment: clinical applications of nanotechnology in thoracic surgery, Thorac Surg Clin. 26(2): 215–228.

Ding Y., Zhu W. H., Xie Y. (2017) Development of ion chemosensors based on porphyrin analogues, Chem Rev. 117(4): 2203–2256.

Dong D., Wang X., Wang H., Zhang X., Wang Y., Wu B. (2015) Elucidating the in vivo fate of nanocrystals using a physiologically based pharmacokinetic model: a case study with the anticancer agent SNX-2112, Int J Nanomedicine. 10: 2521–2535.

Estelrich J., Sánchez-Martín M. J., Busquets M. A. (2015) Nanoparticles in magnetic resonance imaging: from simple to dual contrast agents, Int J Nanomedicine. 10: 1727–1741.

Fang C., Zhang M. (2010) Nanoparticle-based theragnostics: integrating diagnostic and therapeutic potentials in nanomedicine, J Control Release146(1): 2–5.

Fendler W. P., Czernin J., Herrmann K., Beye T. (2016) Variations in PET/MRI operations: results from an international survey among 39 active sites, J Nucl Med. 57(12): 2016–2021.

Ferro-Flores G., Ocampo-García B. E., Santos-Cuevas C. L., Morales-Avila E., Azorín-Vega E. (2014) Multifunctional radiolabeled nanoparticles for targeted therapy, Curr Med Chem. 21(1): 124–138.

Glasgow M. D. K., Chougule M. B. (2015) Recent developments in active tumor targeted multifunctional nanoparticles for combination chemotherapy in cancer treatment and imaging, J Biomed Nanotechnol. 11(11): 1859–1898.

Glaus C., Rossin R., Welch M. J., Bao G. (2010) In vivo evaluation of 64Cu-labeled magnetic nanoparticles as a dual-modality PET/MR imaging agent, Bioconjug Chem. 21: 715–722.

Gobbo O. L., Sjaastad K., Radomski M. W., Volkov Y., Prina-Mello A. (2015) Magnetic nanoparticles in cancer theranostics, Theranostics. 5(11): 1249–1263.

Goins B., Bao A., Phillips W. T. (2017) Techniques for loading technetium-99m and rhenium-186/188 radionuclides into preformed liposomes for diagnostic imaging and radionuclide therapy, Methods Mol Biol. 1522: 155–178.

Golombek S. K., May J.-N., Theek B., Appold L., Drude N., Kiessling F., Lammers T. (2018) Tumor targeting via EPR: strategies to enhance patient responses, Adv Drug Deliv Rev. 130: 17–38.

Greig N. H., Reale M., Tata A. M. (2013) New advances in pharmacological approaches to the cholinergic system: an overview on muscarinic receptor ligands and cholinesterase inhibitors, Recent Pat CNS Drug Discov. 8(2): 123–141.

Hill T. K., Kelkar S. S., Wojtynek N. E., Souchek J. J., Payne W. M., Stumpf K., Marini F. C., Mohs A. M. (2016) Near infrared fluorescent nanoparticles derived from hyaluronic acid improve tumor contrast for image-guided surgery, Theranostics. 6(13): 2314–2328.

Hoshyar N., Gray S., Han H., Bao G. (2016) The effect of nanoparticle size on in vivo pharmacokinetics and cellular interaction, Nanomedicine (Lond). 11(6): 673–692.

Huang Q., Wang Y., Chen X., Wang Y., Li Z., Du S., Wang L., Chen S. (2018) Nanotechnology-based strategies for early cancer diagnosis using circulating tumor cells as a liquid biopsy, Nanotheranostics. 2(1): 21–41.

Jokerst J. V., Lobovkina T., Zare R. N., Gambhir S. S. (2011) Nanoparticle PEGylation for imaging and therapy, Nanomedicine (Lond). 6(4): 715–728.

Kabanov A. V., Gendelman H. E. (2007) Nanomedicine in the diagnosis and therapy of neurodegenerative disorders, Prog Polym Sci. 32(8-9): 1054–1082.

Kang J. H., Jang W. Y., Ko Y. T. (2017) The effect of surface charges on the cellular uptake of liposomes investigated by live cell imaging, Pharm Res. 34: 704–717.

Key J., Leary J. F. (2014) Nanoparticles for multimodal in vivo imaging in nanomedicine, Int J Nanomedicine. 9: 711–726.

Khan I., Saeed K., Khan I. (2017) Nanoparticles: properties, applications and toxicities, Arab J Chem. 12(7): 908–931.

Kleynhans J., Grobler A. F., Ebenhan T., Machaba Sathekge M., Zeevaart J.-R. (2018) Radiopharmaceutical enhancement by drug delivery systems: a review, J Control Release 287: 177–193.

Kouloulias V. E., Dardoufas C. E., Kouvaris J. R. et al. (2002) Liposomal doxorubicin in conjunction with reirradiation and local hyperthermia treatment in recurrent breast cancer, Clin Cancer Res, 8(2): 374–382.

Lacerda L., Bianco A., Prato M., Kostarelos K. (2006) Carbon nanotubes as nanomedicines: from toxicology to pharmacology, Adv Drug Deliv Rev. 58: 1460–1470.

Lagopati N., Tsilibary E.-P., Falaras P., Papazafiri P., Pavlatou E. A., Kotsopoulou E., Kitsiou P. (2014) Effect of nanostructured TiO_2 crystal phase on photoinduced apoptosis of breast cancer epithelial cells, Int J Nanomedicine. 9: 3219–3230.

Lee H.-Y., Li Z., Chen K., Hsu A. R., Xu C., Xie J., Sun S., Chen X. (2008) PET/MRI dual-modality tumour imaging using arginine-glycine-aspartic (RGD)-conjugated radiolabeled iron oxide nanoparticles, J Nucl Med. 49: 1371–1379.

Lee S. G., Gangangari K., Kalidindi T. M., Punzalan B., Larson S. M., Pillarsetty N. V. K. (2016) Copper-64 labeled liposomes for imaging bone marrow, Nucl Med Biol. 43(12): 781–787.

Li S., Goins B., Zhang L., Bao A. (2012) A novel multifunctional theranostic liposome drug delivery system: construction, characterization, and multimodality MR, near-infrared fluorescent and nuclear imaging, Bioconjug Chem. 23: 1322–1332.

Li T., Ao E. C. I., Lambert B., Brans B., Vandenberghe S., Mok G. S. P. (2017) Quantitative imaging for targeted radionuclide therapy dosimetry – technical review, Theranostics. 7(18): 4551–4565.

Lin Z., Jaberi-Douraki M., He C., Jin S., Yang R. S. H., Fisher J. W., Riviere J. E. (2017) Performance assessment and translation of physiologically based pharmacokinetic models from acslX to Berkeley Madonna, MATLAB, and R language: oxytetracycline and gold nanoparticles as case examples, Toxicol Sci. 158(1): 23–35.

Liu D., Auguste D. T. (2015) Cancer targeted therapeutics: from molecules to drug delivery vehicles, J Control Release. 219: 632–643.

Liu M., Zhao Z. Q., Fang W., Liu S. (2017) Novel approach for ^{99m}Tc-labeling of red blood cells: evaluation of 99mTc-4Saboroxime as a blood pool imaging agent, Bioconjug Chem. 28(12): 2998–3006.

Ma J., Waxman D. J. (2008) Combination of anti-angiogenesis with chemotherapy for more effective cancer treatment, Mol Cancer Ther. 7(12): 3670–3684.

Maas M., Rutten I. J. G., Nelemans P. J., Lambregts D. M. J., Cappendijk V. C., Beets G. L., Beets-Tan R. G. H. (2011) What is the most accurate whole-body imaging modality for assessment of local and distant recurrent disease in colorectal cancer? A meta-analysis, Eur J Nucl Med Mol Imaging. 38(8): 1560–1571.

Masarudin M. J., Cutts S. M., Evison B. J., Phillips D. R., Pigram P. J. (2015) Factors determining the stability, size distribution, and cellular accumulation of small, monodisperse chitosan nanoparticles as candidate vectors for anticancer drug delivery: application to the passive encapsulation of [14C]-doxorubicin, Nanotechnol Sci Appl. 8: 67–80.

Müller C., van der Meulen N. P., Benesov M., Schibli R. (2017) Therapeutic radiometals beyond 177Lu and 90Y: production and application of promising a-particle, b2-particle, and auger electron emitters, J Nucl Med. 58(9): 91S–96S.

Muro S. (2012) Challenges in design and characterization of ligand-targeted drug delivery systems, J Control Release. 164(2): 125–137.

Nisini R., Poerio N., Mariotti S., De Santis F., Fraziano M. (2018) the multirole of liposomes in therapy and prevention of infectious diseases, Front Immunol. 9: 155.

Norouzi H., Khoshgard K., Akbarzadeh F. (2018) In vitro outlook of gold nanoparticles in photo-thermal therapy: a literature review, Lasers Med Sci. 33(4): 917–926.

Olsen T. J. J. (2014) Companion diagnostics for targeted cancer drugs – clinical and regulatory aspects, Front Oncol. 4: 105.

Panjnoush M., Kheirandish Y., Mohseni Kashani P., Bashizadeh Fakhar H., Younesi F., Mallahi D. M. (2016) Effect of exposure parameters on metal artifacts in cone beam computed tomography, J Dent (Tehran). 13(3): 143–150.

Peng S. F., Yang M. J., Su C. J., Chen H. L., Lee P. W., Wei M. C., Sung H. W. (2009) Effects of incorporation of poly(gamma-glutamic acid) in chitosan/DNA complex nanoparticles on cellular uptake and transfection efficiency, Biomaterials. 30(9): 1797–1808.

Phillips W. T., Bao A., Brenner A. J., Goins B. A. (2014) Image-guided interventional therapy for cancer with radiotherapeutic nanoparticles, Adv Drug Deliv Rev. 76: 39–59.

Pisal D. S., Kosloski M. P., Balu-Iyer S. V. (2010) Delivery of therapeutic proteins, J Pharm Sci. 99(6): 2557–2575.

Pratt E. C., Shaffer T. M., Grimm J. (2016) Nanoparticles and radiotracers: advances toward radio-nanomedicine, Wiley Interdiscip Rev Nanomed Nanobiotechnol. 8(6): 872–890.

Riahi R., Wang S., Long M., Li N., Chiou P.-Y., Zhang D. D., Kin Wong P. (2014) Mapping photothermally induced gene expression in living cells and tissues by nanorod-locked nucleic acid complexes, ACS Nano. 4: 3597–3605.

Ruggiero A., Villa C. H., Holland J. P., Sprinkle S. R., May C., Lewis J. S., Scheinberg D. A., McDevitt M. R. (2010) Imaging and treating tumour vasculature with targeted radiolabeled carbon nanotubes, Int J Nanomedicine 5: 783–802.

Sartor O. (2004) Overview of samarium Sm 153 lexidronam in the treatment of painful metastatic bone disease, Rev Urol. 6(Suppl 10): S3–S12.

Scheinberg D. A., McDevit M. R. (2011) Actinium-225 in targeted alpha-particle therapeutic applications, Curr Radiopharm. 4(4): 306–320.

Shahbazi R., Ozpolat B., Ulubayram K. (2016) Oligonucleotide-based theranostic nanoparticles in cancer therapy, Nanomedicine (Lond). 11(10):1287–1308.

Shen H., Kreisel D., Robert Goldstein D. (2013) Processes of sterile inflammation, J Immunol. 191(6): 2857–2863.

Signore A., Mather S. J., Piaggio G., Malviya G., Dierckx R. A. (2010) Molecular imaging of inflammation/infection: nuclear medicine and optical imaging agents and methods, Chem Rev. 110(5): 3112–3145.

Singh R., Pantarotto D., Lacerda L., Pastorin G., Klumpp C., Prato M., Bianco A., Kostarelos K. (2006) Tissue biodistribution and blood clearance rates of intravenously administered carbon nanotube radiotracers, PNAS 103: 3357–3362.

Singh R., Lillard J. W. Jr. (2009) Nanoparticle-based targeted drug delivery, Exp Mol Pathol. 86(3): 215–223.

Sneider A., VanDyke D., Paliwal S., Rai P. (2017) Remotely triggered nano-theranostics for cancer applications, Nanotheranostics. 1(1): 1–22.

Sofou S. (2008) Radionuclide carriers for targeting of cancer, Int J Nanomedicine. 3(2): 181–199.

Strauss K. J., Kaste S. C. (2006) The ALARA (as low as reasonably achievable) concept in pediatric interventional and fluoroscopic imaging: striving to keep radiation doses as low as possible during fluoroscopy of pediatric patients—a white paper executive summary, Pediatr Radiol. 36(Suppl 2): 110–112.

Sun X., Li Y., Liu T., Li Z., Zhang X., Chen X. (2017) Peptide-based imaging agents for cancer detection, Adv Drug Deliv Rev. 110–111: 38–51.

Tiwari G., Tiwari R., Sriwastawa B., Bhati L., Pandey S., Pandey P., Bannerjee S. K. (2012) Drug delivery systems: an updated review, Int J Pharm Investig. 2(1): 2–11.

Ulbrich W., Lamprecht A. (2010) Targeted drug-delivery approaches by nanoparticulate carriers in the therapy of inflammatory diseases, J R Soc Interface. 7(Suppl 1): S55–S66.

van Dalen J. A., Vogel W. V., Corstens F. H. M., Oyen W. J. G. (2007) Multi-modality nuclear medicine imaging: artefacts, pitfalls and recommendations, Cancer Imaging. 7(1): 77–83.

Ventola C. L. (2017) Progress in nanomedicine: approved and investigational nanodrugs, P T. 42(12): 742–755.

Wang J., Wang L., Fan Y. (2016) Adverse biological effect of TiO_2 and hydroxyapatite nanoparticles used in bone repair and replacement, Int J Mol Sci. 17(6): 798.

Wen H., Jung H., Li X. (2015) Drug delivery approaches in addressing clinical pharmacology-related issues: opportunities and challenges, AAPS J. 17(6): 1327–1340.

Wong A. D., Ye M., Ulmschneider M. B., Searson P. C. (2015) Quantitative analysis of the enhanced permeation and retention (EPR) effect, PLOS ONE. 10(5): e0123461.

Xing Z.-C., Chang Y., Kang I.-K. (2010) Immobilization of biomolecules on the surface of inorganic nanoparticles for biomedical applications, Sci Technol Adv Mater. 11(1): 014101.

Xing Y., Zhao J., Conti P. S., Chen K. (2014) Radiolabeled nanoparticles for multimodality tumor imaging, Theranostics. 4(3): 290–306.

Yeong C.-H., Cheng M.-h., Ng K.-H. (2014) Therapeutic radionuclides in nuclear medicine: current and future prospects, J Zhejiang Univ Sci B. 15(10): 845–863.

Yingchoncharoen P., Kalinowski D. S., Richardson D. R. (2016) Lipid-based drug delivery systems in cancer therapy: what is available and what is yet to come, Pharmacol Rev. 68(3): 701–787.

Yordanova A., Eppard E., Kürpig S., Bundschuh R. A., Schönberger S., Gonzalez-Carmona M., Feldmann G., Ahmadzadehfar H., Essler M. (2017) Theranostics in nuclear medicine practice, Onco Targets Ther. 10: 4821–4828.

Zhang L., Chen H., Wang L., Liu T., Yeh J., Lu G., Yang L., Mao H. (2010) Delivery of therapeutic radioisotopes using nanoparticle platforms: potential benefit in systemic radiation therapy, Nanotechnol Sci Appl. 3: 159–170.

Zhou M., Zhang R., Huang M. et al . (2010) A chelator free multifunctional [64Cu]CuS nanoparticle platform for simultaneous micro-PET/CT imaging and photothermal ablation therapy, J Am Chem Soc. 132: 1535–15358.

Zhu L., Zhou Z., Mao H., Yang L. (2017) Magnetic nanoparticles for precision oncology: theranostic magnetic iron oxide nanoparticles for image-guided and targeted cancer therapy, Nanomedicine (Lond). 12(1): 73–87.

14 CASE Studies in Nuclear Medicine/ MATLAB Approach

Stella Synefia, Elena Ttofi, and Nefeli Lagopati

CONTENTS

14.1 MYOCARDIUM IMAGING DEFECTS QUANTITATIVE EVALUATION

Estimating the affected area of the myocardium is a very important achievement in patients with coronary artery disease (CAD). The single-photon emission computed tomography (SPECT) study consists of several frames which, together, can sample the blood perfusion of the heart muscle in three dimensions. The quantification of the SPECT studies enables objective interpersonal comparison and objective assessment of cardiac status.

Our approach is based on the generation of 3D images of two distinct phases the patient undergoes in the SPECT study: rest and stress protocol.

Data of two sets (stress-rest) of SPECT slices was used. Thus, the myocardial perfusion was estimated by comparing these slices and suspicion of ischemia was indicated (Figure 14.1.1).

The aim of the two following examples is to first determine the threshold value and then to reconstruct a 3D image according to this value.

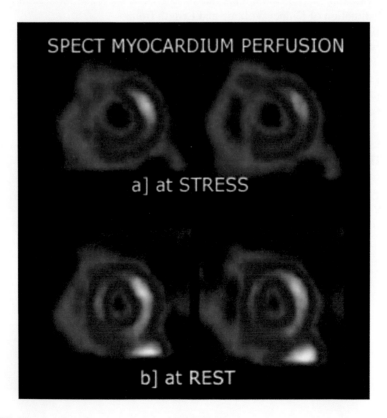

FIGURE 14.1.1   Two sets (stress-rest) of SPECT slices.

14.1.1 Isocontour Surfaces for Threshold Value Determination

Two series (stress-rest) of transverse slices of myocardium images were reconstructed and these images were extracted in a Digital Imaging and Communications in Medicine (DICOM) format. The DICOM file for each phase is imported to MATLAB 7.8 (R2009a). A series of isocontour surfaces are studied in order to identify the appropriate threshold value, which isolates the myocardium surface from the rest area (background) of the image. Example 1 shows the functions used for constructing the isocontour surfaces of the 9th transverse slice of myocardium stress images.

Example 1

SPECT=dicomread('G:\ STRESS_IRNC001_DS.dcm');	*Reads the image data from the compliant Digital Imaging and Communications in Medicine (DICOM) file 'G:\STRESS_IRNC001_DS.dcm', which creates a matrix 'SPECT'.*
I=SPECT(:,:,1,9);	*Creates a matrix 'I' containing the 9th transverse slice of myocardium stress images.*
imagesc(I);	*Displays the data in array 'I'.*
C=contour(I);	*Creates a contour plot 'C' containing the isolines of matrix 'I'.*
clabel(C);	*Labels the current contour plot ('C') with rotated text inserted into each contour line.*

14.1.2 Intensity Volume and 3D Visualization

Volume visualization in Nuclear Medicine is a method of extracting information from volumetric data utilizing and processing a Nuclear Medicine image. In MATLAB, this can be achieved by constructing a 3D surface plot which uses the pixel identities for (x, y) axes. The pixel value is transformed into surface plot height and consequently, color. Apart from that, 3D voxel images can be constructed; SPECT projections are acquired, isocontours are depicted on them including a number of voxels; finally all of them can be added in order to create the desirable image (Figure 14.1.2).

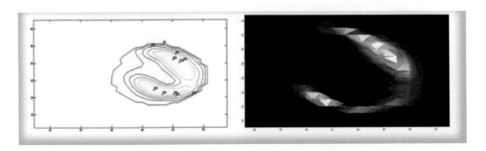

FIGURE 14.1.2 Isocontour surfaces for threshold value determination; REST.

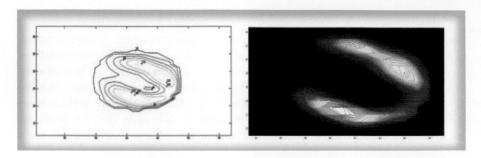

FIGURE 14.1.3   Isocontour surfaces for threshold value determination; STRESS.

In this example, based on the previously calculated threshold value, the myocardium volume is evaluated and reconstructed in a 3D image. Example 2 indicates the functions for the 3D image reconstruction (where 40 is the value of the calculated threshold value) (Figure 14.1.3).

Example 2

mask=logical(zeros(64,64,19));	*Creates an array of logical values*
for k=1:19	*For each element of the 'mask'*
for i=1:64	
for j=1:64	
if (SPECT(i,j,k)>40);	*If the value is more than 40*
mask(i,j,k)=1;	*Fill in the matrix 'mask' the element j, i, k equal to 1*
End	*Evaluates the above expressions*
End	
End	
End	
isosurface(mask,0.9)	*Illustrates the 3D image according to the threshold value 40*

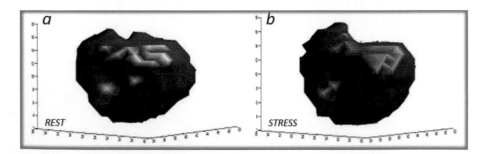

FIGURE 14.1.4   3D representation of the volume of interest at rest (a) and at stress (b) of the same patient as in Figures 14.1.2 and 14.1.3. It is obvious that the total volume is decreased at stress phase due to the general pathogenic condition. Therefore, there is a defect of radionuclide perfusion when compared to the myocardium perfusion at rest.

Stress and rest 3D images are compared and possible differences in voxels are calculated using MATLAB image processing analysis. Thus, quantification and analysis of differences in 3D stress/rest images is permitted. As a result, the location of perfusion defects as well as the myocardium defect volume can be both identified precisely. This results in the quantization of the possible pathogenesis.

REFERENCES

Castellano G, Bonilha L, Li LM, Cendes F 2004, Texture analysis of medical images, Clinical Radiology, 59: 1061–1069.

Lyra M 2009, Single Photon Emission Tomography (SPECT) and 3D Images Evaluation, in Nuclear Medicine, Image Processing, In: Yung-Sheng Chen (Ed). ISBN: 978-953-307-026-1, Chapter 15, pp. 259–286, INTECH, 2009, http://sciyo.com/articles/show/title/single-photon-emission-tomography-spect-and3d-images-evaluation-in-nuclear-medicine

Lyra M, Gavrilelli M, Chatzijiannis C, Skabas N 2009, 3D images quantitative perfusion analysis and myocardium polar index for cardiac scintigraphy improvement, 2009 Annual Scientific Meeting of European Society of Cardiac Radiology, October 8–10, Leipzig, Germany, http://posters.webges.com/get/pdf/escr09/43339/

Lyra M, Sotiropoulos M, Lagopati N, Gavrilleli M 2010, Quantification of myocardial perfusion in 3D SPECT images – stress/rest volume differences, 2010 IEEE International Conference on Imaging Systems and Techniques (IST), Thessaloniki, pp.31–35, DOI: 10.1109/IST.2010.5548486.

Synefia S, Sotiropoulos M, Argyrou M et al 2014, 3D SPECT myocardial volume estimation increases the reliability of perfusion diagnosis, e-Journal of Science & Technology (e-JST), (3), 9, 2014.

Walimbe VBE 2006, Interactive quantitative 3D stress echocardiography and myocardial perfusion SPECT for improved diagnosis of coronary artery disease, dissertation, The Ohio State University, 2006.

14.2   ESTIMATION OF KIDNEY VOLUME AND LENGTH FROM 3D SPECT IMAGES

An assessment of the morphological parameters in renal imaging is an integral point of the evaluation of kidney diseases and organ growth in children. The aim of this study is to create useful pathophysiological tools by MATLAB that will help as to a more accurate and detailed diagnosis in kidney studies.

The clinical value of measuring the size of the two kidneys is widely recognized. Amongst other things, it has made sense in the study of the natural history of certain kidney diseases. Possible enlargement of the kidneys can occur in polycystic disease and in disorders due to fat storage. Renal dysplasia of one of the two kidneys in children leads to the formation of a shrinking kidney and the hypertrophy of the other. Kidney size is also an important factor in the clinical evaluation of patients with diabetes, kidney transplants or renal artery stenosis, and in children with recurrent urinary tract infections or vesicoureteral reflux.

To estimate the volume and length of kidney in conjunction with the usage of 3D volume rendering images, we created algorithms on MATLAB to reconstruct and analyze the data of Technetium-99m dimercaptosuccinic acid (DMSA) scan.

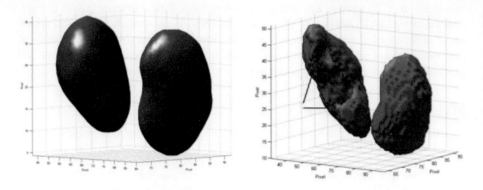

FIGURE 14.2.1 3D volume rendering of the renal surface (transaxial) (a) with smoothing and (b) with renal abnormalities – without smoothing.

14.2.1 MATLAB CODE CREATION

We created a code to manage the extraction of 3D volume rendering image and calculated the volume of the kidneys (Figure 14.2.1). The code starts with an empty matrix which gets binary values (0 or 1) with the help of a counting loop. The loop checks all the values in the three dimensions of the matrix (i, j, k). If the checked value exceeds the threshold, then the number 1 is entered in an empty matrix at corresponding coordinate. For the representation of kidneys in 3D volume rendering the most useful command is *isosurface*. This command extracts isosurface data from volume data at the isosurface value, in the certain case the isovalue is 0.9 so only that organ volume data are displayed. For the data normalization in three dimensions the command *smooth3* is most capable, it smooths the input data.

14.2.1.1 Optimum Threshold Selection

For the complete elimination of the background we used a threshold. It depends on the data which is the optimum threshold. The following code, produced to visualize the kidneys, is flexible to be converted and follow the theory of any threshold. For the data studied, we chose the threshold Otsu based on the intensity histogram of gray (Nobuyuki O., 1979; Mortelmans L. et al., 1986). This threshold is suitable for any medical imaging, as it can check each patient's data separately and export the corresponding threshold. MATLAB enclose the Otsu's theory in function *Multithersh* which returns the single threshold value computed for multilevel image (matrix) using Otsu's method.

14.2.2 QUANTITATIVE DATA – KIDNEY VOLUME

In the same loop except from the visual result, we could extract quantitative results too. If the examiner point is exceeding the threshold, then this point is included in the organ volume and is a voxel of the organ. When the loop is completed, it extracts a value that represents the voxels of the organ.

The pixel size must be known so that to be used for the calculation of the organ volume.

$$\text{Volume(ml)} = \text{Voxel} \cdot (\text{Pixel size}) \cdot (\text{Pixel size}) \cdot (\text{Pixel size}) \cdot 0.001 \quad (14.2.1)$$

```
%3D surface & volume
for k=1:z %changing variable, it depends on the exam slices
        for i=1:x
        for j=1:y
        if (matrix(i,j,k)>(threshold));
        mask(i,j,k)=1;
        Voxel = Voxel + 1; %counting the voxels
        end
        end
        end
end
Voxel
W = smooth3(mask);
isosurface(W,0.9);
```

14.2.3   QUANTITATIVE DATA – KIDNEY LENGTH

Another objective in this study was to find the length of a kidney. In both of codes, we used the sagittal orientation so that the two kidneys can be separated and easily could be found the measurements.

Length can be measured by the maximum distance between the extreme points. The main algorithm begins by checking each slice separately so that we stabilize the third dimension and limit our calculations to two dimensions. Next, it specifies a zero L array into which we enter the non-zero (that is, the value 1 and not 0). Then find the maximum distance of the points for each slice. Finally, we find the maximum distance from all slices, this value representing the length of the kidney.

In the calculation of the maximum two-point distance in the code we used Euclidean distance. According to analytical geometry the distance of two points belonging to the Cartesian coordinate system can be found using the equation of distance. The distance between the points (x_1, y_1) and (x_2, y_2) is given by:

$$d = \sqrt{\left(x_2 - x_1\right)^2 + \left(y_2 - y_1\right)^2} \quad (14.2.2)$$

```
l=0;
for z=1:36    %for the right kidney (for left 36:64)
  L=zeros(20000,2);
  c=1;
  for x=1:128
    for y=1:128
      c=c+1;
      if (m(x,y,z)==1)
```

```
            L(c,1)=x;
            L(c,2)=y;
          end
        end
    end

 L(all(L==0,2),:)=[];
 s=size(L);
 a=s(1);
 l1=0;
 for i=1:a
  for j=1:a
    l2=sqrt(((L(i,1)-L(j,1))^2)+(L(i,2)-L(j,2))^2);
    maximum=max(l1,l2);
    l1=maximum;
  end
end
 l3=max(l1,l);
 l=l3;
end
l
```

To confirm the results of the code, we measured the length manually using the MATLAB tools. Specifically, with the Data Cursor in *Figure*, we traced the coordinates of the points (x_1, y_1, z_1) and (x_2, y_2, z_2), which we considered most extreme points. Then, using the Euclidean distance (Equation 14.2.3) for the three dimensions and we calculated the length of each kidney separately.

$$d = \sqrt{(x_2 - x_1)^2 + (y_2 - y_1)^2 + (z_2 - z_1)^2} \qquad (14.2.3)$$

To validate the code, we performed the above procedure for ten patients and demonstrated that there is no statistical difference between the two values. Undoubtedly, however, the manual method also contains the observer's error, a factor which is eliminated by using the code. Again, with the use of pixel size we converted the result in centimeters (cm).

14.2.4   MATLAB ASSISTANCE IN KIDNEY SIZE CALCULATION/CONCLUSION

So, in the quantitative part of the code, we are able to calculate the length and the total volume of a kidney, two measurements that determine the size of the organ. The quantitative results were compared and confirmed with literature values.

A study of E Ttofi et al. (2016) included 414 children of which 173 were boys and 241 were girls, ageing from 1 month to 18 years old (infants – children – adolescents).

The results of this analysis established that the renal size was larger on the left side than the right side and that there was no significant gender-related difference between them.

Furthermore, it was noted that a rapid renal growth occurs during the first year of life and then this growth rate stabilizes in linear fashion.

From the 3D images and quantitative analysis of the kidneys, we can accomplish to accredit some renal abnormalities, such as damage of renal cortex (scars), distention of the renal pelvis, and congenital abnormalities. These mathematical models compiled on MATLAB can be applied in conjunction with other software applications of Medical Imaging, enabling the extraction of valuable quantitative, morphological, and dosimetry information.

REFERENCES

MathWorks 2019, https://www.mathworks.com/

Mortelmans L, Nuyts J, Van Pamel G, Van den Maegdenbergh V, De Roo M, Suetens P 1986, A new thresholding method for volume determination by SPECT, L. Mortelmans, Dep of Nuclear Medicine, U.Z. Gasthuisberg, Herestraat 49, B-3000 Leuven, Belgiurfl, April 26, 1986.

Nobuyuki O 1979, A threshold selection method from gray-level histograms, IEEE Trans Syst, Men Cybern SMC-9. DOI: 10.1109/TSMC.1979.4310076.

Ttofi E, Lyra M 2016, 3D SPECT reconstruction and analysis of kidney morphology in childhood – use of mathematical models, XIV Mediterranean Conference on Medical and Biological Engineering and Computing 2016, vol. 57 of the series IFMBE Proceedings, pp. 314–319.

14.3   THERAPEUTIC VOXEL DOSIMETRY CASES

Among the advantages of radionuclide treatments is that they allow delivering lethal dose to cancer cells while keeping low dose to healthy tissue (Lyra M. et al., 2013). The precise estimation of the absorbed dose to the tumor and the healthy tissue is very important for the efficiency of the treatment as well as the safety of the patient. It has been proven that there are many fluctuations occurring in absorbed dose between patients that receive radionuclide therapy. Thus, individualized dosimetry is mandatory, since the absorbed dose for each patient is affected by differences in anatomy and in pharmacokinetics. Voxel-based internal dosimetry provides many advantages (Ljungberg M. et al., 2016). Complex activity distributions could be imaged, and inhomogeneous activities can be processed with a high resolution.

14.3.1   MATLAB CALIBRATION FACTOR – ABSORBED DOSE IN TREATMENT BY LU-177, IN-111, I-131, RA-223 AND SR-89

14.3.1.1   Lutetium-177 (Lu-177)

The half time of Lutetium is 6.71 days. It transmits beta particles of an average energy at 498 keV and gamma particles at 208 keV and 113 keV, allowing the imaging of the radiopharmaceutical by γ-camera. It is used in the treatment of neuroendocrine tumors of the gastrointestinal system. It labels with peptides (octreotide) such as DOTANOC, DOTATATE, and various others.

It has been already mentioned that there are many fluctuations occurring in therapeutic absorbed doses between patients. A recent study of our scientific group included ten patients, suffering from neuroendocrine liver cancer that received a therapeutic dose of 4290 MBq Lu-177 Dotatate (Table 14.3.1). The dosimetric data,

TABLE 14.3.1

Administered Dose of Lu-177 Dotatate

#	Initial Radioactivity (MBq)	Gender Female (f)/Male (m)	Age (y)
1	4379.7	F	62
2	4373	M	69
3	3848	F	63
4	4336.4	M	61
5	4351.4	F	62
6	4395	M	69
7	3874	F	63
8	4363.9	M	61
9	4375	F	72
10	4297	F	67

which were obtained through MIRDOSE3.1 software via an individualized protocol, indicated that the maximum observed dose at the tumor area was 21.2 mGy/MBq while the minimum absorbed dose was estimated as 1.2 mGy/MBq (Table 14.3.2); the fluctuations could be attributed to the differences among the tumor size or to the different pharmacokinetics among the patients whose clinical data were exploited at this study.

The personalized dosimetry includes the accurate calculation of the absorbed dose in the tumor and in heathy tissues, the prediction of the therapeutic outcome and the toxicity of the selected therapeutic scheme, the correlation between the dose

TABLE 14.3.2

Absorbed Dose of Lu-177 Dotatate in Tumor Area and in Critical Organs

Number of Patients	Absorbed Dose (mGy/MBq) Liver	Kidneys	Spleen	Pancreas	Tumor
1	0.0008	0.21	0.55	0.0318	2.7
2	0.0021	0.68	0.03	0.0085	21.2
3	0.0015	0.13	0.165	0.072	16.1
4	0.0025	0.1	0.044	0.001	8.5
5	0.003	0.263	0.459	0.045	11.2
6	0.0018	0.087	0.0109	0.03	1.2
7	0.002	0.05	0.26	0.004	5.9
8	0.0024	0.132	0.088	0.047	14.8
9	0.015	0.13	0.16	0.002	17.3
10	0.017	0.087	0.094	0.0047	19.5
Mean Value ± SD	0.002±0.00012	0.13±0.008	0.19±0.0007	0.025±0.0013	13.0± 0.6

and the biological effect, and the comparison among the different clinical techniques and their efficacy.

Thus, a general protocol of internal dosimetry for radionuclide therapies was developed based on an algorithm which was created from data related to the dose per voxel. The first included the calibration of the γ-camera in order to provide quantified scintigraphic images.

14.3.1.2 Calibration of the γ-Camera for the Lutetium-177 – Anterior-Posterior Images

For the calibration of the γ-camera Elcint-APEX SPC4 (NaI-Tl crystal 9.5 mm), at the Aretaieion Hospital of the National and Kapodistrian University of Athens, the acquisition time was 1 minute, the energy window 20% centered at 208 keV and a parallel-hole collimator was used. A rectangular water phantom (0.1 m × 0.3 m × 0.3 m) simulated the patient's body.

At the center of the phantom a 10-mL syringe is placed full of 177Lu-DOTA^0-Tyr3 (66.23 MBq) (Figure 14.3.1). The total number of counts and the count rate were recorded for various values of radioactivity up to 510.23 MBq and a graph of count rate versus radioactivity was created (Figure 14.3.2). A counter Capintec (model CRC15) was utilized to certify the radioactivity values. The mathematical fitting was achieved by Equation (14.3.1):

$$A = 331.24R^2 + 95.383R \qquad (14.3.1)$$

where, R is the count rate (counts/s) and A is the radioactivity (MBq).

14.3.1.3 Calibration of the γ-Camera for the Lutetium-177 – Tomographic Images

In order to correlate the count rate and the radioactivity, a cylindrical phantom with a diameter of 30 cm and width at 10 cm was used. The external shell of the phantom is made of Plexiglas and the internal volume is full of water. In the existed cavities, 10 mL – syringes were placed, filled with 177Lu-DOTA^0-Tyr3 in various radioactivity values. For each value a tomographic image was acquired with the following

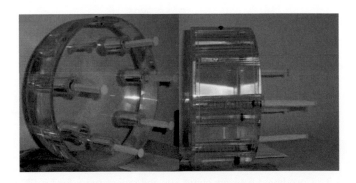

FIGURE 14.3.1   Calibration of the γ-camera for the Lutetium.

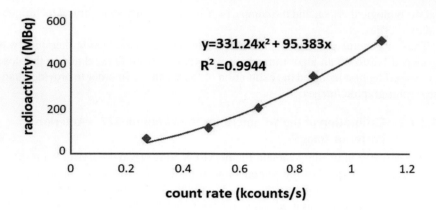

FIGURE 14.3.2   Correlation between count rate and radioactivity for 177Lu anterior-posterior images.

parameters: angle – every 9 degrees, time – 2s, parallel hole collimator – average energy, matrix – 64 × 64, energy window – 20% centered at 208 keV. The mathematical fitting was achieved by Equation (14.3.2):

$$A = 2261R^2 + 511.13R \qquad\qquad (14.3.2)$$

where, R is the count rate (counts/s) and A is the radioactivity (MBq) (Figure 14.3.3).

As it has been previously mentioned, the tomographic data can be inserted in MATLAB and reconstructed as 3D matrices. Each voxel represents the total number of counts happened at this location. The appropriate function is applied in MATLAB (conv routine) to convert the data of the 3D distribution of the count rate in every time point of tomographic acquisition into a 3D distribution of the cumulative radioactivity for different time points. The absorbed dose per voxel is calculated via the

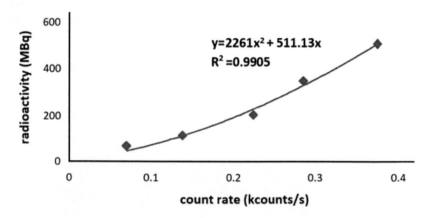

FIGURE 14.3.3   Correlation between count rate and radioactivity for Lu-177 tomographic images.

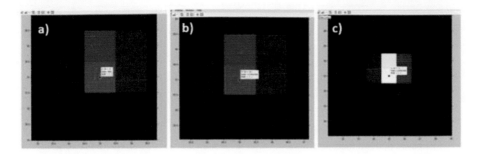

FIGURE 14.3.4 (a) Matrix which represents the counts of the tomographic scintigraphy image, (b) Cumulative radioactivity matrix, and (c) Absorbed dose matrix. (From Vamvakas 2016.)

matrix of cumulative radioactivity through the method of convolution and a new 3 × 3 × 3 matrix of absorbed doses is created (Figure 14.3.4). The method is analytically discussed in Section 11.3.1, in Chapter 11. Then VOIs can be defined and further statistical analysis of the data allows the creation of dose volume histograms (DVHs).

Statistical analysis allows the estimation of the minimum, maximum, and mean value of absorbed dose of Lu-177, through MATLAB (Figure 14.3.5).

The cumulative dose was estimated as 205 GBq·s. Through the convolution method the absorbed dose was calculated as 16.4 Gy and the maximum dose at the central voxel of the contribution as 34.1 Gy. This method can be applied in various studies with different radionuclides.

14.3.1.4 Indium-111 (In-111)

The half-life of Indium is 2.83 days and the physical properties of it are gathered at the Table 14.3.3.

	X
min	555.9
max	5.001e+004
mean	2.064e+004
median	1.589e+004
mode	555.9
std	1.758e+004
range	4.946e+004

FIGURE 14.3.5 Results from statistical analysis Lu-177.

TABLE 14.3.3

Indium Decays

Type of Decay	Energy (keV)	Emission Ratio $(Bq \cdot s)^{-1}$
Photons	150.8	$3 \cdot 10^{-5}$
Photons	171.3	0.906
Photons	245.4	0.941
Electrons (internal transform)	145–170	0.1
Electrons (internal transform)	218–245	0.06
Electrons Auger	19–25	0.16
Electrons Auger	2.6–3.6	1.02
Electrons Auger	0.5	1.91

It is used in neuroendocrine tumors treatment and is labeled with somatostatin analogues. The therapeutic effect of Indium is related to the transmission of Auger electrons. Through the somatostatin receptors the radiopharmaceutical is transferred inside the affected cell, close to the nucleus, in order to destroy it. Since the cancer cells of the neuroendocrine tumors overexpress the somatostatin receptors, the radiopharmaceutical succeeds in entering the cell membrane.

In a previous study of our research group, ten patients were treated for neuroendocrine liver cancer by 4290 MBq In-111 octreotide each. Patient-specific dosimetry protocol with tomographic scintigraphy images was held and the dose calculations were undergone by the MIRDOSE 3.1 software. Significant variation on the absorbed dose at the tumor and critical organs were observed. For the calibration of γ-camera, known activities of In-111 were inserted in a 30-cm diameter cylindrical phantom filled with water. SPECT scintigraphy images were obtained, and a standard imaging protocol was used (Figure 14.3.6). The absorbed dose at every voxel of the cumulative activity matrix was computed with the aforementioned convolution method. The S values of the factors which were required were determined from G Sgouros et al. 2008, for In-111. A cumulative activity of 61.95 GBq·s was calculated and the voxel convolution method showed that mean absorbed dose was 176.3 Gy. Maximum absorbed dose of 763.5 Gy was calculated for the central voxel in the activity.

14.3.1.5　Iodine-131 (I-131)

From the middle of the previous century, Iodine has been widely used for diagnostic and therapeutic applications, related to thyroid gland diseases, in Nuclear Medicine. The half time of Iodine-131 is 8.1 days. It transmits β-particles of an average energy at 0.192 MeV and maximum energy at 0.61 MeV. The average range inside the biological tissues is 0.8 mm. Iodine also transmits γ-particles at 364 keV, allowing the imaging of the radiopharmaceutical by γ-camera. The administration of Iodine is oral in the form of Sodium Iodine. The common dose is 80–200 μCi per gram of thyroid gland mass. For post-operational treatments of cancer of thyroid gland, the typical dose is 2775–5550 mCi, while for non-operational approaches, the dose is increased (5550–7400 mCi). If there is a metastatic area in another organ or tissue

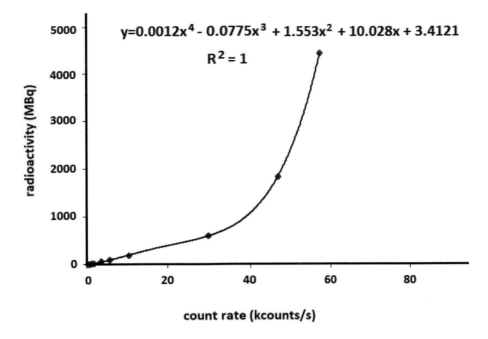

FIGURE 14.3.6   Correlation between count rate and radioactivity for In-111 tomographic images.

the administered dose must be over than 7400 mCi. For accurate dosimetry, acquisition of the data in many time stops is necessary.

In therapeutic applications, it is of crucial importance to protect the sensitive organs by preventing them from over exposure.

14.3.1.5.1   *Voxel-Based Internal Dosimetry during I-131 Radionuclide Therapy*

A dual head Siemens Symbia γ-camera was calibrated in order to obtain quantitative scintigraphy images. This γ-camera has two NaI(Tl) crystals of 9.5 mm thickness each. The high-energy-low-resolution collimator was used, with a 15% energy window, centered at 364 keV. Tomographic images of known activities were acquired, and the count rate was measured. For this reason, an I-131 Theracap sodium iodide capsule was used at two separate time points, when the capsule had activity of 166.5 MBq and 64.01 MBq, respectively. Then, the capsule was inserted into a polymethyl methacrylate PMMA cylindrical phantom (diameter: 16 cm). The capsule was placed 1 cm under the phantom surface for tomographic acquisition to be completed, by matrix 64 × 64 and zoom 1, 23 cm constant circle radius.

The image processing was undergone by Butterworth filter with cut-off 0.4 and order 5 and attenuation correction was held with the Chang method.

The total counts in activity region were measured and the measured counts had to relate to the known activities. Linear fit was applied between data; thus, a first-degree function was determined that converted total acquired counts to activity.

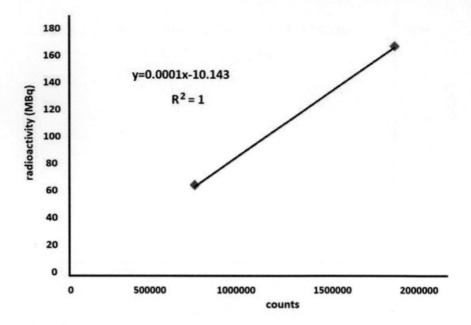

FIGURE 14.3.7 I-131 Counts-Activity diagram. Linear fit was applied between data. According to the linear equation, measured counts were corresponded to activity, for each voxel.

In the next step, the tomographic scintigraphy images were imported in MATLAB as it can perform mathematical calculations by matrices.

So, every axial slice of the scintigraphy image was represented as a 64×64 matrix and every element of this matrix had a numerical index, equal to the number of counts measured for the specific voxel. A 3D matrix, named count matrix, with size $64 \times 64 \times 64$ was obtained from the 64 axial slices of the tomographic scintigraphy image. The voxel size of the matrix was 6.1 mm.

The count matrix was multiplied according to the equation $y = 0.0001x\text{-}10.143$ which was determined from the γ-camera calibration procedure (Figure 14.3.7).

Since, every voxel of the count matrix was calculated according to this equation, the result was a new 3D matrix $64 \times 64 \times 64$ size, named activity matrix. The index of every voxel of the activity matrix was a numerical value, equal to the activity in MBq of each voxel. The activity matrix actually represents the 3D activity distribution as acquired from the tomographic scintigraphy image.

In order to calculate the absorbed dose, cumulative activity must be known. Cumulative activity distribution can be determined if scintigraphy images are acquired at several time points. For each time point distinct activity matrices can be calculated and the cumulative activity matrix can be calculated from the time intervals between image acquisitions, particularly by calculating the area under the activity-time diagram.

In this study, a residence time of 3600 seconds was assumed. Hence, the cumulative activity was calculated by multiplying the activity matrix with 3600. The

TABLE 14.3.4

Central Slice of the Convolution Matrix; the Central Element 0.129 Converts the Cumulative Activity of the Convolved Voxel to Absorbed Dose. The Peripheral Elements Define the Dose Distribution around the Convolved Voxel

0.0001	0.0001	0.0001	0.0001	0.0001
0.0001	0.0029	0.0029	0.0029	0.0001
0.0001	0.0029	0.129	0.0029	0.0001
0.0001	0.0029	0.0029	0.0029	0.0001
0.0001	0.0001	0.0001	0.0001	0.0001

obtained cumulative activity matrix had voxels indexed with numerical values that represented cumulative activity in MBq·s. Every voxel gives a specific value of absorbed dose to itself and it distributes absorbed dose to the adjacent and to the more distant voxels. The cumulative activity matrix containing $64 \times 64 \times 64 = 262144$ voxels, so this is the number of dose distributions which must be determined and summed in order to obtain the total absorbed dose distribution. The absorbed dose distribution for I-131 can be found, published in MIRD 17 in mGy/MBq. A new 3D convolution matrix with size $5 \times 5 \times 5$ was created with 6.1 mm voxel size. Every voxel of the convolution matrix was indexed with absorbed dose per cumulative activity value according to MIRD 17 published data. The values of the central slice of the convolution matrix are presented in Table 14.3.4. The absorbed dose distribution is calculated with the convolution method, by MATLAB and the cumulative activity matrix is convolved with the convolution matrix, resulting to a new 3D matrix with size $64 \times 64 \times 64$, named dose matrix (www.mathworks.com). Every voxel of the dose matrix is indexed with a numeric value that is equal to absorbed dose to voxel in mGy. Furthermore, using MATLAB, regions of interest can be drawn, and dose volume histograms can be generated. Dose statistics as minimum, maximum, and mean dose are available.

A cumulative activity of 230 GBq s was calculated, and the voxel convolution method showed that mean absorbed dose was 20.6 Gy. Maximum dose of 50 Gy was calculated for the central voxel in the activity.

The same methodology was selected in recent studies of our group, for two more approaches applying Ra-223 or Sr-89, in skeletal metastases cases, either for thera-peutic scheme or for palliative treatment. To avoid duplication and unnecessary repetitions, only some information about the properties of the selected radionu-clides is presented.

14.3.1.6   Radium-223 (Ra-223)

The half time of Ra-223 is 11.43 days. It transmits alpha particles at 5.65 MeV and gamma particles at 154 keV, suitable for imaging and quantification of radioactivity by γ-camera. The pharmacokinetics of radium is similar to strontium and for this reason it is entrapped selectively in bones. For this reason, radium is selected for the treatment of skeletal metastasis often follows the breast or prostate cancer. The low range of alpha particles protects the red marrow. A typical administered dose is 50 kBq per kilogram of the patient's body.

14.3.1.6.1   Calibration of the Dose Counter for Therapeutic Applications with $^{223}RaCl_2$

$^{223}RaCl_2$ is used for skeletal cancer due to ideal characteristic of the particles which are emitted. The low range as well as the high LET are efficient factors in destroying the cancer cells, while leaving the normal ones unaffected.

The calibration process was undergone by the counter Capintec (model CRC15). A pre-measured amount of 6000 kBq Ra-223, at a specific time point, was used; based on this measurement the radioactivity at the time of calibration (6 days later) was allowed and estimated at 4220 kBq. The measurement of the background count rate was undergone and afterward the measurement of the standard amount of the 4220 kBq of Ra-223 was followed. Also, the count rate in a half-a-meter-distance from the Geiger detector (Rotem) was measured. The measured value of the 4220 kBq was 700 cpm, thus the conversion factor between count rate and radioactivity was 6.028.

14.3.1.7   Strontium-89 (Sr-89)

Strontium is the most common radionuclide in skeletal metastasis treatment with a physical half-life at 50.5 days. It transmits beta particles of an average energy at 1.46 MeV. The maximum range of the beta particles inside the tissues is 6.7 mm and the average range at 2.4 mm. It is widely used as $^{89}SrCl_2$.

The biokinetics of Strontium is similar to Calcium. The 50% of the administered dose is entrapped in the bones and the other is eliminated by the kidneys and the gastrointestinal system. The typical administered dose is 2.22 MBq/kg.

14.3.2   PATIENT-SPECIFIC ESTIMATION, ABSORBED DOSE TO THE TUMOR VS NORMAL TISSUE

The radionuclide treatments ought to maximize the therapeutic dose in the tumor area, leaving the healthy tissues unaffected, by eliminating the dose in critical organs. Targeted radionuclide therapy (TRNT) is the method that allows this approach to be achievable. TRNT is based on the use of high-affinity molecules as carriers of radionuclides TRNT delivered to tumor cells. The development of novel spatial visualization methods for the accurate estimation of absorbed dose both in tumors and normal tissues is of critical importance (Huizing D et al., 2018). Precise tumor topography and spatial dosimetry make possible the design of an optimal personalized treatment allowing patient convenience.

The fundamental of TRNT is the selective accumulation of the radiopharmaceutical only in tumor area. This strategy requires high-therapeutic index, managing to acquire high efficiency with minimal health risks. The selection of the administered radionuclide is of great importance.

The therapeutic index is a ratio which compares the absorbed dose in tumor and in healthy tissues or in critical organs. The higher this ratio is obtained, the more efficient treatment is considered. Current challenges in the field of radionuclide therapeutic approaches of neoplasms need precise algorithms and accurate calculations.

MATLAB can contribute to this direction, optimizing the personalized therapies.

REFERENCES

Handkiewicz-Junak D, Poeppel TD, Bodei L, Aktolun C, et al 2018, EANM guidelines for radionuclide therapy of bone metastases with beta-emitting radionuclides, Eur J Nucl Med Mol Imaging, 45 (5): 846–859.

Huizing D, de Wit-van der Veen BJ, Verheij M, Stokkel M 2018, Dosimetry methods and clinical applications in peptide receptor radionuclide therapy for neuroendocrine tumours: a literature review, EJNMMI Research, 8 (1): 89.

Jadvar H, Quinn DI 2013, Targeted α-particle therapy of bone metastases in prostate cancer, Clin Nucl Med, 38 (12): 966–971.

Kassis AI 2008, Therapeutic radionuclides: biophysical and radiobiologic principles, therapeutic radionuclides: biophysical and radiobiologic principles, Semin Nucl Med, 38 (5): 358–366.

Kost SD, Dewaraja YK, Abramson RG, Stabin MG 2015, VIDA: a voxel-based dosimetry method for targeted radionuclide therapy using Geant4, Cancer Biother Radiopharm, 30 (1): 16–26.

Lanconelli N, Pacilio M, Lo Meo S, Botta F, Di Dia A, Torres Aroche L A, Coca Pérez M A, Cremonesi M 2012, A free database of radionuclide voxel S values for the dosimetry of nonuniform activity distributions, Phys Med Biol, 57: 517–533.

Li T, Ao ECI, Lambert B, Brans B, Vandenberghe S, Mok GSP 2017, Quantitative imaging for targeted radionuclide therapy dosimetry – technical review, Theranostics, 7 (18): 4551–4565.

Ljungberg M, Sjögreen Gleisner K 2016, Personalized dosimetry for radionuclide therapy using molecular imaging tools, Biomedicines, 4: 25.

Ljungberg M, Hendrik Pretorius P 2018, SPECT/CT: an update on technological developments and clinical applications, Br J Radiol, 91 (1081): 20160402.

Loke KS, Padhy AK, Ng DC, Goh AS, Divgi C 2011, Dosimetric considerations in radioimmunotherapy and systemic radionuclide therapies: a review, World J Nucl Med, 10 (2): 122–138.

Lyra M, Andreou M, Georgantzoglou A et al 2013, Radionuclides used in nuclear medicine therapy-from production to dosimetry, Curr Med Imaging Rev, 9 (1): 51–87.

Macedo F, Ladeira K, Pinho F, Saraiva N, Bonito N, Pinto L, Goncalves F 2017, Bone metastases: an overview, Oncol Rev, 11 (1): 321.

Pereira JM, Stabin MG, Lima F, Guimarães M, Forrester JW 2010, Image quantification for radiation dose calculations – limitations and uncertainties, Health Phys, 99 (5): 688–701.

Sapienza MT, Willegaignon J 2019, Radionuclide therapy: current status and prospects for internal dosimetry in individualized therapeutic planning, Clinics (Sao Paulo), 74: e835.

Sgouros G, Frey E, Wahl R, He B, Prideaux A, Hobbs R 2008, Three-dimensional imaging-based radiobiological dosimetry, Semin Nucl Med, 38 (5): 321–334.

Shevtsova ON, Shevtsova VK 2017, Mathematical simulation of transport kinetics of tumor-imaging radiopharmaceutical 99mTc-MIBI, Comput Math Methods Med, 2017: 2414878.

Vamvakas I, Lyra M 2015, Voxel based internal dosimetry during radionuclide therapy, Hell J Nucl Med, 18 (Suppl. 1): 76–80.

Vamvakas I, Synefia S, Lyra M, Kostakis V, Ttofi E 2016a, MATLAB in voxel internal dosimetry 111In and 177Lu therapy, 5th Balkan & 13th National Congress of Nuclear Medicine BCNM 17-20/06/2016.

Vamvakas IC 2016, Patient specific dosimetry during radionuclide treatment, new techniques, National and Kapodistrian University of Athens, PhD Dissertation. Athens, Greece.

www.mathworks.com

List of Acronyms

A

AAPM	*American Association of Physicists in Medicine*
ACR	*American College of Radiology*
ADCs	*Analog-to-Digital Converters*
ADME	*Absorption, Drug, Elimination and Metabolism*
ALARA	*As Low As Reasonably Achievable*
ANN	*Artificial Neural Network*
ANZSNM	*Australian/New Zealand Standards*
APDs	*Avalanche Photodiodes*
ASICs	*Application-Specific Integrated Circuits*
ASNC	*American Society of Nuclear Cardiology*

B

BGO	*Bismuth Germanate Oxide*
BREP	*Boundary Representation Phantom*
BSS	*Basic Safety Standards*

C

CAD	*Computer Aided Design*
CAD	*Coronary Artery Disease*
CASToR	*Customizable and Advanced Software for Tomographic Reconstruction*
CDR	*Collimator–Detector Response*
CMY	*Cyan, Magenta, and Yellow*
CNN	*Convolutional Neural Network*
CNTs	*Carbon NanoTubes*
COR	*Center of Rotation*
CRL	*Count Rate Loss*
CRP	*Coordinated Research Projects*
CT	*Computed Tomography*
CYP	*Cytochrome P450s*
CZT	*Cadmium Zinc Telluride*

D

DAC	*Digital-to-Analogue Converters*
DAT	*Dopamine Transporter*
DCT	*Discrete Cosine Transform*
DICOM	*Digital Imaging and Communications in Medicine*
DIN	*Deutsches Institut fur Normung*
DMSA	*DimercaptoSuccinic Acid*
DSiPMs	*Digital Silicon Photomultipliers*
DU	*Differential Uniformity*
DVHs	*Dose-Volume Histograms*

E

EANM	*European Association of Nuclear Medicine*
EC	*European Commission*
ECT	*Emission Computed Tomography*
EFOMP	*European Federation of Organizations of Medical Physicists*
EGS	*Electron Gamma Shower*
EM	*Expectation-Maximization algorithm*
EPR	*Enhanced Permeation and Retention*
ESR	*European Society of Radiology*
ET	*Essential Tremor*
EU	*European Union*

F

FBP	*Filtered Back Projection*
FCM	*Fuzzy C-Means*
FDA	*Food and Drug Administration*
FDG	*Fluorodeoxyglucose*
FFT	*Fast Fourier Transform*
FMO	*Flavin-Containing Monooxygenase*
FORTRAN	*Formula Translation*
FOV	*Field of View*
FT	*Fourier Transform*
FWHM	*Full Width at Half Maximum*

G

GAMOS	*GEANT4-based Architecture for Medicine-Oriented Simulation*
GEANT4	*GEometry ANd Tracking*
GSF	*National Research Center for Environment and Health*
GUIDE	*Graphic User Interface Development Environment*
GUI	*Graphical User Interface*

H

HCC	*HepatoCellular Carcinoma*

I

IAEA	*International Atomic Energy Agency*
IC	*Internal Conversion*
ICRP	*International Commission on Radiological Protection*
ICRU	*International Commission of Radiation Units and Measurements*
IDL	*Interactive Data Language*
IEC	*International Electrotechnical Commission*
IOMP	*International Organization for Medical Physics*
IPT	*Image Processing Toolbox*
ISTR	*International Symposium on Trends in Radiopharmaceuticals*
IU	*Integral Uniformity*

L

LDL	*Low-Density Lipoprotein*
LET	*Linear Energy Transfer*
LoG	*Laplacian of Gaussian*
LOR	*Line of Response*
LSD	*Line-Spread Function*

M

MAbs	*Monoclonal Antibodies*
MATLAB	*Matrix-Laboratory*
MC	*Monte Carlo simulation*
MCNP	*Monte Carlo N-Particle Transport*
MCNPX	*Monte Carlo N-Particle eXtended*
MED	*Medical Exposure Directive*
MINC	*Medical Imaging NetCDF*
MIP	*Maximum Intensity Projection*
MIRD	*Medical Internal Radiation Dose*
MIRT	*Michigan Image Reconstruction Toolbox*
MITA	*Medical Imaging and Technology Alliance*
ML-EM	*Maximum Likelihood-Expectation Maximization*
MnMEIO	*Md-doped Magnetism Engineered Iron Oxide*
MRI	*Magnetic Resonance Imaging*
MTF	*Modulation Transfer Function*
MWNT	*Multi-Walled Nano Tubes*

N

NCA	*NonCompartmental Analysis*
NCAT	*NURBS-based CArdiac-Torso (NCAT)*
NCRP	*National Council on Radiation Protection*
NECR	*Noise Equivalent Count Rate*
NEMA	*National Electrical Manufacturers Association*
NHL	*Non-Hodgkin's Lymphoma*
NIfTI	*Neuroimaging Informatics Technology Initiative*
NIH	*National Institutes of Health*
NMI	*Nuclear Medicine Imaging*
NMQC	*Nuclear Medicine-QC*
NURBS	*Non-Uniform Rational B-Spline*

O

OLINDA/EXM	*Organ Level INternal Dose Assessment/EXponential Modeling*
OLINDA	*Organ Level INternal Dose Assessment*
ORNL	*Oak Ridge National Laboratory*
OS-EM	*Ordered Subsets-Expectation Maximization*

P

PACS	*Picture Archiving and Communication System*
PBPK	*Physiologically-Based PharmacoKinetic*

PDE	*Photo-Detection Efficiency*
PD	*Parkinson Disease*
PET	*Positron Emission Tomography*
PHA	*Pulse Height Analyzer*
PK/PD	*PharmacoKinetic/PharmacoDynamic*
PMT	*PhotoMultiplier Tubes*
PSF	*Point Spread Functions*
PVC	*Partial Volume Correction*
PVE	*Partial Volume Effect*

Q

QA	*Quality Assurance*
QC	*Quality Control*
QIN	*Quantitative Imaging Network*
QM	*Quality Management*

R

RADAR	*RAdiation Dose Assessment Resource*
RGB	*Red-Green-Blue*
RIT	*RadioImmunoTherapy*
ROIs	*Region of Interest*

S

SAF	*Specific Absorption Fraction*
SEC	*Size-Exclusion Chromatography*
SIMIND	*Simulation of Imaging Nuclear Detectors*
SiPMs	*Silicon PhotoMultipliers*
SIRF	*Synergistic Image Reconstruction Framework*
SIRT	*Selective Internal Radiation Therapy*
SNMMI	*Society of Nuclear Medicine and Molecular Imaging*
SNM	*Society of Nuclear Medicine*
SNR	*Signal-to-Noise Ratios*
SPECT	*Single Photon Emission Computed Tomography*
SSDL	*Secondary Standard Dosimetry Laboratories*
SSM	*Statistical Shape Model*
SSTR	*Somatostatin Receptors*
STIR	*Software for Tomographic Image Reconstruction*
SUV	*Standard Uptake Value*
SWNT	*Single-Walled Nano Tubes*

T

TAC	*Time-Activity Curve*
TARE	*Trans-Arterial Radio-Embolization*
TCA	*TriChloroacetic Acid*
TDCS	*Transmission-Dependent Convolution Subtraction*
TEW	*Triple-Energy Window*
TOF	*Time-of-Flight*
TRT	*Targeted Radionuclide Therapy*

U
UF   *University of Florida*

V
VIM   *International Vocabulary of Metrology*
VIP   *Vasoactive Intestinal Peptide*
VOI   *Volume of Interest*

W
WHO   *World Health Organization*

X
XCAT   *eXtended CArdiac-Torso*